Springer Theses

Recognizing Outstanding Ph.D. Research

Aims and Scope

The series "Springer Theses" brings together a selection of the very best Ph.D. theses from around the world and across the physical sciences. Nominated and endorsed by two recognized specialists, each published volume has been selected for its scientific excellence and the high impact of its contents for the pertinent field of research. For greater accessibility to non-specialists, the published versions include an extended introduction, as well as a foreword by the student's supervisor explaining the special relevance of the work for the field. As a whole, the series will provide a valuable resource both for newcomers to the research fields described, and for other scientists seeking detailed background information on special questions. Finally, it provides an accredited documentation of the valuable contributions made by today's younger generation of scientists.

Theses are accepted into the series by invited nomination only and must fulfill all of the following criteria

- They must be written in good English.
- The topic should fall within the confines of Chemistry, Physics, Earth Sciences, Engineering and related interdisciplinary fields such as Materials, Nanoscience, Chemical Engineering, Complex Systems and Biophysics.
- The work reported in the thesis must represent a significant scientific advance.
- If the thesis includes previously published material, permission to reproduce this must be gained from the respective copyright holder.
- They must have been examined and passed during the 12 months prior to nomination.
- Each thesis should include a foreword by the supervisor outlining the significance of its content.
- The theses should have a clearly defined structure including an introduction accessible to scientists not expert in that particular field.

More information about this series at http://www.springer.com/series/8790

Laurens Vanderstraeten

Tensor Network States and Effective Particles for Low-Dimensional Quantum Spin Systems

Doctoral Thesis accepted by
the University of Ghent, Ghent, Belgium

Springer

Author
Dr. Laurens Vanderstraeten
Department of Physics and Astronomy
University of Ghent
Ghent
Belgium

Supervisor
Prof. Frank Verstraete
Department of Physics and Astronomy
University of Ghent
Ghent
Belgium

ISSN 2190-5053 ISSN 2190-5061 (electronic)
Springer Theses
ISBN 978-3-319-64190-4 ISBN 978-3-319-64191-1 (eBook)
DOI 10.1007/978-3-319-64191-1

Library of Congress Control Number: 2017948193

Printed on acid-free paper

This Springer imprint is published by Springer Nature
The registered company is Springer International Publishing AG
The registered company address is: Gewerbestrasse 11, 6330 Cham, Switzerland

Supervisor's Foreword

The quantum many-body problem has fascinated many different generations of physicists. This is part due to the fact that interacting systems exhibit extremely interesting and new physical phenomena, and part due to the fact that this is the central and hardest problem in theoretical physics. As realized by the founders of quantum mechanics, the full richness of quantum many-body physics is already witnessed by the intricate physics present in quantum spin systems such as the Heisenberg model. A plethora of analytical and approximate methods has been devised to simulate such effective models, leading to game changing methods such as the renormalization group, diagrammatic methods, quantum Monte Carlo algorithms and the Bethe ansatz.

One-dimensional quantum spin chains are of particular interest as they exhibit the strongest correlations and entanglement. In the case of 1D integrable systems, a pretty complete picture of the ground state and energy spectrum of the related Hamiltonians has been obtained using Bethe ansatz techniques. Luttinger liquids yield a unifying picture for describing critical systems. And for the case of non-integrable systems, the density matrix renormalization group method, originally introduced by Steven White, and related matrix product state techniques have revolutionized the way ground states can be characterized and dynamical properties can be extracted.

One central question that was left open or at least confusing in all those approaches is the nature of the elementary excitations and their interactions. Fundamental results in quantum field theory dictate that those should be understood in terms of localized particles, but neither the Bethe ansatz nor the density matrix renormalization group yield such a local description when the ground state exhibits strong quantum correlations. This problem is precisely the one tackled and solved in this thesis, where Laurens Vanderstraeten develops a full picture for describing elementary excitations and their interactions in terms of localized wave packets encoded by local perturbations of the underlying matrix product states or higher dimensional analogues of it.

In this remarkable thesis, the author develops a framework for describing wave packets of excitations of generic quantum spin chains such as magnons and spinons and completely solves the problem of how those particles interact and scatter. As lattice systems do not exhibit any of the continuous symmetries assumed in standard scattering theory, a novel formalism had to be developed to capture scattering processes in a variational framework. This provides a beautiful connection between S-matrices related to eigenstates and dynamics and opens up novel ways of predicting dynamical structure factors as seen in experiments.

This thesis is a showcase of how a new language, in this case of entanglement theory and of tensor networks, provides novel valuable insights into the quantum many-body problem. Laurens has unlocked the door towards a local description of quantum many-body wavefunctions, in which excitations and particles are described by harmonics of the entanglement degrees of freedom. The challenge is now to lift this beautiful formalism, developed in this thesis for artificial quantum spin chains, to real-world problems involving, e.g., superconductivity and QCD.

Ghent, Belgium
June 2017

Prof. Frank Verstraete

Abstract

In this thesis, we develop the formalism for capturing elementary excitations on top of tensor network ground states for low-dimensional quantum spin systems.

For one-dimensional quantum spin systems, we build on the established framework of matrix product states, which provide an extremely successful variational parametrization for gapped ground states. We introduce a variational ansatz for elementary excitations on top of a matrix product state and show that it reproduces dispersion relations and spectral weights with very high precision. Interpreting these excitations as quasiparticles, we develop the framework for studying their interactions and show how to compute the two-particle S matrix, the two-particle contributions to dynamical correlation functions (spectral functions) and the real-time scattering properties. Moreover, we use the one- and two-particle information to construct an approximate Bethe ansatz for describing the condensation of quasiparticles under a magnetic field. We apply our formalism to some paradigmatic examples such as the spin-1 Heisenberg chain, the two-leg Heisenberg spin ladder, the spin-1/2 XXZ chain and the dimerized $J_1 - J_2$ chain.

For two-dimensional systems, for which tensor network methods are much less developed, we first introduce a variational ground-state optimization algorithm for projected entangled-pair states in the thermodynamic limit. This requires the construction of a new effective environment, combining ideas from existing contraction algorithms. The optimization algorithm is benchmarked on the two-dimensional Heisenberg and Ising models and improves on existing state-of-the-art algorithms. In the last section, we introduce the ansatz for elementary excitations in the two-dimensional setting and compute excitation spectra for frustration-free systems such as the square-lattice AKLT model and the filtered toric-code model.

Acknowledgements

It is my pleasure to thank the people without whom I would never have been able to write this thesis. First of all, I want to thank my supervisor Frank for his confidence, enthusiasm and endless supply of suggestions and ideas. I am grateful to my co-supervisor Jutho, who has never spared any efforts to help me in implementing all these ideas. Without this perfect tandem supporting me, I would never have gotten this far. Of course, I would also like to thank all the people of the Ghent quantum group for the support and, more importantly, for generating the perfect working place. I am grateful to the people from Vienna for our joint projects and, above all, the instructive group meetings.

I am indebted to the members of my doctoral committee for the careful reading of my thesis and especially Philippe Corboz and Fabian Essler for taking the effort of coming to Ghent. I thank Philippe as well as Didier Poilblanc for inviting me to Amsterdam and Toulouse and for the fruitful collaborations that have followed.

Ghent, Belgium
June 2017

Laurens Vanderstraeten

Contents

Chapter 1
Overview

In this dissertation we develop new techniques for simulating the low-energy behaviour of quantum spin systems in one and two dimensions. These techniques arise through the combination of two themes in many-particle physics: (i) the concept of quasiparticles as the effective low-energy degrees of freedom in a condensed-matter system, and (ii) entanglement as the characteristic feature for describing quantum phases of matter. Whereas the former gave rise to the use of effective field theories for understanding many-particle systems, the latter led to the development of tensor network states for describing the distribution of entanglement in quantum ground states. We will use the formalism of tensor network states to derive an effective particle description for one- and two-dimensional spin systems that exhibit strong quantum correlations.

Before giving an overview, a few remarks as to the style of this dissertation are in order. The general motive of theoretical physics is designing an idealized representation of certain parts of nature in terms of mathematical concepts and laws. Yet, in this work we will not make any attempt at mathematical rigour, mainly due to the author's incompetence in that area. Also, the general question regarding the application of the mathematical formalism to real-life experiments will be left rather vague, essentially for the same reason. Instead, we hope that the following does have some meaning in virtue of the physical intuition behind the mathematical concepts and the possibility of implementing the formalism for numerical simulations.

Chapter 2. Introduction to Quantum Many-Body Physics

In the first chapter, we aim at situating and motivating the contents of the next chapters. First we introduce the quantum many-body problem, we show how simple many-body models appear as effective levels of theory, and what the most interesting properties of these models are; also, we indicate shortly how theoretical many-body physics is partly driven by a number of great experiments in the last decades.

L. Vanderstraeten, *Tensor Network States and Effective Particles for Low-Dimensional Quantum Spin Systems*, Springer Theses,
DOI 10.1007/978-3-319-64191-1_1

We go on by identifying entanglement as the crucial ingredient for understanding quantum many-body physics and give some important results in entanglement theory that form the rationale behind our methods. In particular, we show how entanglement drives the quantum effects in a low-dimensional system, and we highlight a few characteristic phenomena of one- and two-dimensional quantum physics. As this dissertation aims at devising new numerical techniques, we also review the most important alternatives for simulating strongly-correlated quantum physics.

Next we give a more detailed overview of three concepts that will play an important role in the following chapters. Firstly, we show how the Bethe ansatz allows to determine the ground state and the low-lying excitation spectrum of one-dimensional integrable systems. Secondly, we give a detailed overview of the field of tensor network states and we explain how they offer both a theoretical tool and a variational parametrization for studying strongly-correlated quantum spin systems. Finally, we shortly discuss the concept of renormalization and the idea of effective particles for capturing the low-energy behaviour of condensed-matter systems.

Chapter 3. Effective Particles in Quantum Spin Chains: The Framework

In this chapter we develop the formalism for treating effective particles in quantum spin chains using the framework of matrix product states (MPS). As a preliminary, we situate our work within a tradition of variational approaches towards elementary excitations and quasiparticles starting with the works of Landau, Anderson, and Feynman. Next we introduce the class of uniform MPS, a class of states that describes translation-invariant ground state of gapped quantum spin chains in the thermodynamic limit.

The next section discusses the ansatz that we will use for capturing the wave function of an elementary excitation as a localized perturbation with definite momentum, living on an MPS background. We show how to compute expectation values of local operators and how to optimize the energy in order to find the elementary dispersion relation. We also show that the ansatz can be extended to the case of topological excitations and bound states. Interpreting the elementary excitations as particles, we introduce the natural ansatz for describing two-particle states, and we explain how to solve the scattering problem. This leads to a natural definition of the two-particle S matrix and we indicate what happens when a bound state forms. At this point, we take a step back and make the connection with different variational methods in many-body physics. In particular, it is explained that our work relies heavily on the concept of the tangent space on the MPS manifold.

After the constructive work, we start discussing how the one- and two-particle energies and wave functions allow us to characterize the properties of quasiparticles in generic spin chains. Firstly, we explicitly show that the S matrix that we have defined earlier, is the same as the one that is introduced in standard scattering theory, and, as such, gives us full access to all two-particle scattering properties. We make the connection with real-time scattering processes, define a scattering length and indicate how the formation of bound states is reflected in the S matrix. Secondly, we consider a finite density of excitations as it is described by the approximate Bethe ansatz. We discuss what happens under the application of a chemical potential for the particles or when the temperature is turned on. Making the connection with Luttinger liquids, we determine the critical properties of a condensed gas of effective particles. Thirdly, the variational wave functions are used to evaluate spectral functions, we trace the origin of divergences, and explain how to artificially add an imaginary frequency. Also, we discuss the extension to the case of non-zero temperature.

In the last two sections, we become more speculative and look at what the possible extensions of the framework are. We give some clues towards the construction of a proper second-quantized effective field theory and note what the difficulties are towards this goal. Also, we discuss to what extent we can understand out-of-equilibrium processes such as quantum quenches and transport in terms of effective particles.

Chapter 4. Effective Particles in Quantum Spin Chains: Applications

The third chapter contains the applications of the framework that was introduced above. The sections discuss resp. the spin-1 Heisenberg chain, the spin-1/2 Heisenberg ladder, the dimerized J_1-J_2 chain, the gapped XXZ chain, and the anisotropic spin-1 Heisenberg chain.

For these models, we typically start by mapping out the elementary excitation spectrum, where the properties of the dispersion relations of the different particles are discussed. Also, the presence of bound states can be observed from the excitation spectrum.

Next we zoom in on the two-particle S matrix that describes all the two-particle interactions in the system. We show how to bring it in diagonal form by going over to the coupled-spin basis, and compute the scattering lengths in the different symmetry sectors. The scattering length also signals the formation of bound states, which we explicitly show for the case of the two-magnon bound state formation in the spin-1 chain and the ladder. For larger relative momenta, the scattering phase loses its linear dependence, which points to higher-energy scattering effects. Another effect is the breaking of Galilean invariance of the S matrix because of the lattice on which the particles live.

The second application of our formalism is towards the computation of spectral functions. Whereas the one-particle excitations contribute δ peaks, the two-particle excitations give a continuous band. Because we target stationary eigenstates directly in the thermodynamic limit, we have perfect resolution in frequency and momentum. We reproduce the generic form of a momentum slice in the two-particle continuum, as well as the divergences whenever a bound state is formed. Our methods are shown to work equally well for two-magnon and two-spinon continuum bands. For the case of bound states that emerge out of the continuum at a certain momentum, we nicely see a divergence at the edge of the two-particle band. In the case of the formation of a bound state as a result of tuning an external parameter, we see that the stabilization of a bound state out of the continuum is signalled by divergences in the spectral function at the band's edge. Together with the information in the S matrix, this gives a physically intuitive tool to investigate bound states.

Thirdly, we investigate the condensation of magnons under a magnetic field or a non-zero temperature. We apply the approximate Bethe ansatz in order to follow the magnetization curve and the Luttinger parameter as a function of magnetic field. The phase transition at the point where the magnetic field crosses the spectral gap and the magnons start condensing is smoothened out when the temperature is turned on. Also, we show preliminary results for the influence of a crystal-field anisotropy on the magnetization process: the C/IC transition is transformed into an Ising transition, which we still seem to capture with our methods.

Finally, we give an overview of the possible future applications of our methods.

Chapter 5. Towards a Particle Theory in Two Dimensions

In the last chapter we add a second dimension. We explain that projected entangled-pair states (PEPS) provide a natural parametrization for the low-energy states of two-dimensional quantum lattice systems, and, after discussing some of the state-of-the-art PEPS algorithms, we formally introduce a PEPS in the thermodynamic limit, as well as a general tangent vector.

Compared to the one-dimensional case, the development of a particle theory is considerably less straightforward because the contraction and optimization of infinite PEPS are already a highly non-trivial computational task. Therefore, we first direct our attention to PEPS ground-state algorithms. We review the different contraction methods that are available, and introduce our own corner environment contraction method as a hybrid version. We discuss how the channel environments allow us to compute the static structure factor and the energy variance. In the next section, we introduce an entirely new ground-state optimization method for PEPS in the thermodynamic limit, based on a gradient-search algorithm of the global energy functional. The gradient is computed with the contraction scheme that we introduced earlier. Our PEPS code has the advantage that it is variational, it has a clear convergence criterion, and it can easily take account of symmetries. We benchmark our new PEPS code on the transverse-field Ising model and the Heisenberg model. In both cases, we show significant improvement over state-of-the-art methods.

Only after we have found a reliable ground-state optimization method, we introduce the excitation ansatz for two-dimensional systems and discuss how the eigenvalue problem can be solved efficiently. We apply it to (i) the two-dimensional AKLT model, for which we compute the single-mode approximation dispersion relation, the spectral gap, the magnon velocity and the spectral weight in the magnon mode, and (ii) the filtered toric code, where we compute the dispersion of the charge and flux excitations, and we illustrate the closing of the gap in one of the two sectors if the model is tuned to a topological phase transition.

We close the chapter by discussing the further directions that should be pursued in the future, based on the algorithms that we have just laid out.

Appendix. Technical Details

In the main text we have omitted some of the formulas and derivations, because of considerations of length and readability. These additional details are taken up in the Appendix and should only be read in utter despair.

Chapter 2
Introduction to Quantum Many-Body Physics

In the introduction to his 1929 paper on many-electron systems [1], Dirac envisioned that the two problems facing quantum mechanics were "in connection with the exact fitting in of the theory with relativity ideas" on the one hand, and the fact that "the exact application of these laws leads to equations much too complicated to be soluble" on the other. These two problems would dominate theoretical physics throughout the twentieth century: Whereas the former led to the formulation of relativistic quantum field theories and, ultimately, the standard model of particle physics, the latter posed the central problem for the field of quantum many-body physics.

In a way that Dirac could not foresee, the quantum many-body problem continues to puzzle physicists. Although a set of theoretical and computational methods has been developed to solve most problems in atomic and nuclear physics, there is still a class of problems that defy any satisfactory solution. These are the problems for which the quantumness of the physical world is most pressing, making any treatment with classical concepts and computers intractable. For the same reasons, these are the most interesting problems for a physicist wanting to explore new physical phenomena.

In 1935 Schrödinger already noted that *entanglement* is the "characteristic trait of quantum mechanics, the one that enforces its entire departure from classical lines of thought" [2], a trait that was problematized by Einstein in the same year [3]. Yet, it is only in the last decades that entanglement in quantum many-body systems has been considered as the defining feature of so-called quantum phases of matter. This has led to an explosion of new insights into the quantum properties of systems with many degrees of freedom, where quantum correlations are no longer seen as a difficulty for simulating the system, but rather explored as a resource for exotic physical phenomena.

In this introductory chapter, we will give a short overview of quantum many-body physics, with an emphasis on the different ways in which entanglement determines

L. Vanderstraeten, *Tensor Network States and Effective Particles for Low-Dimensional Quantum Spin Systems*, Springer Theses,
DOI 10.1007/978-3-319-64191-1_2

the quantum behaviour of systems with many degrees of freedom. In the three last sections, we will discuss integrability, tensor networks and effective particle descriptions in more detail, as we will need these concepts in the following chapters.

2.1 Defining the Problem

In this dissertation we investigate the low-energy properties of systems consisting of a large number of quantum-mechanical degrees of freedom that interact locally. In this section, we explain how this is modelled mathematically, we give a physical motivation and we discuss what the interesting properties are.[1]

2.1.1 *The Basics*

The building block of a quantum many-body system is a d-level Hilbert space with basis vectors $|s\rangle$ ($s = 1, \ldots, d$). These elementary degrees of freedom or "spins" are brought together as parts of a large physical system, for which the Hilbert space is obtained as the direct product of all the individual spins

$$\mathbb{H} = \bigotimes_{i \in \mathcal{L}} \mathbb{H}_i .$$

The direct-product structure is a direct consequence of the superposition principle and allows for entanglement between different spins. It also implies that the size of the Hilbert grows exponentially with the number of spins, a fact that makes the quantum many-body problem notoriously difficult to deal with. In physical systems the spins are typically arranged in some spatial structure or lattice $\mathcal{L}$. In this dissertation, we are alway interested in infinitely large lattices, so that only the physics of the bulk is taken into account.[2]

A wave function for a system of N spins can be written as a superposition

$$|\Psi\rangle = \sum_{s_1,\ldots,s_N} c_{s_1,\ldots,s_N} |s_1\rangle \otimes \cdots \otimes |s_N\rangle ,$$

where the $c_{s_1,\ldots,s_N}$ are complex coefficients. The wave function describes a physical state of the system, for which all physical information should in the end be obtained as the expectation value of a well-chosen operator O

[1]We assume the basic postulates and formalism of quantum mechanics [4, 5] to be in place, as well as all mathematical concepts and techniques. We refer to Ref. [6] for a modern overview.

[2]We will always ignore the mathematical difficulties of defining this thermodynamic limit.

$$\langle O \rangle = \langle \Psi | O | \Psi \rangle.$$

More generally, every state of the system is described by a density matrix ρ, acting on $\mathbb{H}$, which has expectation values

$$\langle O \rangle = \mathrm{Tr}(O\rho).$$

A model is further defined by its Hamiltonian H, modelling the interactions between the spins. We will be interested in local interactions, such that the Hamiltonian can be written as a sum of operators h_i that only act non-trivially on a small patch around site i in the lattice, i.e.

$$H = \sum_{i \in \mathcal{L}} h_i.$$

The Hamiltonian is of central importance for the physical properties of the system because

- it is associated with the time-dependence of a state $|\Psi(t)\rangle$ via the Schrödinger equation
$$i \frac{\partial}{\partial t} |\Psi(t)\rangle = H |\Psi(t)\rangle,$$
- it determines the Gibbs distribution of a system at inverse temperature β as
$$\rho = \mathrm{e}^{-\beta H}.$$

In condensed-matter systems, where one typically works at very low temperatures, one is mainly interested in the sector of Hilbert space that corresponds to the lowest eigenvalues or energies of the Hamiltonian. The eigenstate with the lowest energy is called the ground state, which determines all the static properties of the system at zero temperature, whereas the next few eigenstates or excitations determine the dynamical properties of the system.

2.1.2 Elementary Models

In classical statistical mechanics a system is often not very interesting at zero temperature, but quantum-mechanically the ground state can be a complicated state with very non-trivial properties. This can be traced back to a basic property of the Hamiltonian, viz. that the different local terms do not commute. This implies that they cannot be minimized simultaneously and that the ground state is a complicated superposition of different configurations that minimize the global energy of the state. Interestingly, the local interactions don't need to be complex in order to give rise to interesting physics; throughout this dissertation, it will become clear that surprisingly simple Hamiltonians can give rise to an extreme variety of physical behaviour.

One paradigmatic example is the Hubbard model, which describes interacting fermions on a lattice [7]. As we will see, the Hubbard model will form the basis for a lot of models that we consider in this dissertation, so it proves worthwhile to understand how it arises as an effective model for electrons in a metal [8–10]. First of all, the assumption is made that the heavy ions in the metal's crystal form a static lattice and only the electrons are taken into account as degrees of freedom. Next, we determine one-electron wave functions in an optimal way, in the sense that the interactions between the electrons are treated as much as possible on the mean-field level. These wave functions are, because of the periodic lattice, delocalized Bloch functions that describe electrons moving independently (uncorrelated) through a potential generated by the ions and the other electrons. For some metals, this mean-field description (band theory) is not accurate enough and it proves necessary to include interactions further. This can be done by first defining Wannier functions, which are one-particle wave functions centred around a lattice site, and rewriting the second-quantized Hamiltonian in this basis as

$$H = \sum_{i,j} \sum_{\alpha,a} t_{ij} c^{\dagger}_{\alpha i,a} c_{\alpha j,a} + \frac{1}{2} \sum_{\alpha \ldots \delta, i \ldots l} U^{\alpha\beta\gamma\delta}_{ijkl} \sum_{ab} c^{\dagger}_{\alpha i,a} c^{\dagger}_{\beta j,b} c_{\gamma k,b} c_{\delta l,a},$$

where $c_{\alpha i,a}$ annihilates an electron in a Wannier orbital of band α, centred around site i with spin a. The simplest version of the Hubbard model is obtained by further assuming that (i) the Fermi surface lies inside a single conduction band and we are interested in low enough energies such that the other bands only determine the parameters t and U, (ii) the Wannier functions are strongly localized around one lattice site, such that only hopping matrix elements t_{ij} between nearest-neighbour lattice sites $\langle ij \rangle$ are retained, and (iii) the range of the interaction matrix elements $U^{\alpha\beta\gamma\delta}_{ijkl}$ is very small, such that only intra-atomic interactions have to be taken into account. Under these assumptions, the Hubbard model reads

$$H = -t \sum_{\langle ij \rangle} c^{\dagger}_{i,a} c_{j,a} + U \sum_{i} n_{i,\uparrow} n_{i,\downarrow},$$

where we have introduced the particle number operators $n_{i,a} = c^{\dagger}_{i,a} c_{i,a}$ and the diagonal hopping t_{ii} has been omitted. Despite its simplicity, the Hubbard model has been studied intensively because it captures the essential physical features of general strongly-correlated quantum systems.

Another elementary model that exhibits strongly-correlated quantum behaviour is the Heisenberg model [11], introduced to explain ferro- and antiferromagnetism in solids. It describes the behaviour of a collection of spins that are placed on fixed sites in a lattice, and is defined by the Hamiltonian

$$H = J \sum_{\langle ij \rangle} \vec{S}_i \cdot \vec{S}_j,$$

where $\vec{S}_i = (S_i^x, S_i^y, S_i^z)$ are the three spin operators. The most fundamental is the spin-1/2 model, but higher spin representations are also possible. In magnetic materials the Heisenberg interaction arises via (super-) exchange processes between effective spin degrees of freedom, where e.g. an antiferromagnetic interaction ($J > 0$) is obtained because anti-aligned spins can virtually hop to the same site and lower their kinetic energy [12]. The Heisenberg model can also be derived as an effective Hamiltonian for the Hubbard model when $U/t \gg 1$ and there is one electron per site (half-filling); indeed, in that case, the charge degrees of freedom are frozen and only the spins of the electrons remain to be taken into account.

2.1.3 Statics and Dynamics

Finding the lowest-lying eigenstates of a many-body Hamiltonian is not a goal in itself, but serves rather as a means to compute the physical, low-energy properties of the system. So what are the interesting low-energy properties of a quantum-many body system?[3]

First of all, we will be interested in thermodynamic properties such as the ground state energy density e or the average magnetization m (in e.g. the z direction)[4]

$$e = \frac{1}{|\mathcal{L}|}\langle\psi_{\mathrm{gs}}|H|\psi_{\mathrm{gs}}\rangle, \qquad m = \frac{1}{|\mathcal{L}|}\langle\psi_{\mathrm{gs}}|\sum_{i\in\mathcal{L}} S_i^z|\psi_{\mathrm{gs}}\rangle.$$

Different static susceptibilities quantify the change in these observables as a result of external perturbations such as a magnetic field or a change in temperature.

Also, we will be interested in correlation functions with respect to a given operator O

$$c(i, j) = \langle\psi_{\mathrm{gs}}|O_i^\dagger O_j|\psi_{\mathrm{gs}}\rangle,$$

quantifying the quantum correlations that are present in the ground state. If the operator O has a zero-temperature expectation value, it is rather the connected part of the correlation function that determines the non-trivial correlations, i.e.

$$c(i, j)_{\mathrm{con}} = \langle\psi_{\mathrm{gs}}|O_i^\dagger O_j|\psi_{\mathrm{gs}}\rangle - \langle\psi_{\mathrm{gs}}|O_i^\dagger|\psi_{\mathrm{gs}}\rangle\langle\psi_{\mathrm{gs}}|O_j|\psi_{\mathrm{gs}}\rangle.$$

In typical systems, these correlation functions decay exponentially in the distance between the operators; the correlation length ξ is defined in terms of this long-distance behaviour[5]

---

[3] The concepts in this section can be found in more detail in standard textbooks on quantum many-body systems [9, 10, 13–15].

[4] The number of lattice sites is denoted as $|\mathcal{L}|$, which, in the thermodynamic limit, is sent to infinity.

[5] We denote $\vec{n}_i$ as the lattice vector of site i.

$$c(i,j)_{\text{con}} \to \exp\left(-\frac{|\vec{n}_i - \vec{n}_j|}{\xi}\right), \qquad |\vec{n}_i - \vec{n}_j| \gg 1,$$

and can be interpreted as an effective length scale in the system. For a translation-invariant Hamiltonian in the thermodynamic limit it makes sense to look at the momentum representation of a correlation function, the so-called static structure factor

$$s(\vec{q}) = \sum_{i\in\mathcal{L}} \mathrm{e}^{-i\vec{q}\cdot\vec{n}_i} \langle \psi_{\text{gs}} | O_i^\dagger O_0 | \psi_{\text{gs}} \rangle,$$

for which δ-peaks signal the presence of long-range order in the ground state (the wave vector $\vec{q}$ determines the periodicity of the order).

More interesting are the dynamical properties of the system such as the dynamical correlation function

$$C(i,t) = \langle A_i(t) A_0(0) \rangle$$

with $A_i(t) = \mathrm{e}^{iHt} A_i \mathrm{e}^{-iHt}$ the operator A working at site i, time-evolved in the Heisenberg picture. Upon Fourier transforming this space-time correlation function, we obtain the frequency-momentum resolved dynamical correlation function or spectral function $S(\vec{q},\omega)$

$$S(\vec{q},\omega) = \sum_{i\in\mathcal{L}} \int_{-\infty}^{+\infty} \mathrm{d}t \; C(i,t) \mathrm{e}^{-i(\vec{q}\cdot\vec{n}_i - \omega t)}.$$

This spectral function[6] can be rewritten as

$$\begin{aligned} S(\vec{q},\omega) &= \sum_{i\in\mathcal{L}} \int_{-\infty}^{+\infty} \mathrm{d}t \; \mathrm{e}^{-i\vec{q}\cdot\vec{n}_i} \mathrm{e}^{i\omega t} \langle \psi_{\text{gs}} | \mathrm{e}^{iHt} A_i \mathrm{e}^{-iHt} \left(\sum_\alpha |\alpha\rangle\langle\alpha| \right) A_0(0) | \psi_{\text{gs}} \rangle \\ &= \sum_\alpha \left| \langle \alpha | A(\vec{q}) | \psi_{\text{gs}} \rangle \right|^2 2\pi\delta(\omega - E_\alpha) \end{aligned} \tag{2.1}$$

where $\sum_\alpha$ represents a sum over all excited states of the system with excitation energies E_α and $A(\vec{q}) = \sum_{i\in\mathcal{L}} \mathrm{e}^{i\vec{q}\cdot\vec{n}_i} A_i$. Typically, the operator A will only excite the first low-lying states, so that the spectral function essentially contains the momentum-frequency resolved information on the low-lying excitation spectrum of the system. A crucial observable is the gap Δ, the energy difference between the ground state and the first excited state, such that $S(\vec{q},\omega) = 0$ if $\omega < \Delta$. In the case of periodic boundary conditions and a translation-invariant Hamiltonian, all excited states can be

[6]Note that the static structure factor is regained from the dynamic correlations as

$$s(\vec{q}) = \int \mathrm{d}\omega \; S(\vec{q},\omega).$$

labeled by a momentum quantum number $\vec{p}$, defined as the phase of the eigenvalue of the translation operator $T_{\vec{e}}$

$$T_{\vec{e}}|\vec{p}\rangle = \mathrm{e}^{i\vec{p}\cdot\vec{e}}|\vec{p}\rangle.$$

Typically, a Hamiltonian has a few low-lying momentum-energy curves in its excitation spectrum; one such curve is often interpreted as a particle with a characteristic dispersion relation.[7]

Another set of observables is related to the sudden application of non-local perturbations to the system. Suppose we apply an external time- and space-dependent perturbation $h(i, t)$ that couples to the operator A at site i, such that the Hamiltonian changes abruptly as

$$H \to H - \sum_{i\in\mathcal{L}} h(i, t)A_i.$$

The shift away from the ground state value for a given observable O at site i is given by

$$\delta\langle O_i\rangle(t) = \sum_{j\in\mathcal{L}} \int_{-\infty}^{+\infty} \mathrm{d}t'\ \chi(\vec{n}_i - \vec{n}_j, t - t')h(j, t'),$$

where the dynamic susceptibility $\chi(\vec{n}, t)$ characterizes the response of the system. Again, it is most interesting to consider the momentum-frequency resolved version,

$$\chi(\vec{q}, \omega) = \sum_{i\in\mathcal{L}} \int_0^{\infty} \mathrm{d}t\ \mathrm{e}^{-i(\vec{q}\cdot\vec{n}_i - \omega t)}\chi(\vec{n}_i, t).$$

Interestingly, for $A = O$, the spectral function is related to the imaginary part of the susceptibility as ($\omega > 0$)

$$\mathrm{Im}\chi(\vec{q}, \omega) = \frac{1 - \mathrm{e}^{-\beta\omega}}{2} S(\vec{q}, \omega),$$

in the regime where linear-response theory is valid.

2.1.4 Quantum Phases and Phase Transitions

One of the unifying goals of quantum many-body systems is the identification of all possible quantum phases of matter. A quantum phase is loosely defined as a collection of Hamiltonians, and corresponding ground states, that are all adiabatically connected, i.e. can be transformed into each other by a smooth interpolation of the parameters of the Hamiltonian without any drastic (non-analytic) changes in e.g.

[7] We refer to Chap. 3 for a worked-out version of the particle interpretation of elementary excitations.

the ground state energy. All ground states within the same phase have the same qualitative features such as similar correlation functions.

In contrast to classical phase transitions, a quantum phase transition [15] takes place at $T = 0$, and is driven by the competition between different terms in the Hamiltonian. In the Landau paradigm of phase transitions [16], a phase is characterized by the way the symmetries of the model are reflected in the ground state subspace and can be characterized by a local order parameter. As a phase transition is approached, the (quantum) fluctuations in this order parameter will diverge, explaining the system's non-analytic behaviour. The corresponding divergence of the correlation length ξ implies that there is no characteristic length scale in the system, giving rise to scale invariance.

The paradigmatic example of a second-order quantum phase transition is provided by the one-dimensional transverse-field Ising model [17], a chain of spins-1/2 with a Hamiltonian

$$H = -\sum_{\langle ij \rangle} \sigma_i^x \sigma_j^x - \lambda \sum_i \sigma_i^z,$$

where $\sum_{\langle ij \rangle}$ denotes a sum over nearest-neighbour pairs. At $\lambda = 1$, the model exhibits a phase transition between a single polarized state ($\lambda > 1$) and a twofold degenerate, ordered (i.e. symmetry-broken) ground state ($\lambda < 1$); the order parameter is $\langle \sigma^x \rangle$. This is reflected in the excitation spectrum as a closing of the gap at the transition; the new ground state that emerges after the transition point is formed as an extensive number of excitations have collapsed onto the old ground state. These vague notions will be made more precise in the following chapters.

Whereas the classical picture of Landau captures many second-order quantum phase transitions, a range of quantum phases has recently been discovered that cannot be characterized by any local order parameter. Indeed, these topological phases are characterized by global order parameters, related to long-range quantum correlations in the ground-state subspace – see Sect. 2.3.

2.1.5 The Link with Experiment

Although the problems that we try to solve have a clear mathematical meaning without any connection to an external reality, it is interesting to note that these simple many-body models can be realized in real-life experiments. In fact, the field of quantum many-body physics is characterized by a fruitful interplay between theory and experiment, so that a short overview of the experimental work is appropriate.[8]

[8]In the following chapters we will also mention a few experimental realizations of the specific models that we will study.

First of all, we have previously identified the Hubbard model as a paradigmatic example for theoretical strongly-correlated quantum physics, but the model seems to have experimental relevance as well. Indeed, in transition metal oxides the electrons are mostly localized around the lattice ions, and are at the same time interacting strongly. In these systems, metal-insulator phase transitions [18] can be observed [19]. The physics of the Hubbard model has received increased attention as it might explain the origin of high-T_c superconductivity [20].

Quantum spin systems can be realized in a number of different compounds [21], but in order to display non-trivial quantum physics, these systems should have a reduced dimensionality and/or frustrated interactions; in this way, a non-classical many-body state can be a stable ground state at $T \to 0$. The thermodynamic properties such as the magnetic susceptibility and the specific heat can be measured, and using e.g. inelastic neutron scattering [22] the spectral function can be probed directly.[9]

Recently, elementary quantum many-body models were artificially realized in experiments with ultracold gases. After the experimental realization of Bose–Einstein condensation [24, 25] in 1995, a new research field of artificially creating strongly-correlated quantum many-body physics has emerged [26–28]. Famously, the quantum phase transition of a bosonic gas in a lattice between a superfluid and a Mott insulator, as proposed theoretically [29, 30], was observed in the lab [31], as well as the one-dimensional Tonks–Girardeau gas [32–34] or quantum spin models [35, 36]. Alternatively, the ideas of using trapped ions [37] to simulate quantum spin models [38], or photons [39] to study e.g. entanglement in spin systems [40] have been implemented.

The overarching theme and goal in these recent developments is the realization of a universal quantum simulator [41, 42], a device that can simulate the quantum behaviour of generic many-body systems [43]. Part of the motivation for this line of research is the development of a universal quantum computer that would realize an immense speed-up for certain computational tasks as compared to classical computers [44].

2.2 Entanglement in the Low-Energy Subspace

The fundamental ingredient that characterizes a quantum-mechanical treatment of a many-body system is entanglement; it is because spins are entangled in a quantum-mechanical ground state that they can have non-trivial correlations even at temperature zero. Moreover, it is through entanglement that phenomena such as quantum phase transitions [46, 200] and topological order can exist [47–49]. From this perspective, it is crucial to have a clear understanding of how entanglement is distributed in a physical system. Nonetheless, it is only recently that the study of entanglement – originally the subject of quantum information theory, where entanglement has been

[9] We refer to Ref. [23] for all details on magnetic materials.

studied as a resource for non-classical computation [44, 50] – has found its way in quantum many-body physics [51] and has led to a number of great advances in both theoretical understanding and numerical simulation.[10]

The first step is taken by quantifying entanglement in a many-body state. The idea is that we take a state of a large system, define the reduced density matrix ρ_A of a small region A by tracing over all the other spins in the lattice, and compute the Von Neumann entropy associated with this density matrix as

$$S(\rho_A) = -\mathrm{Tr}\left(\rho_A \log \rho_A\right).$$

This quantity expresses the amount of quantum correlations between region A and its environment. If the region is completely uncorrelated with its environment, the entanglement entropy is zero. For generic quantum states, where every spin is correlated with every other spin, the entanglement scales with the size of the considered region – as one would expect for an entropy measure.

For the low-energy states of physical systems, one expects that neither of these scenarios will be realized; the system's interactions will generate some quantum correlations but, because the interactions are local, entanglement will not spread out through the system over arbitrary distances. In fact, it has been observed that the entanglement entropy for low-energy states of local Hamiltonians typically scales as the perimeter of the considered region, instead of the volume. This *area law* of entanglement entropy [53] has led to the realization that the entanglement in physical states is distributed in a special way, and that a clear understanding of the entanglement structure is crucial for solving the quantum many-body problem.

In fact, many properties of ground states of local Hamiltonians have been rigorously proven. These proofs often start from Lieb–Robinson bounds [54], which state that the speed of information propagation in quantum spin systems is bounded by some characteristic velocity. These bounds play the role that strict causality plays in relativistic quantum field theory, and allow to translate results from relativistic quantum field theory to the setting of quantum lattice systems. One such result is that the ground state of a gapped local Hamiltonian has exponential decay of correlations [55], i.e.

$$|\langle O_A O_B \rangle - \langle O_A \rangle \langle O_B \rangle| < \mathrm{e}^{-d(A,B)/\xi}$$

where $d(A, B)$ is the distance between the operators O_A and O_B acting on separate regions A and B. This stands in contrast to critical models, where the ground state correlations generically decay algebraically with the distance.

Although the area law for the entanglement entropy is believed to be valid for all gapped systems, this has only been rigorously proven in one dimension. There it has been shown [56] that the entanglement entropy of a region is bounded – the area of a one-dimensional region is a constant – for ground states of gapped systems. For critical one-dimensional ground states, it scales logarithmically with the size of the subsystem (which is still very small compared to arbitrary quantum states), as

[10] The results mentioned in this section are laid out in exquisite detail in Ref. [52].

suggested by conformal field theory [57]. For higher-dimensional system, no rigorous results have been obtained with the same generality.

The proof of the area law for gapped one-dimensional systems confirmed the belief that their ground states can be approximated by matrix product states [56, 58], a class of states with a number of parameters that scales polynomially with the size of the system – see Sect. 2.6. Moreover, it is guaranteed that this approximation can also be found with a polynomial-time algorithm [59]. These results imply that the classical simulation of quantum many-body system is not impossible, but does require that entanglement is treated in an efficient way.

2.3 Low-Dimensional Quantum Physics

This work is focused on the low-energy physics of quantum systems of reduced dimensionality for two reasons: (i) the methods that we develop only work for this low-dimensional setting, and (ii) low-dimensional quantum matter can exhibit physical phenomena that are not observed for their three-dimensional counterparts. Whereas the former is of course not an argument for studying these systems, in this section we will argue for the latter and show why low-dimensional quantum physics is worth considering.

A first point in case is the fact that quantum correlations are stronger in lower-dimensional settings. The reason is the so-called monogamy property of entanglement [60, 61], implying that there is a limited resource of entanglement that a quantum degree of freedom can share with other ones. In particular, if a qubit A is maximally entangled with B, it can not be entangled with a third party C. In an extended quantum spin system, every spin will try to spread its limited amount of entanglement in an optimal way with all neighbouring spins, leading to strongly-correlated states. Now the number of neighbours clearly depends on the coordination number and dimensionality of the lattice; the entanglement will be strongest in low-dimensional lattices. On the other hand, in the limit for infinite-dimensional lattices where every spin has an infinite number of neighbours, the ground state of a translation-invariant model will have no entanglement at all and mean-field theory can be safely applied.

A second important result is the Mermin–Wagner theorem [62, 63], stating that one- and two-dimensional systems of statistical mechanics do not break continuous symmetries at finite temperature. The quantum version of this theorem was proven by Coleman [64], and states that one-dimensional quantum systems do not break continuous symmetries, even at zero temperature. The physical reason is that, if such a symmetry would be broken, the resulting Goldstone bosons would restore this symmetry due to the strong quantum fluctuations in one dimension. These results imply that low-dimensional systems have a stronger tendency to be disordered, i.e. no long-range order will develop because of strong quantum ($T = 0$) or thermal ($T > 0$) fluctuations.

The Heisenberg antiferromagnet serves as the best example to illustrate the effects of dimensionality. Classically (in a bipartite lattice, where classical frustration is absent), the ground state is the Néel state, where every spin is anti-aligned with all its nearest neighbours. This staggered magnetization survives on the two-dimensional square lattice at zero temperature, although it is highly suppressed by quantum fluctuations; any finite temperature immediately destroys long-range order [65, 66]. For the one-dimensional chain, the Néel order is completely destroyed by quantum fluctuations and the ground state is disordered [67, 68].

This picture can be understood from spin-wave theory [69, 70], a third way to illustrate the connection between dimensionality and quantum fluctuations. The idea is to perturb from the large-s limit where the spins become essentially classic, and look at (quantum) $1/s$ perturbations in terms of "spin waves". To this end the model is mapped to an effective boson model by the Holstein–Primakoff transformation [71]

$$S^+ = \left(\sqrt{2s - b^\dagger b}\right) b, \qquad S^- = b^\dagger \left(\sqrt{2s - b^\dagger b}\right), \qquad S^z = s - b^\dagger b.$$

The square-root factors can be expanded in $1/s$, such that the Heisenberg interaction ($J > 0$) is written as a bosonic Hamiltonian

$$H = -\frac{|\mathcal{L}|}{2} J s^2 z + J s z \sum_{\vec{k}} \left(b^\dagger_{\vec{k}} b_{\vec{k}} + \frac{\gamma_{\vec{k}}}{2} \left(b^\dagger_{\vec{k}} b^\dagger_{-\vec{k}} + b_{\vec{k}} b_{-\vec{k}} \right) \right) + \dots$$

with $\gamma_{\vec{k}} = \sum_{\langle ij \rangle} \mathrm{e}^{i\vec{k}\cdot(\vec{n}_i - \vec{n}_j)}$ and z the coordination number of the lattice; the dots represent higher order terms in $1/s$ such as quartic terms in the bosonic operators, representing spin wave interactions [72]. The quadratic part is readily solved by a Bogoliubov transformation, yielding the spin-wave spectrum

$$\omega(\vec{k}) = J s z \sqrt{1 - \gamma^2_{\vec{k}}}.$$

In the case of a cubic lattice in any dimension, the spectrum becomes gapless around the points $\vec{k} = 0$ and $\vec{k} = \vec{\pi}$ with a linear dispersion. The Bogoliubov transformation introduces Holstein–Primakoff bosons in the ground state, which reduces the zero-temperature Néel order of the antiferromagnet. In fact, the zero'th-order correction to the staggered magnetization upon the classical Néel ground state diverges in one dimension for every value of s, whereas, in two dimensions, it diverges at finite temperature. In correspondence with the Mermin–Wagner theorem, it is the long wavelength or low-energy ($\omega(\vec{k}) \to 0$) modes that destroy the long-range order.

2.3.1 One Dimension

In one dimension, quantum fluctuations and disorder are strongest, which makes them exhibit unexpected quantum phenomena. Let us review the most important results.[11]

[11] See also Ref. [73].

Interestingly, a number of analytic results are known for one-dimensional quantum systems. The first one is the Lieb–Schultz–Mattis theorem [74], stating that the spin-s Heisenberg chain is gapless if s is half-integer. This result was extended [75] to the claim that basically any reasonable local Hamiltonian for a half-integer spin chain either has a unique ground state with gapless excitations or degenerate symmetry-broken ground states.

The theorem does not make any predictions on the case of integer spin, and, in fact, it appears that the opposite is true. It was Haldane [76, 77] who first made the conjecture that generic Hamiltonians for integer-spin models have a unique ground state with a gap to the first excited state that survives in the thermodynamic limit. The conjecture was based on a mapping of the Heisenberg model onto the non-linear sigma model, which captures the long-wavelength behaviour of the Heisenberg model for large spin but remains qualitatively correct for small spin as well [78]. Although the conjecture came as a shock initially, soon a spin-1 chain model was introduced [79, 80] for which an excitation gap can be proven, and numerical simulations quickly settled the issue in favour of a gap for the spin-1 Heisenberg model as well [81].

A third result that has shaped the physics of one-dimensional quantum systems is the Luttinger-liquid concept [73, 82, 83], which provides a phenomenological description of the low-energy excitations of critical one-dimensional quantum liquids [84, 85] and spin chains [86]. It maps these systems onto the Tomonaga–Luttinger model [87, 88], a model for describing a gas of electrons with a linear dispersion, which can be exactly solved by bosonization even in the presence of density-density interactions. This universal mapping implies that any critical quantum system in one dimension has an excitation spectrum consisting of bosonic density waves with a linear spectrum, and different correlation functions have a universal form depending on essentially two parameters [73].

Finally, the presence of a number of non-trivial symmetries and conserved quantities has dramatic effects on the physics of the class of integrable [Sect. 2.5] and/or conformal models [73, 89]. Whereas the former give rise to the absence of thermalization or ballistic transport, a conformal symmetry has great implications on the correlations and excitations of critical systems.

These results have led to a unified understanding of the physics of one-dimensional quantum systems in terms of the elementary excitations. Still, there is an extremely rich variety in the properties of these excitations that can be traced back to a purely quantum-mechanical origin. One important point is that elementary excitations can have non-trivial topological properties, a feature that was observed e.g. in the spin-1/2 Heisenberg model [90] where the elementary excitations can only be created in pairs because of their topological nature. In addition, elementary excitations in one dimension often have fractionalized quantum numbers, which points again to their collective nature. A third characteristic is that the statistics of the excitations or quasiparticles is not very well-defined [91], as it is impossible to clearly isolate a particle's statistics from its interactions and topological properties. This has led to the realization that bosonic and fermionic descriptions often lead to complementary descriptions of the same physics, and that mappings between the two can give additional physical insights [92].

We refer to Chaps. 3 and 4 for a more detailed and idiosyncratic view on the physics of one-dimensional quantum systems, and, in particular, the properties of elementary excitations.

2.3.2 Two Dimensions

Whereas the Heisenberg antiferromagnet on the square lattice develops long-range order in the ground state and can be treated, at least qualitatively, using spin-wave theory, in the last years a range of two-dimensional models have been show to remain disordered at zero temperature. These so-called quantum spin liquids [93–95] exhibit massive amounts of entanglement and, therefore, cannot be connected to any classical phase of matter. The easiest way to construct a quantum spin liquid is in terms of valence bonds, i.e. pairs of spins that form maximally entangled singlet states. If all spins are partitioned in specific valence bonds – a so-called valence bond solid or crystal – the system is non-magnetic. It does break translational symmetry, however, and has no long-range entanglement, so these states cannot be true quantum spin liquids. Instead, we should allow for quantum fluctuations between different valence bond configurations, giving rise to a resonating valence-bond state [96, 97]. These states restore all symmetries and exhibit long-range entanglement. They (possibly) appear as ground states of quite simple model Hamiltonians such as the nearest-neighbour antiferromagnet on the kagomé lattice [98–100] or the next-nearest-neighbour antiferromagnet on the triangular lattice [101, 102]. Interestingly, quantum spin liquids typically support excitations with fractional quantum numbers, pictured as localized excitations with a non-local string running through the lattice. These strings correspond to local rearrangements of the spins and have no associated energy cost, and imply that the excitations can have non-local properties.

The second reason for studying two-dimensional quantum physics is the observation of phases of matter that are not detectable by any local order parameter, but are instead characterized by topological order [103, 104]. Although this concept is lacking a clear-cut definition, topologically ordered systems have a few characteristic properties [104] such as (i) a ground-state degeneracy depending on the topology of the surface on which it is defined [105], (ii) emergent quasiparticle excitations with anyonic statistics [106–109] (iii) gapless boundary excitations [110–112], and (iv) the presence of long-range entanglement in the ground state [47, 48]. Topological order is realized in fractional quantum Hall states, which appear as ground states of two-dimensional electrons under strong magnetic fields [113, 114], and has recently attracted a lot of attention as it might be used as a resource for quantum memories [115] and computations [116].

Gapped quantum spin liquids provide examples of topological order, so both phenomena are intimately connected; the unifying feature seems to be the non-trivial entanglement structure of two-dimensional quantum ground states. As we have seen, the study of entanglement in many-body physics is a relatively new topic, so that its theoretical understanding is all but complete at this point. In addition, numerical techniques for capturing these exotic quantum phases are less developed than in one dimension, so that a lot of interesting physics remains unexplored.

2.4 Numerical Techniques

One of the most attractive features of the domain of quantum many-body physics is the interplay between analytic techniques and numerical simulations, an interplay that is deemed essential to obtain physical insight into quantum materials. The numerical techniques that have dominated twentieth-century physics and that are based on mean-field techniques or independent particle models – think of Hartree–Fock [117–119] and density functional theory [120, 121] – are inaccurate for the purpose of simulating strongly-correlated systems. Moreover, since we are primarily interested in quantum phases that are not connected to some trivial (classical or free) limit, it seems that all standard forms of perturbation theory such as diagrammatic techniques [122] are excluded. In this section we briefly review some numerical algorithms that *can* capture strong quantum correlations in order to know what the methods of this dissertation are competing against.

The most obvious method for capturing the low-energy physics of simple models is by numerically diagonalizing the Hamiltonian on small system sizes, a method known as exact diagonalization [123, 124]. At low temperatures the interest is only in the low-energy states, such that only a small number of extremal eigenvalues and eigenvectors is wanted; in that case, iterative eigensolvers with fast convergence properties such as the Lanczos algorithm [125] can be applied. The major drawback of the method is the exponential scaling of the Hilbert space as a function of system size; for a system of N spins with d internal levels, the size of the Hilbert space is d^N. This complexity can be scaled down significantly by exploiting physical symmetries, but even then the present-day limit seems to be around 40-50 sites (in the case of a spin-1/2 model) [126].

Larger systems can be simulated using quantum Monte Carlo methods, based on the Monte Carlo technique [127] for simulating systems of classical statistical mechanics. As the original method is based on sampling a probability distribution, the challenge for quantum Monte Carlo is to map a quantum-mechanical problem to a representation with probabilistic interpretation. A popular approach is based on a stochastic series expansion [128], an expansion of the partition function in powers of inverse temperature and in "operator strings". In all quantum Monte Carlo methods, these mappings can lead to negative signs in the probability distributions, which can be formally circumvented, but lead to large errors for large system sizes and low temperatures. In special cases, this sign problem [129] can be eliminated, but for generic quantum systems with frustration or fermions, it appears to be an insurmountable obstacle for any quantum Monte Carlo approach.

A third set of methods comprises of different perturbative series expansions [130], which allow to, starting from a certain well-understood limit, trace how properties change under certain perturbations. A slightly different approach is a continuous unitary transformation [131–133], where a given Hamiltonian is continuously deformed into an effective model Hamiltonian for which the physical properties can be understood. As these methods are inherently perturbative, they are not well-suited to phase transitions or strongly-correlated phases that are not straightforwardly connected to a trivial limit.

Lastly, there is the variational approach to strongly-correlated quantum physics. In this approach, a variational wave function or ansatz is proposed for the ground state or low-lying excited states for a given model Hamiltonian, possibly with a number of variational parameters that can be numerically optimized. Famous examples of this approach include Hartree–Fock theory, for which the variational wave function is a Slater determinant [134, 135], the Gross–Pitaevskii equation for the interacting Bose gas [136, 137], the BCS theory of superconductivity [138, 139], the resonating valence bond state for two-dimensional antiferromagnets [96, 97], and the Laughlin wave function for the fractional quantum Hall effect [114]. Recently, the variational approach has been combined with quantum Monte Carlo techniques [140, 141], where a Gutzwiller-projected wave function is used to describe a quantum spin liquid [142]. In any case, the variational approach potentially works for every phase and has the advantage that it provides physical insight into the problem at hand, but, as it always involves an ansatz, the accuracy of the method is not completely under control.

2.5 Integrability and the Bethe Ansatz

In one-dimensional quantum many-body physics a special place is taken by the so-called integrable models [143–145]. These are discrete or continuous models for which the Hamiltonian has some special symmetry that allows to write down the exact ground state and excited states. The history goes back to Bethe's solution of the spin-1/2 Heisenberg chain [68], and later on the "Bethe ansatz" provided exact solutions of simple spin chain models such as the XXZ [146–148] and XYZ [149] chain, as well as the one-dimensional Hubbard model [8, 150] and the Lieb–Liniger gas [151, 152]. In this section we show how the coordinate Bethe ansatz works, and how it determines the low-energy spectrum. Afterwards, we explain how this spectrum is characterized and how the low-energy dynamics of an integrable model can be understood.

2.5.1 Solving the Heisenberg Model

Let us first follow the Bethe solution for the antiferromagnetic Heisenberg spin-1/2 chain, defined by the Hamiltonian

$$H = \sum_{n=1}^{N} \vec{S}_n \cdot \vec{S}_{n+1} = \sum_{n=1}^{N} \frac{1}{2}\left(S_n^+ S_{n+1}^- + S_n^- S_{n+1}^+\right) + S_n^z S_{n+1}^z$$

where we identify the spins 1 and $N+1$ (periodic boundary conditions). An obvious eigenstate is the spins-up state

$$|\Omega\rangle = |\ldots \uparrow\uparrow\uparrow \ldots\rangle$$

with an energy $E_\Omega = N/4$. Upon this reference state, we can now build magnon states, which are obtained by flipping one spin and making a momentum superposition

$$|1\rangle = \frac{1}{\sqrt{N}} \sum_{n=1}^{N} e^{ipn} S_n^- |\Omega\rangle,$$

with dispersion relation

$$E(p) - E_\Omega = \cos p - 1.$$

Adding a second magnon is less straightforward because of the magnon-magnon interactions; generally, an eigenstate in the two-magnon sector can be written down as

$$|2\rangle = \sum_{n_1 < n_2 = 1}^{N} a(n_1, n_2) S_{n_1}^- S_{n_2}^- |\Omega\rangle,$$

The magnons only interact when they are on neighbouring sites, so let us take the following wave function for the two-magnon state

$$a(n_1, n_2) = e^{ip_1 n_1} e^{ip_2 n_2} + A(p_1, p_2) e^{ip_2 n_1} e^{ip_1 n_2},$$

which obeys the eigenvalue equation for $n_2 > n_1 + 1$ with an energy

$$E(p_1, p_2) - E_\Omega = (\cos(p_1) - 1) + (\cos(p_2) - 1).$$

When the particles are next to each other ($n_2 = n_1 + 1$), the eigenvalue equation reduces to

$$2[E - E_\Omega] a(n_1, n_2) = 2a(n_1, n_2) - a(n_1 - 1, n_2) - a(n_1, n_2 + 1),$$

or the equation for the coefficient $A(p_1, p_2)$

$$A(p_1, p_2) = -e^{i\theta} = -\frac{e^{i(p_1+p_2)} + 1 - 2e^{ip_2}}{e^{i(p_1+p_2)} + 1 - 2e^{ip_1}}.$$

This is a pure phase factor capturing the interactions between the magnons.

At this point, the boundary conditions are imposed. For the one-magnon state, the trivial condition $a(n+N) = a(n)$ leads to the momentum quantization

$$p = \frac{2\pi j}{N}, \qquad j = 0, \ldots, N-1.$$

For the two-magnon state the condition is $a(n_2, n_1 + N) = a(n_1, n_2)$, which results in

$$\begin{cases} -e^{ip_1 N} e^{i\theta} = 1 \\ e^{ip_2 N} = -e^{i\theta} \end{cases} \quad \rightarrow \quad \begin{cases} Np_1 = 2\pi j_1 - \tilde{\theta}, & j_1 = 0, \ldots, N-1 \\ Np_2 = 2\pi j_2 + \tilde{\theta}, & j_2 = 0, \ldots, N-1 \end{cases}$$

with $\tilde{\theta} = \theta + \pi$. Through the non-trivial scattering phase $\theta(p_1, p_2)$, the magnon interactions determine the momentum quantization and, consequently, the energy of the two-magnon state.

Going to higher magnon numbers seems to be an intractable task, because a priori multi-particle scattering processes should be taken into account. It appears, however, that for this specific model, these processes always have zero amplitude and all magnon interactions can be decomposed into two-particle scattering processes, described by the two-magnon scattering phase shift $\theta(p_1, p_2)$. This implies that an M-magnon wave function can be written down explicitly as

$$|M\rangle = \sum_{n_1 < \cdots < n_M = 1}^{N} a(n_1, \ldots, n_M) S^-_{n_1} \ldots S^-_{n_M} |\Omega\rangle,$$

with the Bethe ansatz wave function for the coefficients

$$a(n_1, \ldots, n_M) = \sum_{\mathcal{P}} (-1)^{\mathcal{P}} \left(\prod_{\langle ij \rangle \in \mathcal{P}} e^{i\theta(\lambda_i, \lambda_j)} \right) e^{i\lambda_{\mathcal{P}(1)} n_1} \ldots e^{i\lambda_{\mathcal{P}(M)} n_M},$$

where the sum runs over all permutations of the M different λ's. Again imposing periodic boundary conditions leads to the Bethe equations

$$\lambda_j = \frac{2\pi j}{N} - \frac{1}{N} \sum_{k=1}^{M} \theta(\lambda_j, \lambda_k),$$

for a set of integers $\{j\}$. It can be checked that this M-magnon state is indeed an eigenstate with an energy equal to the sum of the one-magnon energies

$$E(\lambda_1, \ldots, \lambda_M) = \sum_{j=1}^{M} (\cos(\lambda_j) - 1).$$

The ground state of the antiferromagnetic chain is obtained as the state with $N/2$ magnons; the corresponding Bethe equations can be solved numerically.

2.5.2 Coordinate Bethe Ansatz: General Set-Up

Let us take a more general perspective now: we consider a gas of one-dimensional bosons in the continuum that interact locally. The bosons have a dispersion relation $\epsilon_0(p)$, which is defined as the energy of one boson with momentum p in an infinite system. The only information on the interactions that is needed is the two-body S matrix $S(p_1, p_2)$, which can be defined as the coefficient that pops up in the asymptotic part of the stationary two-body wave function,[12]

$$\Psi_{p_1,p_2}(x_1, x_2) \to \mathrm{e}^{ip_1x_1}\mathrm{e}^{ip_2x_2} + S(p_1, p_2)\mathrm{e}^{ip_2x_1}\mathrm{e}^{ip_1x_2}, \qquad x_1 \ll x_2.$$

Based on this information, the coordinate Bethe ansatz wave function for a collection of N particles can be written down as

$$\Psi_{\{\lambda_i\}}(x_1, \dots, x_N) = \sum_{\mathcal{P}} A(\mathcal{P})\mathrm{e}^{i(\lambda_{\mathcal{P}1}x_1+\cdots+\lambda_{\mathcal{P}N}x_N)}$$

where $\sum_{\mathcal{P}}$ is a sum over permutations of the λ's and $A(\mathcal{P})/A(\mathcal{P}') = S(\lambda_i, \lambda_j)$ if the permutations $\mathcal{P}$ and $\mathcal{P}'$ differ by the interchange of the momenta λ_i and λ_j. The wave function is determined by the set of N pseudo-momenta $\{\lambda_i\}$.

Imposing periodic boundary condition again leads to the Bethe equations

$$L\lambda_j + \sum_{k=1}^{N} \theta(\lambda_j, \lambda_k) = 2\pi n_j \qquad \forall j \tag{2.2}$$

where θ is the phase of the S-matrix, i.e. $S(\lambda_i, \lambda_j) = -\mathrm{e}^{i\theta(\lambda_i,\lambda_j)}$, and the n_j are (half) integers, depending on the number of particles being odd or even. Every solution of the Bethe equations is an eigenstate of our system and any eigenstate can be uniquely parametrized with a set of pseudo-momenta obeying Eq. (2.2). The total energy and momentum are given by

$$E_N = \sum_{j=1}^{N} \epsilon_0(\lambda_j) \qquad \text{and} \qquad P_N = \sum_{j=1}^{N} \lambda_j,$$

the sum of the bare energies and momenta. That the energy simply is the sum of the bare energies of all particles reflects the fact that all interactions have already been incorporated in the Bethe ansatz wave function. In fact, we do not even have to know how the particles interact or how the first-quantized Hamiltonian looks like; only

[12] This definition of the two-particle S matrix will be discussed in full detail in Chap. 3.

the one-particle dispersion and the two-particle S matrix are needed to extract all thermodynamic information about the system. The reason for this is of course that in an integrable system, the complete S matrix (three-, four-, ...particle scattering) can be determined from the two-particle S matrix.

We now take the thermodynamic limit, i.e. we take $L, N \to \infty$ but keeping the density $D = N/L$ finite. The pseudo-momenta will form a dense set, so we define a density of pseudo-momenta $\rho(\lambda)$ describing a Fermi-sea filled up to a certain level q, such that $\rho(\lambda) = 0$ for $|\lambda| > q$. The Bethe equations are transformed into a linear integral for $\rho(\lambda)$ [151, 152]

$$\rho(\lambda) - \frac{1}{2\pi} \int_{-q}^{q} K(\lambda, \mu)\rho(\mu)\mathrm{d}\mu = \frac{1}{2\pi} \tag{2.3}$$

where the kernel of the integral equation is given by the derivative of the scattering phase, i.e. $K(\lambda, \mu) = \partial_\lambda \theta(\lambda, \mu)$. The Fermi level q is not known in advance; it is determined by the density as

$$D = \int_{-q}^{q} \rho(\lambda)\mathrm{d}\lambda.$$

When the density is not fixed, we have to implicitely determine the Fermi level q through the function $\epsilon(\lambda)$, the dressed energy of a pseudo-particle. This function obeys

$$\epsilon(\lambda) - \frac{1}{2\pi} \int_{-q}^{q} K(\lambda, \mu)\epsilon(\mu)\mathrm{d}\mu = \epsilon_0(\lambda).$$

The Fermi level q is determined by imposing that $\epsilon(q) = 0$, which implies that we have to solve the system for the function $\epsilon(\lambda)$ and the value for q, given a certain $\epsilon_0(\lambda)$ self-consistently.

Elementary excitations are easily characterized in terms of the pseudo-particles of the Bethe ansatz. We can construct two types of excitations: either we take one particle with momentum $|\lambda| < q$ out of the Fermi sea (hole excitation) or we add one particle with momentum $|\lambda| > q$ (particle excitation). The energy of these excitations has two contributions: the bare energy $\epsilon_0(\lambda)$ and the dressing from all the other particles. These two contributions are contained within the function $\epsilon(\lambda)$ so the energy of an excitation is given by the above equation. Analogously, the physical momentum of an excitation has two contributions:

$$p(\lambda) = \pm\lambda \pm \int_{-q}^{q} \theta(\lambda, \mu)\rho(\mu)\mathrm{d}\mu \tag{2.4}$$

where the plus, resp. minus, sign is for a particle, resp. hole. Here it is assumed that λ is the bare momentum of the particle.

2.5.3 Thermodynamic Bethe Ansatz

At zero temperature, the coordinate Bethe ansatz describes an integrable system in its ground state by filling up a Fermi sea of quasi-momentum states; its excitations are holes and particles above this Fermi sea. When a finite temperature T is applied, these particles and holes will have finite distribution densities. By associating an entropy to these distributions and minimizing the free energy, one arrives at the Yang–Yang equation [153]

$$\epsilon(\lambda) = \epsilon_0(\lambda) - \frac{T}{2\pi} \int_{-\infty}^{+\infty} K(\lambda, \mu) \ln \left(1 + \mathrm{e}^{-\epsilon(\mu)/T}\right) \mathrm{d}\mu,$$

a non-linear integral equation for the dressed energy $\epsilon(\lambda)$ of the quasi momentum states; the equation can be solved by iteration [154]. The density of occupied vacancies $\rho(\lambda)$ is given by

$$\theta(\lambda) = \frac{\rho(\lambda)}{\rho_v(\lambda)} = \frac{1}{1 + \mathrm{e}^{\epsilon(\lambda)/T}}$$

with $\rho_v(\lambda)$ the density of all (occupied and empty) vacancies. Through this equation the density of occupied vacancies satisfies the integral equation

$$\rho(\lambda) = \frac{\theta(\lambda)}{2\pi} \left(1 + \int_{-\infty}^{+\infty} K(\lambda, \mu)\rho(\mu)\mathrm{d}\mu\right),$$

such that the total density can be calculated as

$$D = \int_{-\infty}^{+\infty} \rho(\lambda)\mathrm{d}\lambda.$$

2.5.4 Dynamics and Out-of-Equilibrium

If one wishes to extract the physical properties of integrable systems the construction of the ground state is only the first step. Due to the special symmetries of integrable systems, however, the low-energy spectrum has an equally special structure. Indeed, due to the particular conservation laws in integrable models, the elementary excitations can be treated as "particles" scattering off each other according to two-body S matrices, where all particle production processes are absent and particle number is conserved [155].

S Matrix and Faddeev–Zamolodchikov Algebra

We can describe this low-lying spectrum formally with the so-called Faddeev-Zamolodchikov operators [155–157], that create the exact elementary excitations out of the ground state $|\psi_{\text{gs}}\rangle$ of the interacting theory:

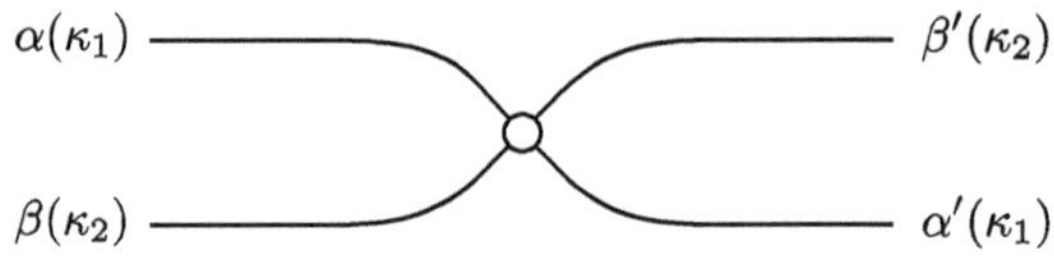

Fig. 2.1 Graphical representation of the two-particle S matrix $S_{\alpha\beta}^{\alpha'\beta'}(\kappa_1, \kappa_2)$. Momentum and energy conservation in one dimension imply that the outgoing momenta are equal to the incoming ones

$$|\kappa\rangle_\alpha = Z_\alpha^\dagger(\kappa)|\psi_{\text{gs}}\rangle$$

where $|\kappa_\alpha\rangle$ is an elementary excitation of type α with momentum κ. The S matrix $S_{\alpha\beta}^{\alpha'\beta'}(\kappa_1, \kappa_2)$ describes the scattering amplitude and phase for the process where two particles $\{\alpha, \beta\}$ with momenta κ_1 and κ_2 scatter into two, potentially different, particles $\{\alpha', \beta'\}$ (the individual momenta are preserved); it is visualized in Fig. 2.1. The S matrix determines the commutation relations of the Faddeev-Zamolodchikov operators

$$\begin{aligned}
Z_\alpha(\kappa_1)Z_\beta(\kappa_2) &= S_{\alpha\beta}^{\alpha'\beta'}(\kappa_1, \kappa_2)Z_{\beta'}(\kappa_2)Z_{\alpha'}(\kappa_1) \\
Z_\alpha^\dagger(\kappa_1)Z_\beta^\dagger(\kappa_2) &= S_{\alpha\beta}^{\alpha'\beta'}(\kappa_1, \kappa_2)Z_{\beta'}^\dagger(\kappa_2)Z_{\alpha'}^\dagger(\kappa_1) \\
Z_\alpha(\kappa_1)Z_\beta^\dagger(\kappa_2) &= 2\pi\delta(\kappa_1 - \kappa_2)\delta_{\alpha\beta} + S_{\beta\alpha'}^{\beta'\alpha}(\kappa_1, \kappa_2)Z_{\beta'}^\dagger(\kappa_2)Z_{\alpha'}(\kappa_1).
\end{aligned}$$

The hallmark of integrability now is the fact that all scattering processes can be decomposed into two-particle processes described by the two-particle S matrix $S_{\alpha\beta}^{\alpha'\beta'}(\kappa_1, \kappa_2)$. The three-body S matrix is decomposed into consecutive two-body S matrices in two ways [see Fig. 2.2]

$$\begin{aligned}
S_{\alpha\beta\gamma}^{\alpha'\beta'\gamma'}(\kappa_1, \kappa_2, \kappa_3) &= S_{\alpha\beta}^{ij}(\kappa_1, \kappa_2)S_{i\gamma}^{\alpha'k}(\kappa_1, \kappa_3)S_{jk}^{\beta'\gamma'}(\kappa_2, \kappa_3) \\
&= S_{\beta\gamma}^{jk}(\kappa_2, \kappa_3)S_{\alpha k}^{i\gamma'}(\kappa_1, \kappa_3)S_{ij}^{\alpha'\beta'}(\kappa_1, \kappa_2).
\end{aligned}$$

These two decompositions should be equal, which leads to a consistency condition on the two-body S matrix; this equation is called the Yang–Baxter equation [149, 158]. This elastic scattering implies that all low-lying excitations of the fully-interacting system can be pictured as many-particle states on top of the strongly-correlated ground state. Labelling these states by their particle numbers, all low-lying states are created by the Faddeev-Zamolodchikov operators as

$$|\kappa_1, \ldots, \kappa_m\rangle_{\alpha_1,\ldots,\alpha_m} = Z_{\alpha_1}^\dagger(\kappa_1)\ldots Z_{\alpha_m}^\dagger(\kappa_m)|\psi_{\text{gs}}\rangle.$$

Dynamical Correlations

This structure of the low-lying spectrum can be used to compute dynamical correlation functions or spectral functions of integrable systems. We have seen [Eq. (2.1)]

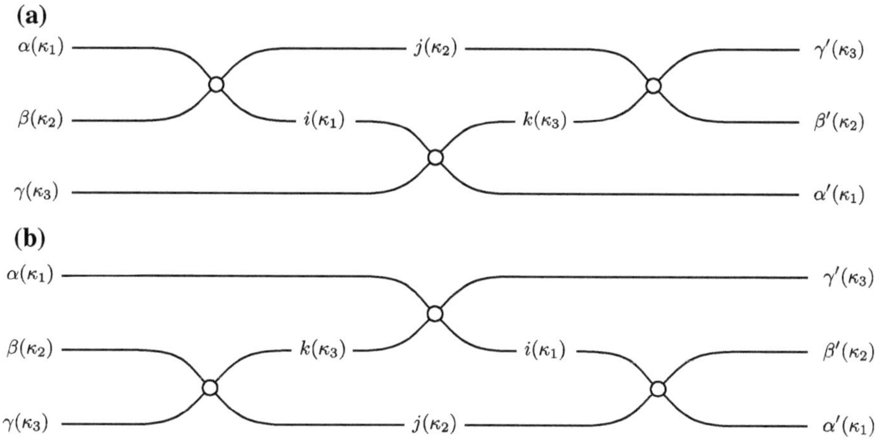

Fig. 2.2 Two ways in which the three-particle S matrix $S_{\alpha\beta\gamma}^{\alpha'\beta'\gamma'}(\kappa_1, \kappa_2, \kappa_3)$ can be decomposed in two-particle scattering processes

that the spectral function can be written as a sum of matrix elements of local operators in the eigenstate basis (form factors). The computation is, because of the unwieldy form of the Bethe ansatz wave functions, a rather tedious task, but a lot of progress has been made recently [159–162]. The success of this approach is explained by the fact that it is only the few-particle excitations (as created by the FZ operators) that contribute to the low-energy spectral functions. For a gapped system this is obvious: because every particle carries a minimum energy of Δ (the gap), an m-particle state has an energy larger than $m\Delta$. At low enough energies, only a few particles will contribute to the spectral function and it can be computed exactly by summing over all of them. Even for critical models, the spectral weight in higher-particle states seems to decrease rapidly.

For the case of gapped systems, this same approach can be applied to the computation of dynamical correlation functions at finite temperature [155, 163–168]. Indeed, the factor $e^{-n\Delta/T}$ provides a suppression factor for the contribution of n-particle states, which allows for an efficient summation of the form factors at low enough temperatures.

Real-Time Evolution

The absence of three-particle scattering has another interesting consequence in the behaviour of integrable systems out of equilibrium. Since the ground-breaking "quantum Newton's cradle" experiment, showing the absence of thermalization of a quasi one-dimensional many-body system [34], there has been a number of theoretical studies on the effect of integrability on thermalization processes. In particular, it has been shown that, instead of relaxing to a thermal Gibbs ensemble, integrable systems relax to a so-called generalized Gibbs ensemble [169–171]. The reason is that integrable systems have, in addition to the energy, an extensive number of local [143] or quasi-local [172, 173] conserved quantities $\{I_m\}$ that determine the reduced density matrix at infinite times as

$$\rho_{\text{GGE}} = \frac{1}{Z}\exp\left(-\sum_m \lambda_m I_m\right),$$

similar to the way in which ensembles are constructed in statistical mechanics [174, 175].

We can again make sense of this in terms of the particle structure of the low-lying excitations and the Faddeev-Zamolodchikov operators. Indeed, because they correspond to exact excitations of the fully-interacting Hamiltonian and they are preserved in all scattering processes, the particle numbers

$$N_\alpha(\kappa) = Z_\alpha^\dagger(\kappa) Z_\alpha(\kappa)$$

are integrals of motion, such that a representation of the generalized Gibbs ensemble can be written down in terms of them [176–178]. In the case of free theories, quenching from the one Hamiltonian to the other amounts to doing a Bogoliubov rotation in the space of the particle creation operators, such that the particle occupation numbers $n_\alpha(\kappa) = \langle N_\alpha(\kappa)\rangle$ can be computed exactly [179]. In the case of interacting integrable models, this approach seems to be harder to follow through [180].

Another approach for constructing the generalized Gibbs ensemble is through the so-called generalized thermodynamic Bethe ansatz [181–183]. In this approach, the thermodynamic Bethe ansatz is extended to account for the additional conserved quantities. Again, in the thermodynamic limit one can find a "saddle-point" root density $\rho(\lambda)$ that describes the generalized Gibbs ensemble. Additionally, in Ref. [181] it was shown that a representative eigenstate $|\Phi_s\rangle$ can be found (i.e. go over to the micro-canonical ensemble) and that the dynamics at large enough times are determined by the non-extensive "excitations" on top of this state – these excitations can have positive and negative excitation energies, since the reference state is not the ground state. This implies that, for quenched integrable models, one has an excitation spectrum (with infinitely long-lived excitations) on top of the generalized Gibbs ensemble that determines the dynamics at long times. More specifically, one can observe the light-cone dynamics in quenched integrable models as a result of this quasiparticle spectrum [184].

2.5.5 *Breaking Integrability*

Although the Bethe ansatz gives rise to a number of concepts that are applicable to non-integrable one-dimensional systems as well – a large portion of this dissertation is a case in point – the question remains how generic the physics of integrability is for one-dimensional quantum many-body systems.

A number of studies has been performed that track the influence of small non-integrable terms in the Hamiltonian on the thermalization process. Between Hamiltonians far away from integrability, which seem to thermalize rather quickly, and integrable ones, which don't thermalize at all, there seems to be a class of Hamiltonians with small non-integrable terms that lead to so-called prethermalization plateaux [179, 185–188]. These plateaux are characterized by a generalized Gibbs ensemble with the "almost" conserved quantities, and only after longer times do these systems relax to thermal Gibbs state.

2.6 Tensor Networks

In the last twenty-five years another set of methods was developed for simulating strongly-correlated quantum systems grouped under the name of "tensor network methods".[13] The unifying feature of these methods is their ability to capture the specific entanglement structure of low-energy states, and as such to provide an extremely accurate variational parametrization for describing and simulating quantum many-body physics. We have seen that quantum correlations in ground states of local Hamiltonians are distributed in a very special way, and the idea is to design a network of tensors that mimics this special distribution. In this way, tensor network states parametrize the "tiny corner of Hilbert space", where all the relevant physics is going on.

In order to see how a network of tensors can describe a many-body state, take a general state of N spins on an arbitrary lattice

$$|\Psi\rangle = \sum_{i_1,\ldots,i_N} c_{i_1,\ldots,i_N} |i_1, \ldots, i_N\rangle.$$

The coefficients $c_{i_1,\ldots,i_N}$ are complex numbers for every input of the indices; alternatively, they can be interpreted as defining a tensor with N indices. If we take the basis states $|i_1, \ldots, i_N\rangle$ as a given, we can say that the tensor describes the state. Graphically, the state can be depicted as

$|\Psi\rangle =$ 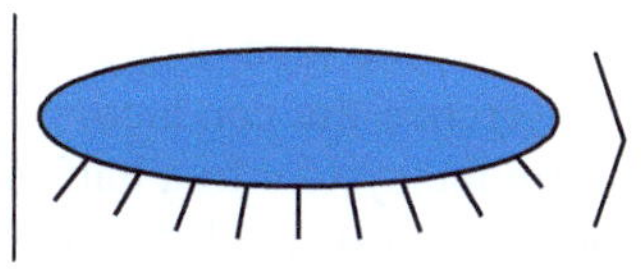 ,

where a geometric figure always represents a tensor with the legs corresponding to the tensor's indices. Now, a tensor network should be pictured as a decomposition of this one N-leg tensor as a contraction of a number of smaller tensors:

[13]This section should be more or less self-contained, but the reader is referred to review papers [189–191] for more details.

$$|\Psi\rangle_{\text{TNS}} = \Bigg|$$

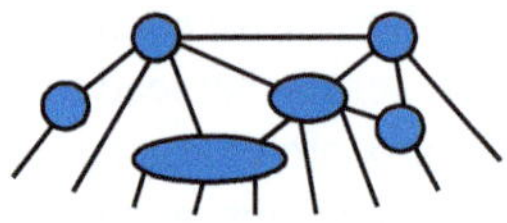

$$\Bigg\rangle .$$

In this expression, we have introduced the graphical notation for a tensor network contraction: (i) whenever two tensors are connected, the connected indices are identified and summed over, and (ii) unconnected legs remain as indices of the resulting new tensor (if all legs are connected the diagram corresponds to a scalar). These graphical representations will be very useful throughout this dissertation.

The most important feature of tensor network states is that the dimensions of the virtual indices in these contractions will not be very large, so that the number of parameters that describe these states is small. This implies that such a low-rank tensor-network decomposition of a quantum state is generally not possible, but, of course, it will prove to be the case that tensor network states exactly target the low-energy subspace of physical systems. For the rest of this section, we will explain why different tensor networks indeed have the right entanglement properties and, most importantly, how this opens up the possibility to simulate strongly-correlated quantum many-body physics.

2.6.1 The Density-Matrix Renormalization Group

The story of tensor network states takes off with a few seemingly unrelated breakthroughs in different fields of physics; the first (not in a chronological sense) is the invention of the density-matrix renormalization group by S. White in 1992.

The first numerical implementation of the renormalization group was done by K. Wilson in his seminal work [192] on the Kondo problem. The Kondo model [193–195] describes a magnetic impurity coupled to the conduction band of a nonmagnetic metal; the crucial question, unsolvable by perturbation theory, is the low-temperature behaviour of this impurity spin. The idea of the numerical renormalization group [196] by Wilson is to (i) discretize the conduction band to discrete energy levels logarithmically, (ii) transform the system to a half-infinite spin chain with the first spin representing the impurity, and (iii) solving this spin-chain system iteratively. Starting from the impurity spin in every iteration a new degree of freedom is added to the system and, in order to keep the size of the Hilbert space tractable, the number of states is truncated by only keeping the lowest-energy states of the Hamiltonian for the current part of the chain.

The application of this numerical renormalization method to strongly-correlated spin chains proved to work rather poorly. White and Noack [197] showed that the truncation procedure of only keeping the low-energy states of a certain region of a spin chain is inaccurate for a real-space version of the renormalization group. Instead, White proposed [198, 199] to embed this region in a larger environment and use the

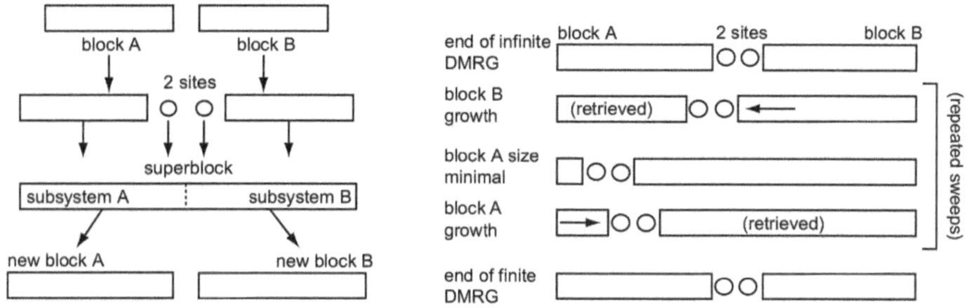

Fig. 2.3 Infinite-system (*left*) and finite-system (*right*) DMRG [201] (figure taken from Ref. [202]). In the infinite-system algorithm the system (block A) and the environment (block B) are grown by adding a site in every iteration. For the superblock the ground state $|\psi\rangle$ is found by a sparse eigensolver, after which a new reduced basis for block A is found by only keeping the m leading eigenvectors of the density matrix on block A, $\rho_A = \mathrm{Tr}_B(|\psi\rangle\langle\psi|)$; similarly for block B. The Hamiltonian and other observables are updated in every iteration by projecting on the kept subspace. The finite-system DMRG algorithm is applied to increase the accuracy: now the system size is kept to a fixed length L, and the growing of block A is at the cost of a shrinkage of block B and vice versa. The truncation procedure is only applied to the block that grows in size. If a boundary of the system is reached, the growing/shrinking is reversed. This algorithm converges to a good ground state after a number of "sweeps"

region's density matrix as the natural basis for truncating the number of states kept in every iteration. Suppose that the wave function of a block A and its environment E is given by ψ, and is approximated as $\tilde{\psi}$ by only keeping a basis of m states on the block A,

$$|\tilde{\psi}\rangle = \sum_{\alpha=1}^{m} \sum_{j=1}^{N_E} a_{\alpha j} |\alpha\rangle |j\rangle.$$

The error on the wave function $\epsilon = \| |\psi\rangle - |\tilde{\psi}\rangle \|$ is minimized by choosing the $|\alpha\rangle$ to be the eigenvectors corresponding to the m largest eigenvalues of the density matrix ρ_A, obtained by tracing out the environment degrees of freedom,

$$\rho_A = \mathrm{Tr}_E \left(|\psi\rangle\langle\psi| \right).$$

This truncation scheme explicitly maximizes the entanglement between the block A and its environment [45, 200]. This interpretation reveils the basic motive of the DMRG prodecure as shifting the renormalization flow towards the entanglement degrees of freedom, instead of the original energy/momentum rescaling of Wilson.

The success of DMRG was immediate; a first illustration of its accuracy was given in Ref. [203] where the ground-state energy of the spin-1 Heisenberg antiferromagnet was calculated at almost machine precision, whereas the Haldane gap [Sect. 2.3] was estimated with an accuracy of five digits. In the following years, DMRG has evolved in the standard for obtaining static properties of one-dimensional quantum lattice models [204, 205] – in Fig. 2.3 the general strategy of DMRG is presented schematically.

Yet, the success of DMRG has for a long time been something of a miracle; its power for simulating quantum spin chains has only been fully understood by identifying it as a variational method that optimizes over a specific class of tensor network states. Indeed, it was realized [206–208] that the finite-system DMRG algorithm leads to a ground state in the form of a *matrix product state*. Only some years later this development was taken seriously and the power of matrix product states investigated systematically. As a result, it was understood how DMRG efficiently captures the entanglement distribution in one-dimensional quantum lattice models and in what ways DMRG can be extended in order to capture e.g. dynamical and finite-temperature behaviour [202, 209].

2.6.2 Finitely Correlated States

A few years before the conception of DMRG, however, matrix product states had already appeared under the name of valence bond solids or finitely correlated states.

The former were introduced by Affleck, Kennedy Lieb and Tasaki (AKLT) in 1987 [79, 80] as a class of translation-invariant quantum states with exponential decay of correlations. They generalized the work of Majumdar and Ghosh [210–212], who showed that the two states with purely nearest-neighbour valence bonds are the exact ground states of the spin-1/2 antiferromagnet with equal nearest-neighbour and next-nearest-neighbour coupling. From such a dimerized state, a translation-invariant state is obtained by projecting spin-1/2 pairs onto the symmetric spin-1 subspace. The AKLT state on the spin-1 chain that is thus obtained has no broken symmetries and is the unique ground state of the simple Hamiltonian

$$H = \sum_i \left(\frac{1}{2}(\vec{S}_i \cdot \vec{S}_{i+1}) + \frac{1}{6}(\vec{S}_i \cdot \vec{S}_{i+1})^2 + \frac{1}{3} \right),$$

where every term is nothing more than a projector onto the spin-2 subspace of every nearest-neighbour pair of spins. In Ref. [80] it was also proven that this Hamiltonian has a finite excitation gap, a result that showed for the first time the existence of a gapped spin-1 antiferromagnet without broken symmetries in correspondence with Haldane's conjecture [Sect. 2.3]. The AKLT construction can be extended to higher spin models and higher dimensions, but the provability of the gap, although highly expected in some cases [see Sect. 5.6], is lost.

The one-dimensional AKLT state was later generalized [213] to the so-called finitely correlated states, a class of translation-invariant states for quantum spin chains in the thermodynamic limit. These states have, under some conditions, exponential decay of correlations and they are the ground states of local Hamiltonians with a finite gap.

In retrospect, finitely correlated states can be identified as uniform matrix product states in the thermodynamic limit, a class of states that will be extensively worked with in this dissertation. Although the use of finitely correlated states as variational trial states is already mentioned by the authors of Ref. [213], the connection with DMRG was, unfortunately, only noticed many years later. Also, it took another number of breakthroughs until the class of uniform matrix product states was used directly as a variational class for simulating spin chains in the thermodynamic limit directly.

2.6.3 Matrix Product States

Matrix product states (MPS) have appeared as subject of extensive research only in the years 2003–04, by the combination of ideas in quantum information theory and many-body physics. There are a few complementary ways of arriving at an MPS, all of which we will review in this section.

Construction of MPS

Consider a quantum state on a spin chain with N sites

$$|\Psi\rangle = \sum_{i_1,\dots,i_N} c_{i_1,\dots,i_N} |i_1,\dots,i_N\rangle.$$

The number of coefficients $c_{i_1,\dots,i_N}$ scales exponentially in the number of sites and provides a complete yet awkward parametrization of the state. The challenge is to find a representation that has a more local structure. Consider thereto a partition of the spin chain into two blocks A and B, with a corresponding Schmidt decomposition

$$|\Psi\rangle = \sum_{\alpha=1}^{D} \lambda_\alpha |\Phi_\alpha^A\rangle \otimes |\Phi_\alpha^B\rangle \tag{2.5}$$

where the Schmidt vectors $|\Phi_\alpha^{A,B}\rangle$ are orthonormal and the Schmidt numbers λ_α are arranged in decreasing order. The reduced density matrices on the two blocks are given by

$$\rho_{A/B} = \sum_{\alpha=1}^{D} \lambda_\alpha^2 |\Phi_\alpha^{A/B}\rangle\langle\Phi_\alpha^{A/B}|$$

such that the entanglement entropy between blocks A and B is given by

$$\begin{aligned} S_{A|B} &= -\mathrm{Tr}\left(\rho_A \log \rho_A\right) = -\mathrm{Tr}\left(\rho_B \log \rho_B\right) \\ &= -\sum_{\alpha=1}^{D} \lambda_\alpha^2 \log \lambda_\alpha^2. \end{aligned}$$

The advantage of the Schmidt decomposition is that the entanglement properties of low-energy states can be exploited: if the entanglement entropy of the region A is bounded, it is expected that the Schmidt spectrum falls off quickly and that the sum in Eq. (2.5) can be truncated efficiently. Indeed, an approximate wave function $|\tilde{\Psi}\rangle$ is obtained by only including the $D' < D$ leading Schmidt vectors, giving rise to a truncation error of

$$\| |\Psi\rangle - |\tilde{\Psi}\rangle \|^2 = \sum_{\alpha=D'+1}^{D} \lambda_\alpha^2. \tag{2.6}$$

Let us now use this Schmidt decomposition successively for all partitions of the spin chain [214, 215]. Isolating the first spin results in

$$|\Psi\rangle = \sum_{\alpha_1} \lambda_{\alpha_1} |\Phi_{\alpha_1}^{[1]}\rangle \otimes |\Phi_{\alpha_1}^{[2\ldots N]}\rangle = \sum_{i_1,\alpha_1} \Gamma_{\alpha_1}^{[1],i_1} \lambda_{\alpha_1} |i_1\rangle \otimes |\Phi_{\alpha_1}^{[2\ldots N]}\rangle,$$

where in the second step $|\Phi_{\alpha_1}^{[1]}\rangle$ was written out in the local spin basis $|i_1\rangle$. Next we make a partition after the second spin and apply the same procedure,

$$|\Psi\rangle = \sum_{i_1,\alpha_1} \sum_{i_2,\alpha_2} \Gamma_{\alpha_1}^{[1],i_1} \lambda_{\alpha_1} \Gamma_{\alpha_1,\alpha_2}^{[2],i_2} \lambda_{\alpha_2} |i_1, i_2\rangle \otimes |\Phi_{\alpha_2}^{[3\ldots N]}\rangle.$$

By iterating, the state is rewritten as

$$\begin{aligned} |\Psi\rangle &= \sum_{i_1,\ldots,i_N} \sum_{\alpha_1,\ldots,\alpha_{N-1}} \Gamma_{\alpha_1}^{[1],i_1} \lambda_{\alpha_1} \Gamma_{\alpha_1,\alpha_2}^{[2],i_2} \lambda_{\alpha_2} \ldots \Gamma_{\alpha_{N-2},\alpha_{N-1}}^{[N-1],i_{N-1}} \lambda_{\alpha_{N-1}} \Gamma_{\alpha_{N-1}}^{[N],i_N} |i_1, \ldots, i_N\rangle \\ &= \sum_{i_1,\ldots,i_N} \Gamma^{i_1}[1]\Lambda[1]\Gamma^{i_2}[2]\Lambda[2]\ldots\Gamma^{i_{N-1}}[N-1]\Lambda[N-1]\Gamma^{i_N}[N] |i_1, \ldots, i_N\rangle \end{aligned} \tag{2.7}$$

where the simplified notation was obtained by identifying the sums over indices as matrix multiplications and introducing the diagonal matrices $\Lambda[n]$. The matrices in this decomposition now have a physical interpretation: the Γ's describe some local description of the spin degrees of freedom, whereas the Λ's contain the Schmidt spectrum corresponding to a bipartition of the chain. An exact representation of any quantum state $|\Psi\rangle$ would require the size of these matrices to scale as $\mathcal{O}(\exp N)$, but for states with a bound on the bipartite entanglement entropy the matrices can be safely truncated with an error that is given by Eq. 2.6; the maximal size of the matrices is called the bond dimension D. We can now represent this *matrix product state* pictorially as

$|\Psi\rangle =$ 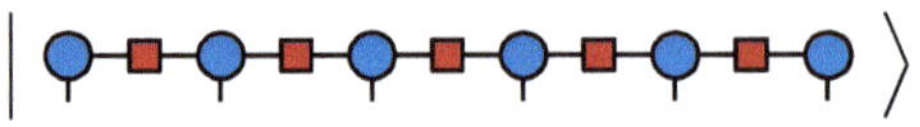, where every blue circle represents a Γ tensor – the index i is represented as the middle "physical" leg – and every red square is a Λ matrix.

Another way of arriving at a matrix product state is in the spirit of the AKLT construction above. In a so-called valence bond solid [216] two ancilla degrees of freedom $|\alpha\rangle_{a,i}$ and $|\alpha\rangle_{b,i}$ are attached to every physical site i, and ancillas from neighbouring sites are brought into a maximally entangled state

$$|\phi\rangle_{i,i+1} = \sum_{\gamma} |\gamma\rangle_{b,i} \otimes |\gamma\rangle_{a,i+1}.$$

A physical state $|\Psi\rangle$ is obtained by projecting the two ancillas onto a physical spin at every site i with the projection operator

$$P_i = \sum_{s=1}^{d} \sum_{\alpha,\beta} A^{[i],s}_{\alpha,\beta} |s\rangle \left(\langle\alpha|_{a,i} \otimes \langle\beta|_{b,i} \right).$$

If the system has open boundary conditions, two ancillas remain unpaired. The advantage of this construction, however, is that the case of periodic boundary conditions is easily considered by placing these two unpaired ancillas in an entangled state as well [217]. The state that is obtained in this way is given by

$$\begin{aligned} |\Psi\rangle &= \sum_{i_1,\dots,i_N} \sum_{\gamma_1,\dots,\gamma_N} A^{[1],i_1}_{\gamma_N,\gamma_1} A^{[2],i_2}_{\gamma_1,\gamma_2} \cdots A^{[N-1],i_{N-1}}_{\gamma_{N-2},\gamma_{N-1}} A^{[N],i_N}_{\gamma_{N-1},\gamma_N} |i_1,\dots,i_N\rangle \\ &= \sum_{i_1,\dots,i_N} \mathrm{Tr}\left(A^{[1],i_1} A^{[2],i_2} \cdots A^{[N-1],i_{N-1}} A^{[N],i_N} \right) |i_1,\dots,i_N\rangle \end{aligned}$$

where in the second line we have again introduced a matrix-product notation. Note that the projection transforms a product state of valence bonds into a state with finite-range entanglement. In fact, this form is equivalent to the state in Eq. 2.7 up to the boundary conditions, so that we have again obtained an MPS, now represented as

$|\Psi\rangle =$ 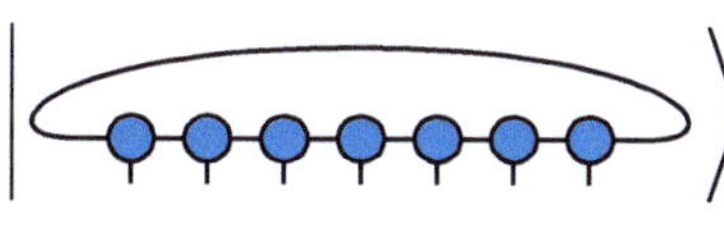 ,

with every blue circle an MPS tensor A, again with the physical index i in the middle.

A third interpretation of the class of MPS is in terms of the holographic principle. An MPS can be thought of as being generated by a D-level auxiliary system that interacts sequentially with the different spins on the lattice [218–220]. The spatial correlations of the state that is generated is directly related to the temporal correlations of the auxiliary system. This shows that an MPS can be described by the dissipative dynamics of a lower-dimensional auxiliary system, providing a holographic interpretation that can be extended to higher-dimensional settings as well.

Properties of MPS

The viability of MPS as a variational class was clear from the outset because of the success of DMRG, but it proved that investigating the properties of MPS in a more formal fashion would open up a whole new range of applications.

For our purposes, it suffices to consider the set of translation-invariant MPS on a chain of N sites with periodic boundary conditions

$$|\Psi(A)\rangle = \sum_{i_1,\ldots,i_N} \mathrm{Tr}\left(A^{i_1}A^{i_2}\cdots A^{i_N}\right)|i_1,\ldots,i_N\rangle. \tag{2.8}$$

In this form, the same $D \times d \times D$ tensor A is repeated on every site, so it is translation invariant by construction; it can be shown that every translation-invariant MPS can be brought into this form [221]. Closely connected to this MPS is the completely positive map $\mathcal{E}$ acting on the space of $D \times D$ matrices as

$$\mathcal{E}(X) = \sum_i A^i X (A^i)^\dagger.$$

In the generic[14] case [221] this map has a leading eigenvalue equal to one, and all other eigenvalues within the unit circle. The leading eigenvector or fixed point of $\mathcal{E}$ is guaranteed [222] to be a positive matrix Λ.

Now it is clear that there is a redundancy in the parametrization of the state $|\Psi(A)\rangle$ in terms of the tensor A because the gauge transformation

$$A^i \to X^{-1}A^iX \qquad \forall i$$

leaves the state invariant. This gauge freedom leaves the possibility to define a normal or canonical form [221] for the tensor A by imposing that the fixed point of the cp map is the unit matrix, i.e.

$$\sum_i A^i (A^i)^\dagger = \mathbb{1}.$$

Now the fundamental theorem of MPS states that this canonical form is unique up to an additional unitary gauge transformation for any generic MPS [221, 223]. This implies that if two MPS $|\Psi(A_1)\rangle$ and $|\Psi(A_2)\rangle$ are equal there should exist a gauge transform that links the two MPS tensors A_1 and A_2.

From the form in Eq. (2.7) it is clear that the bipartite entanglement entropy of an MPS is bounded by $\log D$ and, if the bond dimension is kept finite, obeys the area law of entanglement entropy. Therefore, it seems plausible that an MPS can

[14]The generic case excludes the cases where the MPS can be written as (i) the superposition of multiple translation-invariant MPSs with smaller bond dimension or (ii) the superposition of p p-periodic states each of which can be written as an MPS. The condition under which an MPS is generic is related to the injectivity of the MPS, which means that by concatenating enough A tensors the map from the virtual space to the physical one becomes injective, and to the fact that there is only eigenvalue of $\mathcal{E}$ on the unit circle [221].

approximate any state with an area law, such as ground states of local Hamiltonians [58, 224]. This statement was made rigorous in Ref. [56] where it was indeed proven that ground states of local gapped Hamiltonians can be approximated by MPS. This implies that the class of matrix product states provides a dense parametrization of all gapped phases in one dimension. For ground states of critical Hamiltonians, for which the entanglement entropy of a block scales logarithmically with the size, an MPS approximation needs a bond dimension that scales polynomially in the system size [225].

This description of all gapped ground states in one dimension can be lifted to the level of Hamiltonians. As we have seen, matrix product states are generalizations of the valence bond solids, which were proven to be the unique ground states of local Hamiltonians. It appears that this property is general: any MPS is the unique ground state of a local gapped Hamiltonian, called the parent Hamiltonian [221]. Since we know that the ground state of a given Hamiltonian can be well approximated by an MPS, this allows to replace this original Hamiltonian by the parent Hamiltonian of the MPS. This replacing can be done without closing the gap, so that we can represent all gapped Hamiltonians within a certain phase by a well-chosen parent Hamiltonian. In this way, the classification of all gapped quantum phases can be done by tracing paths of MPS and their parent Hamiltonians. If one imposes that certain physical symmetries have to be preserved on these paths – these physical symmetries are reflected in the MPS language as projective representations on the MPS's virtual degrees of freedom [226] – it appears that there are so-called symmetry-protected topological phases that do not connect to a trivial phase without breaking these symmetries [227–229].

Implicit in all these considerations is the assumption that the number of sites N should be large, such that the MPS describes the bulk physics of a spin chain. In the original DMRG algorithm numerical resources scale linearly in the system size, but it is clear that for the translation-invariant form of Eq. (2.8) the limit to infinitely large systems can easily be taken [213, 230–232].

2.6.4 *Simulating Spin Chains with MPS*

The success of MPS for simulating spin chains is mostly based on the fact that expectation values and correlation functions of an MPS can be computed in an efficient way.[15] Importantly, an MPS always has exponential decay of correlations, which shows again that they are not the right variational class to simulate critical models. In this section, we review the different MPS algorithms for computing the properties of spin chains.

[15]We refer to the reviews [189, 191, 202] for a detailed account.

Approximating Ground States

We have seen that the ground state of every (gapped) local Hamiltonian H can be represented as an MPS $|\Psi(\{A_i\})\rangle$, but the challenge remains of finding this optimal state efficiently within the MPS manifold. According to the variational principle, this problem of optimizing the variational parameters $\{A_i\}$ can be written as a minimization problem for the energy functional,

$$\min_{\{A_i\}} \frac{\langle\Psi(\{A_i\})|H|\Psi(\{A_i\})\rangle}{\langle\Psi(\{A_i\})|\Psi(\{A_i\})\rangle}.$$

Finding the optimal set of tensors $\{A_i\}$ is a highly non-linear optimization problem, and is infeasible to solve in one stretch. Luckily, there is a way out of this by realizing that the energy functional is quadratic for every tensor A_i separately, so that the optimization for one tensor can be easily solved by a generalized eigenvalue problem. An efficient algorithm is then obtained by sweeping back and forth through the chain and optimizing every tensor separately at each iteration. Moreover, by fixing the gauge of the MPS tensors in a smart way, the normalization for each tensor can be made trivial, such that the on-site optimization reduces to an ordinary eigenvalue problem. This algorithm is variational in the sense that the energy is lowered at every step in the algorithm. In fact, it appears that this algorithm is almost identical to the one-site version of DMRG [202, 233], a connection that also explains the success of standard DMRG implementations.

For translation-invariant MPS in the thermodynamic limit, where the MPS $|\Psi(A)\rangle$ is parametrized by one single tensor A, this DMRG procedure obviously no longer works. There are essentially three different methods for optimizing infinite MPS:

- Infinite DMRG. This algorithm is a translation of the original infinite-system DMRG [198, 199] into the MPS language. The central idea is the growing of the system by iteratively inserting new sites in the middle and variationally optimizing the MPS tensors on the newly inserted sites. This procedure is repeated until the state has converged in the middle to some description of the bulk physics of the system. The MPS reformulation has led to a tremendous improvement on the original DMRG algorithm through a so-called state prediction, which provides an initial guess for the optimization of the inserted tensors based on the previous iteration [234], see also [235].
- Infinite time-evolving block decimation. This method finds the ground state by evolving a well-chosen initial state with the imaginary-time evolution operator $e^{-H\tau}$; in the limit $\tau \to \infty$ this results in a projection on the ground state. The time evolution is implemented with a Trotter–Suzuki decomposition and, after each time step, truncating the MPS to a certain bond dimension [230]. This truncation is done by considering the truncation of one bond in the MPS and substituting these locally optimal MPS tensors throughout the infinite chain – a procedure that is not guaranteed to be optimal. Because imaginary-time evolution is non-unitary, the MPS should in principle be orthonormalized in every iteration [231].

- Tangent-space methods. In this set of methods the manifold structure of the variational class of MPS is taken seriously, and implementations typically reduce to operations with vectors in the tangent space on the manifold [236]. In particular, imaginary time-evolution is performed in the MPS manifold by projecting the time evolution onto the tangent space, according to the time-dependent variational principle [232]. Alternatively, one could implement a conjugate-gradient optimization within the MPS manifold, where, in every iteration of the optimization, the gradient of the energy functional is a vector in the tangent space [237]; more advanced optimization strategies can take into account the non-trivial geometry of the MPS manifold [238]. Tangent-space methods have the advantage of relying on globally optimal routines only, but often at a higher numerical cost.

Dynamical Properties

Ground-state algorithms can be straightforwardly extended to also capture excited states. In the case where the ground state and the first excitations of a system live in different sectors characterized by different quantum numbers, then an energy minimization in the right sector will lead to these excited states. If the sectors are the same, the excitations have to be computed iteratively, where one particular excitation is targeted by, throughout the algorithm, orthonormalizing with respect to all lower-lying states that were found in a previous run. This implies that a simple adaptation of the DMRG algorithm allows to compute e.g. the gap of generic spin chains.

Typically, however, real interest lies in capturing excitations with well-defined momentum such that dispersion relations can be computed. Therefore, systems with periodic boundary conditions should be treated, which are a lot harder to simulate with DMRG [217, 239], especially if a translation-invariant ground state is wanted [240]. Although the targeting of excitations within a particular momentum sector has been implemented in Ref. [241], the scope of dynamical DMRG simulations has shifted to other approaches.

One of the main interests is the simulation of quantum quenches, i.e. the time evolution of a system after a sudden change of a parameter in the Hamiltonian. All standard approaches for simulating real-time evolution rely on the Trotter–Suzuki decomposition of the evolution operator e^{-iHt}, often in the form of a matrix product operator [242], and subsequent truncations of the bond dimension [202, 214, 215, 243–245]. Other approaches are based on the time-dependent variational principle [232, 246]. The time evolution after a local perturbation on a translation-invariant background can be simulated by introducing a finite window that is grown as time increases [247–249].

Equally important is the dynamical spin structure factor or spectral function

$$S(q,\omega) = \sum_n \mathrm{e}^{-iqn} \int_{-\infty}^{+\infty} \mathrm{e}^{i\omega t} \, \langle 0| S_n^\alpha(t) S_0^\alpha(0) |0\rangle,$$

and different schemes have been developed within the DMRG framework for its evaluation:

- Lanczos algorithms. The first attempt was made by Hallberg [250] by using a continued-fraction expansion based on a Lanczos algorithm. The central idea in this approach is the construction of a number of Lanczos vectors by a multitarget DMRG, where the DMRG basis is optimized for the ground state and Lanczos states simultaneously. This method was improved recently [251] by calculating the Lanczos vectors adaptively, i.e., only the three last states are kept in the recursion. This method was recently much improved by implementing it in the MPS/MPO language [252].
- Correction vector method. The spectral function at a certain energy can be computed as the spectral weight of a so-called correction vector, which can be computed with conventional DMRG methods [253–255] or variational MPS methods [256]. This method is quite expensive as this correction vector has to be computed for every ω separately, involving an operator inversion problem.
- Real-time evolution. The time-domain correlator can be computed using real-time evolution algorithms, so that a Fourier transform gives direct access to the spectral function. The time-dependence can be further interpolated using linear prediction techniques [257–261]. This method is limited by the growth of entanglement during real-time evolution.
- Chebyshev expansion. Spectral functions can be represented approximately by a Chebyshev expansion, which can be computed efficiently with MPS methods [262].

Although these methods differ a lot in implementation and efficiency, they share the same basic strategy: one starts from the action of a local operator on the ground state and captures different operations on this state within the space of MPS. The action of this local operator, however, generates not only the low-lying physics that one is interested in, but also excites higher-energy eigenstates of the system. These higher-energy modes are only eliminated at a final stage, when one looks only at the low-energy part of the spectral function.

From the perspective that we take in this dissertation, this is a clear disadvantage: it is a lot more natural to search for an efficient parametrization of the low-energy subspace first, such that the action of the local operator can be projected on this low-energy space from the start. This approach has the additional advantage that one can target non-extensive properties on a ground state in the thermodynamic limit, which is impossible if one stays within the original space of matrix product states. Instead, we will show in this dissertation that the low-lying dynamics above an MPS ground state are captured within the tangent space, rather than the original manifold itself – see Sect. 3.5.

Finally, it should be noted that also finite-temperature simulations have come within the reach of MPS methods [245, 263, 264], which led to algorithms for computing spectral properties [259] and real-time evolution [265].

2.6.5 Projected Entangled-Pair States

In principle any quantum state can be decomposed as a matrix product state, irrespective of the dimensionality or geometry of the lattice on which it is defined – as long as the bond dimension is chosen large enough. This implies that DMRG and other MPS methods can also be applied to e.g. two-dimensional quantum spin systems. Yet, despite some success [98, 266, 267], it should be obvious that the class of MPS is not a correct variational state for capturing ground states of two-dimensional systems, because MPS do not have the correct entanglement structure for describing a two-dimensional ground state.

A more natural choice is the class of projected entangled-pair states (PEPS), which are easily obtained by generalizing the valence-bond construction to two- and higher-dimensional geometries [216, 268]. For the two-dimensional square lattice, we put four ancilla degrees of freedom on every physical site, put neighbouring ancillas into maximally entangled states and project them onto the physical spins. A PEPS is thus represented as

$$|\Psi\rangle = \sum_{i_{11}, i_{12}, \ldots} C_2(A^{i_{11}} A^{i_{12}} \cdots) |i_{11}, i_{12}, \ldots\rangle$$

where a part of the contraction can be represented as

$$C_2(A^{i_{11}} A^{i_{12}} \cdots) =$$

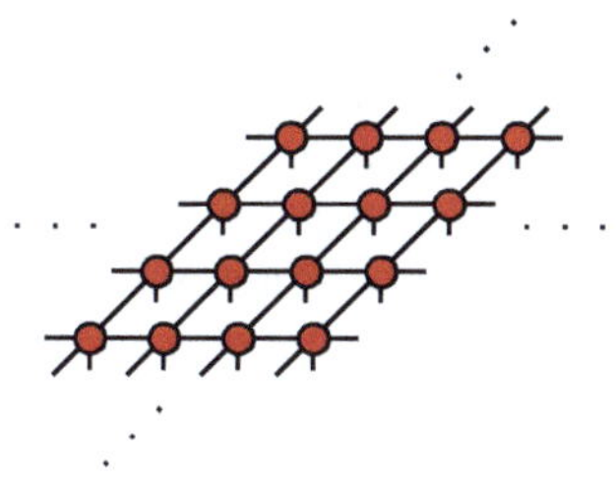

In this definition, A is now a 5-legged tensor with four virtual legs, carrying the quantum correlations in the state, and one physical leg corresponding to the physical spins of the system.

From the above form it is clear that a PEPS has a built-in area law for the entanglement entropy, a property that is shared by many ground states of local Hamiltonians. In contrast to the one-dimensional case, this cannot be rigorously proven in the higher-dimensional setting. Still, every PEPS has a parent Hamiltonian and, if the PEPS is injective, it is the unique ground state [270]. There a few interesting two-dimensional quantum states that have an exact PEPS representation, viz. the two-dimensional AKLT states [79, 80], resonating valence-bond states [97, 269, 271], the toric-code state [116, 269], and, more generally, string nets [272, 273].

Interestingly, whereas MPS could only exhibit exponential decay of correlations, a PEPS can also contain power-law decay of correlations [269]. The way to construct a "critical" PEPS is by noting that to every system of statistical mechanics a quantum state can be associated for which the expectation values of quantum observables reduce to the classical expectation values. If the classical system is now tuned to a critical point – the easiest example is the two-dimensional Ising model [269] – the corresponding PEPS has critical correlations. It seems, however, that the correlations in this state have a classical origin, and, consequently, this property does not imply that two-dimensional quantum critical systems can be captured accurately with PEPS. Another crucial difference is that the norm or expectation values of PEPS cannot be computed efficiently in principle; the cost of contracting the two-dimensional network corresponding to the norm of a PEPS scales exponentially in the number of sites of the lattice [274]. In practice, numerical algorithms still allow to do this efficiently by representing subregions of the lattice as tensor networks. This allows to determine effective environments for local regions in the PEPS such that local expectation values can be evaluated. In Chap. 5 we will review the different algorithms in more detail.

Despite the progress in numerical algorithms, the contraction of a PEPS is still a computationally demanding task, which makes PEPS simulations a rather tedious endeavour. The challenge of finding an optimal PEPS representation for the ground state of a two-dimensional quantum spin system is, in absence of a two-dimensional analogue of DMRG, still in the stage of finding efficient algorithms. All state-of-the-art methods are based on imaginary-time evolution with two-dimensional versions of the time-evolving block decimation, a method that is non-variational and, as such, not completely under control as it is not guaranteed to reach a variational minimum. In Chap. 5 we will consider these different algorithms and present our own variational method for optimizing the PEPS representation of generic ground states.

Next to the viability of PEPS as a variational ansatz for approximating a ground state of a two-dimensional quantum system, they also offer interesting theoretical insights into the entanglement structure of two-dimensional quantum systems. Firstly, they capture bulk-boundary correspondences in a natural way; the unpaired virtual degrees of freedom that appear as a PEPS is cut in half, can be identified with the entanglement spectrum of the subregions, allowing to determine boundary Hamiltonians in a natural way [275, 276]; in addition, one can follow the effect of perturbations in the bulk on the edge physics [277]. Secondly, PEPS allow to consider topologically ordered states in a natural way. Indeed, it was observed that topological order of a PEPS is manifested as a symmetry on the virtual level of the PEPS tensors and that the non-trivial properties of the anyonic excitations can be understood from strings living on the virtual level of the PEPS [278–281]. Also, topological phase transitions can be modelled by perturbing these PEPS in a specific way [276, 282, 283]. It still remains a big question, however, to what extent PEPS can classify all two-dimensional quantum phases.

2.6.6 *Significant Other*

Both MPS and PEPS succeed in capturing the physics of low-energy states of local Hamiltonians by modeling their entanglement structure in the form of a tensor network. In recent years, this fundamental idea has been extended to other quantum many-body states with different entanglement properties. Neglecting any attempt at being exhaustive, let us highlight two other types of tensor networks.

Firstly, we have repeatedly mentioned that MPS fail to capture the properties of critical spin chains, because of the logarithmic scaling of the entanglement entropy and the algebraic decay of correlations. Therefore, in order to capture the ground state of critical spin chain models, the multi-scale entanglement renormalization ansatz has been introduced [284]. In this ansatz, the idea of renormalization group transformations is explicitly modelled, such that, in contrast to the case of MPS [285], it is tailored to sustain a renormalization flow along a newly introduced scale dimension of the network. As such, it can give rise to scale-invariant states at criticality, with the correct entanglement and correlation properties. The same idea has been applied to impurity problems [286], making the connection with the numerical renormalization group of Wilson, and has led to a better understanding of the transfer matrix of an MPS [287, 288].

Secondly, the idea of MPS can be extended to the description of ground states of one-dimensional quantum field theories, by taking the appropriate continuum limit [220, 289–291]. These continuous matrix product states have been successfully applied to capture the ground state, excitations, and real-time dynamics of Bose gases [292–294] and has led to a non-commutative version of the Gross–Pitaevskii equation [295].

2.7 Renormalization and Effective Particles

One concept that we haven't really touched upon is the renormalization group [192, 296], which embraces a set of ideas and procedures that are designed to eliminate irrelevant degrees of freedom in order to arrive at a low-energy effective description for a many-body system. Both in high-energy particle physics [297] and condensed-matter physics [14, 298], such an effective description is typically given in terms of a quantum field theory; for the latter, the fields describe the long-wavelength fluctuations that occur in a quantum material.

One such effective-field construction is Fermi-liquid theory [299, 300], describing the low-energy behaviour of a system of interacting fermions in terms of effective quasiparticles [301, 302]. These quasiparticles are defined starting from the free electrons in the non-interacting limit, and, by adiabatically turning on the interactions, the electrons gain renormalized masses, charges and lifetimes. Close to the Fermi surface, the quasiparticle lifetime diverges and the dressed electrons remain well-defined modes of the system, so that their dynamics explain the low-energy behaviour of a Fermi-liquid.

This idea of capturing the low-energy physics of quantum many-body systems in terms of effective particle-like degrees of freedom is all-around in condensed-matter physics. In the field of strongly-correlated quantum matter, however, they are not defined starting from a non-interacting limit, but they appear as collective modes on top of a strongly-correlated ground state. The long-ranged quantum correlations (entanglement) in the ground state allow for these particles to have exotic emergent properties such as fractional quantum numbers and non-conventional statistics [see Sect. 2.3]. They come under the names of phonons, magnons, spinons, holons, solitons, visons, anyons, etc.

In the following chapters effective particles will play a central role. The way we will arrive at them, however, will be quite different, without field-theory methods such as functional integration. Instead, we will start from the ground state of a microscopic Hamiltonian, on which we build the elementary excitations variationally. Only in a next step we will interpret the excitations as "particles" and start building an effective theory.

References

1. P.A.M. Dirac, Quantum mechanics of many-electron systems. Proc. R. Soc. Lond. Ser. A **123**, 714 (1929). doi:10.1098/rspa.1929.0094
2. E. Schrödinger, Discussion of probability relations between separated systems. Math. Proc. Camb. Philos. Soc. **31**, 555 (1935). doi:10.1017/S0305004100013554
3. A. Einstein, B. Podolsky, N. Rosen, Can quantum-mechanical description of physical reality be considered complete? Phys. Rev. **47**, 777 (1935). doi:10.1103/PhysRev.47.777
4. P.A.M. Dirac, Note on exchange phenomena in the Thomas atom. Math. Proc. Camb. Philos. Soc. **26**, 376 (1930). doi:10.1017/S0305004100016108
5. J. von Neumann, *Mathematische Grundlagen der Quantenmechanik* (Springer, Berlin, 1932)
6. J.J. Sakurai, *Modern Quantum Mechanics* (The Benjamin/Cummings Publishing Company Inc, San Francisco, 1985)
7. J. Hubbard, Electron correlations in narrow energy bands. Proc. R. Soc. A Math. Phys. Eng. Sci. **276**, 238 (1963). doi:10.1098/rspa.1963.0204
8. F.H.L. Essler, H. Frahm, F. Göhmann, A. Klümper, V.E. Korepin, *The One-Dimensional Hubbard Model* (Cambridge University Press, Cambridge, 2005)
9. A. Auerbach, *Interacting Electrons and Quantum Magnetism* (Springer, Berlin, 1994)
10. A. Altland, B. Simons, *Condensed Matter Field Theory* (Cambridge University Press, Cambridge, 2006)
11. W. Heisenberg, Zur Theorie des Ferromagnetismus. Zeitschrift für Physik **49**, 619 (1928). doi:10.1007/BF01328601
12. P.W. Anderson, Antiferromagnetism. theory of superexchange interaction. Phys. Rev. **79**, 350 (1950). doi:10.1103/PhysRev.79.350
13. A.L. Fetter, J.D. Walecka, *Quantum Theory of Many-Particle Systems* (McGraw-Hill, San Francisco, 1971)
14. A.M. Tsvelik, *Quantum Field Theory in Condensed Matter Physics* (Cambridge University Press, Cambridge, 1995)
15. S. Sachdev, *Quantum Phase Transitions* (Cambridge University Press, Cambridge, 2011)
16. L. Landau, On the theory of phase transitions. Zh. Eksp. Teor. Fiz. **7**, 19 (1937)
17. P. Pfeuty, The one-dimensional Ising model with a transverse field. Ann. Phys. **57**, 79 (1970). doi:10.1016/0003-4916(70)90270-8

18. N.F. Mott, The basis of the electron theory of metals, with special reference to the transition metals. Proc. Phys. Soc. Sect. A **62**, 416 (1949). doi:10.1088/0370-1298/62/7/303
19. M. Imada, A. Fujimori, Y. Tokura, Metal-insulator transitions. Rev. Mod. Phys. **70**, 1039 (1998). doi:10.1103/RevModPhys.70.1039
20. J.G. Bednorz, K.A. Müller, Possible high T_c superconductivity in the Ba-La-Cu-O system. Zeitschrift für Physik B Condensed Matter **64**, 189 (1986). doi:10.1007/BF01303701
21. C. Broholm, G. Aeppli, Dynamic correlations in quantum magnets, *Strong Interactions in Low Dimensions* (Springer, Netherlands, 2004), pp. 21–61. doi:10.1007/978-1-4020-3463-3_2
22. S.T. Bramwell, Neutron scattering in highly frustrated magnetism, in *Introduction to Frustrated Magnetism*, ed. by C. Lacroix, F. Mila, P. Mendels (Springer, Berlin, 2011), pp. 45–78. doi:10.1007/978-3-642-10589-0
23. C. Lacroix, P. Mendels, F. Mila, *Introduction to Frustrated Magnetism* (Springer, Berlin, 2011)
24. M.H. Anderson, J.R. Ensher, M.R. Matthews, C.E. Wieman, E.A. Cornell, Observation of Bose-Einstein Condensation in a dilute atomic vapor. Science **269**, 198 (1995). doi:10.1126/science.269.5221.198
25. K.B. Davis, M.O. Mewes, M.R. Andrews, N.J. van Druten, D.S. Durfee, D.M. Kurn, W. Ketterle, Bose-Einstein condensation in a gas of sodium atoms. Phys. Rev. Lett. **75**, 3969 (1995). doi:10.1103/PhysRevLett.75.3969
26. I. Bloch, Ultracold quantum gases in optical lattices. Nat. Phys. **1**, 23 (2005). doi:10.1038/nphys138
27. I. Bloch, W. Zwerger, Many-body physics with ultracold gases. Rev. Mod. Phys. **80**, 885 (2008). doi:10.1103/RevModPhys.80.885
28. I. Bloch, J. Dalibard, S. Nascimbène, Quantum simulations with ultracold quantum gases. Nat. Phys. **8**, 267 (2012). doi:10.1038/nphys2259
29. M.P.A. Fisher, P.B. Weichman, G. Grinstein, D.S. Fisher, Boson localization and the superfluid-insulator transition. Phys. Rev. B **40**, 546 (1989). doi:10.1103/PhysRevB.40.546
30. D. Jaksch, C. Bruder, J.I. Cirac, C. Gardiner, P. Zoller, Cold bosonic atoms in optical lattices. Phys. Rev. Lett. **81**, 3108 (1998). doi:10.1103/PhysRevLett.81.3108
31. M. Greiner, O. Mandel, T. Esslinger, T.W. Hänsch, I. Bloch, Quantum phase transition from a superfluid to a Mott insulator in a gas of ultracold atoms. Nature **415**, 39 (2002). doi:10.1038/415039a
32. B. Paredes, A. Widera, V. Murg, O. Mandel, S. Folling, J.I. Cirac, G.V. Shlyapnikov, T.W. Hänsch, I. Bloch, Tonks–Girardeau gas of ultracold atoms in an optical lattice. Nature **429**, 277 (2004), http://www.nature.com/nature/journal/v429/n6989/suppinfo/nature02530_S1.html
33. T. Kinoshita, T. Wenger, D.S. Weiss, Observation of a one-dimensional Tonks-Girardeau gas. Science **305**, 1125 (2004). doi:10.1126/science.1100700
34. T. Kinoshita, T. Wenger, D.S. Weiss, A quantum Newton's cradle. Nature **440**, 900 (2006), http://www.nature.com/nature/journal/v440/n7086/suppinfo/nature04693_S1.html
35. J. Simon, W.S. Bakr, R. Ma, M.E. Tai, P.M. Preiss, M. Greiner, Quantum simulation of antiferromagnetic spin chains in an optical lattice. Nature **472**, 307 (2011). doi:10.1038/nature09994
36. J. Struck, C. Olschlager, R. Le Targat, P. Soltan-Panahi, A. Eckardt, M. Lewenstein, P. Windpassinger, K. Sengstock, Quantum simulation of frustrated classical magnetism in triangular optical lattices. Science **333**, 996 (2011). doi:10.1126/science.1207239
37. R. Blatt, C.F. Roos, Quantum simulations with trapped ions. Nat. Phys. **8**, 277 (2012). doi:10.1038/nphys2252
38. D. Porras, J.I. Cirac, Effective quantum spin systems with trapped ions. Phys. Rev. Lett. **92**, 207901 (2004). doi:10.1103/PhysRevLett.92.207901
39. A. Aspuru-Guzik, P. Walther, Photonic quantum simulators. Nat. Phys. **8**, 285 (2012). doi:10.1038/nphys2253
40. X.-S. Ma, B. Dakic, W. Naylor, A. Zeilinger, P. Walther, Quantum simulation of the wavefunction to probe frustrated Heisenberg spin systems. Nat. Phys. **7**, 399 (2011). doi:10.1038/nphys1919

41. J.I. Cirac, P. Zoller, Goals and opportunities in quantum simulation. Nat. Phys. **8**, 264 (2012). doi:10.1038/nphys2275
42. I. Buluta, F. Nori, Quantum simulators. Science **326**, 108 (2009). doi:10.1126/science.1177838
43. R.P. Feynman, Simulating physics with computers. Int. J. Theor. Phys. (1982), http://www.springerlink.com/index/T2X8115127841630.pdf
44. M.A. Nielsen, I.L. Chuang, *Quantum Computation and Quantum Information* (Cambridge University Press, Cambridge, 2000)
45. T.J. Osborne, M.A. Nielsen, Entanglement, quantum phase transitions, and density matrix renormalization. Quantum Inf. Process. **1**, 45 (2002). doi:10.1023/A:1019601218492
46. G. Vidal, J.I. Latorre, E. Rico, A. Kitaev, Entanglement in quantum critical phenomena. Phys. Rev. Lett. **90**, 227902 (2003). doi:10.1103/PhysRevLett.90.227902
47. A. Kitaev, J. Preskill, Topological entanglement entropy. Phys. Rev. Lett. **96**, 110404 (2006). doi:10.1103/PhysRevLett.96.110404
48. M. Levin, X.-G. Wen, Detecting topological order in a ground state wave function. Phys. Rev. Lett. **96**, 110405 (2006). doi:10.1103/PhysRevLett.96.110405
49. X. Chen, Z.-C. Gu, X.-G. Wen, Local unitary transformation, long-range quantum entanglement, wave function renormalization, and topological order. Phys. Rev. B **82**, 155138 (2010). doi:10.1103/PhysRevB.82.155138
50. J. Preskill, Quantum Information and Computation (Lecture Notes) (2016), http://www.theory.caltech.edu/people/preskill/ph229/
51. J. Preskill, Quantum information and physics: some future directions. J. Mod. Opt. **47**, 127 (2000). doi:10.1080/09500340008244031
52. M.B. Hastings, Locality in Quantum Systems (2010), arXiv:1008.5137
53. J. Eisert, M. Cramer, M.B. Plenio, Colloquium: area laws for the entanglement entropy. Rev. Mod. Phys. **82**, 277 (2010). doi:10.1103/RevModPhys.82.277
54. E.H. Lieb, D.W. Robinson, The finite group velocity of quantum spin systems. Commun. Math. Phys. **28**, 251 (1972). doi:10.1007/BF01645779
55. M.B. Hastings, T. Koma, Spectral gap and exponential decay of correlations. Commun. Math. Phys. **265**, 781 (2006). doi:10.1007/s00220-006-0030-4
56. M.B. Hastings, An area law for one-dimensional quantum systems. J. Stat. Mech. Theory Exp. **2007**, P08024 (2007). doi:10.1088/1742-5468/2007/08/P08024
57. P. Calabrese, J. Cardy, Entanglement entropy and quantum field theory. J. Stat. Mech. Theory Exp. **2004**, P06002 (2004). doi:10.1088/1742-5468/2004/06/P06002
58. F. Verstraete, J.I. Cirac, Matrix product states represent ground states faithfully. Phys. Rev. B **73**, 094423 (2006). doi:10.1103/PhysRevB.73.094423
59. Z. Landau, U. Vazirani, T. Vidick, A polynomial-time algorithm for the ground state of 1D gapped local Hamiltonians (2013), arXiv:1307.5143
60. V. Coffman, J. Kundu, W.K. Wootters, Distributed entanglement. Phys. Rev. A **61**, 052306 (2000). doi:10.1103/PhysRevA.61.052306
61. T.J. Osborne, F. Verstraete, General monogamy inequality for bipartite qubit entanglement. Phys. Rev. Lett. **96**, 220503 (2006). doi:10.1103/PhysRevLett.96.220503
62. N.D. Mermin, H. Wagner, Absence of ferromagnetism or antiferromagnetism in one- or two-dimensional isotropic Heisenberg models. Phys. Rev. Lett. **17**, 1133 (1966). doi:10.1103/PhysRevLett.17.1133
63. P.C. Hohenberg, Existence of long-range order in one and two dimensions. Phys. Rev. **158**, 383 (1967). doi:10.1103/PhysRev.158.383
64. S. Coleman, There are no Goldstone bosons in two dimensions. Commun. Math. Phys. **31**, 259 (1973). doi:10.1007/bf01646487
65. P.W. Anderson, An approximate quantum theory of the antiferromagnetic ground state. Phys. Rev. **86**, 694 (1952). doi:10.1103/PhysRev.86.694
66. S. Chakravarty, B.I. Halperin, D.R. Nelson, Two-dimensional quantum Heisenberg antiferromagnet at low temperatures. Phys. Rev. B **39**, 2344 (1989). doi:10.1103/PhysRevB.39.2344

67. H. Bethe, Zur Theorie der Metalle. I. Eigenwerte und Eigenfunktionen der linearen Atomkette. Zeitschrift für Physik **71**, 205 (1931). doi:10.1007/BF01341708
68. L. Hulthén, Über das Austauschproblem eines Kristalles. Arkiv för matematik, astronomi och fysik **26A**, 11 (1938)
69. R. Kubo, The spin-wave theory of antiferromagnetics. Phys. Rev. **87**, 568 (1952). doi:10.1103/PhysRev.87.568
70. T. Oguchi, Theory of spin-wave interactions in ferro- and antiferromagnetism. Phys. Rev. **117**, 117 (1960). doi:10.1103/PhysRev.117.117
71. T. Holstein, H. Primakoff, Field dependence of the intrinsic domain magnetization of a ferromagnet. Phys. Rev. **58**, 1098 (1940). doi:10.1103/PhysRev.58.1098
72. F.J. Dyson, General theory of spin-wave interactions. Phys. Rev. **102**, 1217 (1956). doi:10.1103/PhysRev.102.1217
73. T. Giamarchi, *Quantum Physics in One Dimension* (Oxford University Press, Oxford, 2004)
74. E.H. Lieb, T. Schultz, D. Mattis, Two soluble models of an antiferromagnetic chain. Ann. Phys. **16**, 407 (1961). doi:10.1016/0003-4916(61)90115-4
75. I. Affleck, E.H. Lieb, A proof of part of Haldane's conjecture on spin chains. Lett. Math. Phys. **12**, 57 (1986). doi:10.1007/BF00400304
76. F.D.M. Haldane, Continuum dynamics of the 1-D Heisenberg antiferromagnet: identification with the O(3) nonlinear sigma model. Phys. Lett. A **93**, 464 (1983). doi:10.1016/0375-9601(83)90631-X
77. F.D.M. Haldane, Nonlinear field theory of large-spin Heisenberg antiferromagnets: semiclassically quantized solitons of the one-dimensional easy-axis nNeel state. Phys. Rev. Lett. **50**, 1153 (1983). doi:10.1103/PhysRevLett.50.1153
78. I. Affleck, Quantum spin chains and the Haldane gap. J. Phys. Condens. Matter **1**, 3047 (1989), http://cat.inist.fr/?aModele=afficheN&cpsidt=6916826
79. I. Affleck, T. Kennedy, E.H. Lieb, H. Tasaki, Rigorous results on valence-bond ground states in antiferromagnets. Phys. Rev. Lett. **59**, 799 (1987). doi:10.1103/PhysRevLett.59.799
80. I. Affleck, T. Kennedy, E.H. Lieb, H. Tasaki, Valence bond ground states in isotropic quantum antiferromagnets. Commun. Math. Phys. **115**, 477 (1988). doi:10.1007/BF01218021
81. M. Nightingale, H. Blöte, Gap of the linear spin-1 Heisenberg antiferromagnet: a Monte Carlo calculation. Phys. Rev. B **33**, 659 (1986). doi:10.1103/PhysRevB.33.659
82. F.D.M. Haldane, Errata. Phys. Lett. A **81**, 545 (1981). doi:10.1016/0375-9601(81)90464-3
83. A. Imambekov, T.L. Schmidt, L.I. Glazman, One-dimensional quantum liquids: beyond the Luttinger liquid paradigm. Rev. Mod. Phys. **84**, 1253 (2012). doi:10.1103/RevModPhys.84.1253
84. F.D.M. Haldane, Effective harmonic-fluid approach to low-energy properties of one-dimensional quantum fluids. Phys. Rev. Lett. **47**, 1840 (1981). doi:10.1103/PhysRevLett.47.1840
85. F.D.M. Haldane, Demonstration of the Luttinger liquid character of Bethe-ansatz-soluble models of 1-D quantum fluids. Phys. Lett. **81**, 153 (1981), http://www.sciencedirect.com/science/article/pii/0375960181900499
86. F.D.M. Haldane, General relation of correlation exponents and spectral properties of one-dimensional fermi systems: application to the anisotropic S = 1/2 Heisenberg chain. Phys. Rev. Lett. **45**, 1358 (1980). doi:10.1103/PhysRevLett.45.1358
87. S.-I. Tomonaga, Remarks on Bloch's method of sound waves applied to many-fermion problems. Prog. Theor. Phys. **5**, 544 (1950). doi:10.1143/ptp/5.4.544
88. J.M. Luttinger, An exactly soluble model of a many-fermion system. J. Math. Phys. **4**, 1154 (1963). doi:10.1063/1.1704046
89. J. Cardy, *Scaling and Renormalization in Statistical Physics* (Cambridge University Press, Cambridge, 1996)
90. L.D. Faddeev, L.A. Takhtajan, What is the spin of a spin wave? Phys. Lett. A **85**, 375 (1981). doi:10.1016/0375-9601(81)90335-2
91. F.D.M. Haldane, Fractional statistics in arbitrary dimensions: a generalization of the Pauli principle. Phys. Rev. Lett. **67**, 937 (1991). doi:10.1103/PhysRevLett.67.937

92. M.A. Cazalilla, R. Citro, T. Giamarchi, E. Orignac, M. Rigol, One dimensional bosons: from condensed matter systems to ultracold gases. Rev. Mod. Phys. **83**, 1405 (2011). doi:10.1103/RevModPhys.83.1405
93. L. Balents, Spin liquids in frustrated magnets. Nature **464**, 199 (2010). doi:10.1038/nature08917
94. C. Lhuillier, G. Misguich, Introduction to quantum spin liquids, in *Introduction to Frustrated Magnetism*, ed. by C. Lacroix, F. Mila, P. Mendels (Springer, Berlin, 2011), pp. 23–44. doi:10.1007/978-3-642-10589-0
95. L. Savary, L. Balents, Quantum Spin Liquids (2016), arXiv:1601.03742
96. P. Anderson, Resonating valence bonds: a new kind of insulator? Mater. Res. Bull. **8**, 153 (1973). doi:10.1016/0025-5408(73)90167-0
97. P.W. Anderson, The resonating valence bond state in La_2CuO_4 and superconductivity. Science **235**, 1196 (1987). doi:10.1126/science.235.4793.1196
98. S. Yan, D.A. Huse, S.R. White, Spin-liquid ground state of the S = 1/2 Kagome Heisenberg antiferromagnet. Science **332**, 1173 (2011). doi:10.1126/science.1201080
99. H.-C. Jiang, Z. Wang, L. Balents, Identifying topological order by entanglement entropy. Nat. Phys. **8**, 902 (2012). doi:10.1038/nphys2465
100. Y. Iqbal, F. Becca, S. Sorella, D. Poilblanc, Gapless spin-liquid phase in the kagome spin-1/2 Heisenberg antiferromagnet. Phys. Rev. B **87**, 060405 (2013). doi:10.1103/PhysRevB.87.060405
101. Z. Zhu, S.R. White, Spin liquid phase of the S = 1/2 J_1-J_2 Heisenberg model on the triangular lattice. Phys. Rev. B **92**, 041105 (2015). doi:10.1103/PhysRevB.92.041105
102. W.-J. Hu, S.-S. Gong, W. Zhu, D.N. Sheng, Competing spin-liquid states in the spin-1/2 Heisenberg model on the triangular lattice. Phys. Rev. B **92**, 140403 (2015). doi:10.1103/PhysRevB.92.140403
103. X.G. Wen, Topological order in rigid states. Int. J. Mod. Phys. B **04**, 239 (1990). doi:10.1142/S0217979290000139
104. B. Zeng, X. Chen, D.-L. Zhou, X.-G. Wen, Quantum Information Meets Quantum Matter – From Quantum Entanglement to Topological Phase in Many-Body Systems (2015), arXiv:1508.02595
105. X.G. Wen, Vacuum degeneracy of chiral spin states in compactified space. Phys. Rev. B **40**, 7387 (1989). doi:10.1103/PhysRevB.40.7387
106. D.P. Arovas, J.R. Schrieffer, F. Wilczek, Fractional statistics and the quantum hall effect. Phys. Rev. Lett. **53**, 722 (1984). doi:10.1103/PhysRevLett.53.722
107. B.I. Halperin, Statistics of quasiparticles and the hierarchy of fractional quantized Hall states. Phys. Rev. Lett. **52**, 1583 (1984). doi:10.1103/PhysRevLett.52.1583
108. X.G. Wen, Non-Abelian statistics in the fractional quantum Hall states. Phys. Rev. Lett. **66**, 802 (1991). doi:10.1103/PhysRevLett.66.802
109. G. Moore, N. Read, Nonabelions in the fractional quantum Hall effect. Nucl. Phys. B **360**, 362 (1991). doi:10.1016/0550-3213(91)90407-o
110. B.I. Halperin, Quantized Hall conductance, current-carrying edge states, and the existence of extended states in a two-dimensional disordered potential. Phys. Rev. B **25**, 2185 (1982). doi:10.1103/PhysRevB.25.2185
111. A.H. MacDonald, Edge states in the fractional-quantum-Hall-effect regime. Phys. Rev. Lett. **64**, 220 (1990). doi:10.1103/PhysRevLett.64.220
112. X.G. Wen, Gapless boundary excitations in the quantum Hall states and in the chiral spin states. Phys. Rev. B **43**, 11025 (1991). doi:10.1103/PhysRevB.43.11025
113. D.C. Tsui, H.L. Stormer, A.C. Gossard, Two-dimensional magnetotransport in the extreme quantum limit. Phys. Rev. Lett. **48**, 1559 (1982). doi:10.1103/PhysRevLett.48.1559
114. R.B. Laughlin, Anomalous quantum Hall effect: an incompressible quantum fluid with fractionally charged excitations. Phys. Rev. Lett. **50**, 1395 (1983). doi:10.1103/PhysRevLett.50.1395
115. E. Dennis, A. Kitaev, A. Landahl, J. Preskill, Topological quantum memory. J. Math. Phys. **43**, 4452 (2002). doi:10.1063/1.1499754

116. A. Kitaev, Fault-tolerant quantum computation by anyons. Ann. Phys. **303**, 2 (2003). doi:10.1016/S0003-4916(02)00018-0
117. D.R. Hartree, The wave mechanics of an atom with a non-coulomb central field. Part I. Theory and methods. Math. Proc. Camb. Philos. Soc. **24**, 89 (1928). doi:10.1017/S0305004100011919
118. D.R. Hartree, The wave mechanics of an atom with a non-coulomb central field. Part II. Some results and discussion. Math. Proc. Camb. Philos. Soc. **24**, 111 (1928). doi:10.1017/S0305004100011920
119. V. Fock, Näherungsmethode zur Lösung des quantenmechanischen Mehrkörperproblems. Zeitschrift für Physik **61**, 126 (1930). doi:10.1007/BF01340294
120. P. Hohenberg, W. Kohn, Inhomogeneous electron gas. Phys. Rev. **136**, B864 (1964). doi:10.1103/PhysRev.136.B864
121. W. Kohn, L.J. Sham, Self-consistent equations including exchange and correlation effects. Phys. Rev. **140**, A1133 (1965). doi:10.1103/PhysRev.140.A1133
122. R.D. Mattuck, *A Guide to Feynman Diagrams in the Many-Body Problem* (Courier Dover Publications, New York, 1976)
123. N. Laflorencie, D. Poilblanc, Simulations of pure and doped low-dimensional spin-1/2 gapped systems, in *Quantum Magnetism*, ed. by U. Schollwöck, J. Richter, D.J.J. Farnell, R.F. Bishop (Springer, Berlin, 2004), pp. 227–252. doi:10.1007/BFb0119595
124. A.M. Läuchli, Numerical simulations of frustrated systems, in *Introduction to Frustrated Magnetism*, ed. by C. Lacroix, P. Mendels, F. Mila (Springer, Berlin, 2011), pp. 481–511. doi:10.1007/978-3-642-10589-0
125. C. Lanczos, An iteration method for the solution of the eigenvalue problem of linear differential and integral operators. J. Res. Natl. Bur. Stand. **45**, 255 (1950)
126. A.M. Läuchli, An exact diagonalization perspective on the S = 1/2 Kagome Heisenberg antiferromagnet, in *KITP Program: Frustrated Magnetism and Quantum Spin Liquids: From Theory and Models to Experiments*, 13 August–9 November 2012 (2012), http://online.kitp.ucsb.edu/online/fragnets12/laeuchli/
127. N. Metropolis, A.W. Rosenbluth, M.N. Rosenbluth, A.H. Teller, E. Teller, Equation of state calculations by fast computing machines. J. Chem. Phys. **21**, 1087 (1953). doi:10.1063/1.1699114
128. A.W. Sandvik, J. Kurkijärvi, Quantum Monte Carlo simulation method for spin systems. Phys. Rev. B **43**, 5950 (1991). doi:10.1103/PhysRevB.43.5950
129. M. Troyer, U.-J. Wiese, Computational complexity and fundamental limitations to fermionic quantum Monte Carlo simulations. Phys. Rev. Lett. **94**, 170201 (2005). doi:10.1103/PhysRevLett.94.170201
130. M.P. Gelfand, R.R.P. Singh, High-order convergent expansions for quantum many particle systems. Adv. Phys. **49**, 93 (2000). doi:10.1080/000187300243390
131. F. Wegner, Flow-equations for Hamiltonians. Annalen der Physik **506**, 77 (1994). doi:10.1002/andp.19945060203
132. C. Knetter, A. Bühler, E. Müller-Hartmann, G.S. Uhrig, Dispersion and symmetry of bound states in the Shastry–Sutherland model. Phys. Rev. Lett. **85**, 3958 (2000). doi:10.1103/PhysRevLett.85.3958
133. S. Kehrein, *The Flow Equation Approach to Many-Particle Systems* (Springer, Berlin, 2006)
134. J.C. Slater, The theory of complex spectra. Phys. Rev. **34**, 1293 (1929). doi:10.1103/PhysRev.34.1293
135. J.C. Slater, Note on Hartree's method. Phys. Rev. **35**, 210 (1930). doi:10.1103/PhysRev.35.210.2
136. E.P. Gross, Structure of a quantized vortex in boson systems. Il Nuovo Cimento **20**, 454 (1961). doi:10.1007/BF02731494
137. L.P. Pitaevskii, Vortex lines in an imperfect Bose gas. Sov. Phys. JETP **40** (1961)
138. J. Bardeen, L.N. Cooper, J.R. Schrieffer, Microscopic theory of superconductivity. Phys. Rev. **106**, 162 (1957). doi:10.1103/PhysRev.106.162

139. J. Bardeen, L.N. Cooper, J.R. Schrieffer, Theory of superconductivity. Phys. Rev. **108**, 1175 (1957). doi:10.1103/PhysRev.108.1175
140. S. Sorella, Wave function optimization in the variational Monte Carlo method. Phys. Rev. B **71**, 241103 (2005). doi:10.1103/PhysRevB.71.241103
141. F. Becca, L. Capriotti, A. Parola, S. Sorella, Variational wave functions for frustrated magnetic models, in *Introduction to Frustrated Magnetism*, ed. by C. Lacroix, P. Mendels, F. Mila (Springer, Berlin, 2011), pp. 379–406. doi:10.1007/978-3-642-10589-0
142. Y. Iqbal, D. Poilblanc, F. Becca, Spin-1/2 Heisenberg J_1-J_2 antiferromagnet on the kagome lattice. Phys. Rev. B **91**, 020402 (2015). doi:10.1103/PhysRevB.91.020402
143. V.E. Korepin, N.M. Bogoliubov, A.G. Izergin, *Quantum Inverse Scattering Method and Correlation Functions* (Cambridge University Press, Cambridge, 1997)
144. M. Gaudin, *The Bethe Wavefunction* (Cambridge University Press, Cambridge, 2014)
145. B. Sutherland, *Beautiful Models: 70 Years of Exactly Solved Quantum Many-body Problems* (World Scientific, Singapore, 2004)
146. C.N. Yang, C.P. Yang, One-dimensional chain of anisotropic spin-spin interactions. I. Proof of Bethe's hypothesis for ground state in a finite system. Phys. Rev. **150**, 321 (1966). doi:10.1103/PhysRev.150.321
147. C.N. Yang, C.P. Yang, One-dimensional chain of anisotropic spin-spin interactions. II. Properties of the ground-state energy per lattice site for an infinite system. Phys. Rev. **150**, 327 (1966). doi:10.1103/PhysRev.150.327
148. C.N. Yang, C.P. Yang, One-dimensional chain of anisotropic spin-spin interactions III. Applications. Phys. Rev. **151**, 258 (1966). doi:10.1103/PhysRev.151.258
149. R.J. Baxter, One-dimensional anisotropic Heisenberg chain. Ann. Phys. **70**, 323 (1972). doi:10.1016/0003-4916(72)90270-9
150. E.H. Lieb, F. Wu, Absence of Mott transition in an exact solution of the short-range, one-band model in one dimension. Phys. Rev. Lett. **20**, 1445 (1968). doi:10.1103/PhysRevLett.20.1445
151. E.H. Lieb, W. Liniger, Exact analysis of an interacting bose gas. I. The general solution and the ground state. Phys. Rev. **130**, 1605 (1963). doi:10.1103/PhysRev.130.1605
152. E.H. Lieb, Exact analysis of an interacting bose gas II. The excitation spectrum. Phys. Rev. **130**, 1616 (1963). doi:10.1103/PhysRev.130.1616
153. C.N. Yang, C.P. Yang, Thermodynamics of a one-dimensional system of Bosons with repulsive delta-function interaction. J. Math. Phys. **10**, 1115 (1969). doi:10.1063/1.1664947
154. M. Takahashi, *Thermodynamics of One-Dimensional Solvable Models* (Cambridge University Press, Cambridge, 2005)
155. F.H.L. Essler, R.M. Konik, Application of massive integrable quantum field theories to problems in condensed matter physics, *From Fields to Strings: Circumnavigating Theoretical Physics* (World Scientific, Singapore, 2005), pp. 684–830. doi:10.1142/9789812775344_0020
156. A.B. Zamolodchikov, Exact two-particle S-matrix of quantum sine-Gordon solitons. Commun. Math. Phys. **55**, 183 (1977). doi:10.1007/BF01626520
157. G. Mussardo, *Statistical Field Theory: An Introduction to Exactly Solved Models in Statistical Physics* (Oxford University Press, Oxford, 2009)
158. C.N. Yang, Some exact results for the many-body problem in one dimension with repulsive delta-function interaction. Phys. Rev. Lett. **19**, 1312 (1967). doi:10.1103/PhysRevLett.19.1312
159. N.A. Slavnov, Calculation of scalar products of wave functions and form factors in the framework of the algebraic Bethe ansatz. Theor. Math. Phys. **79**, 502 (1989). doi:10.1007/BF01016531
160. N.A. Slavnov, Nonequal-time current correlation function in a one-dimensional Bose gas. Theor. Math. Phys. **82**, 273 (1990). doi:10.1007/BF01029221
161. N. Kitanine, J. Maillet, V. Terras, Form factors of the XXZ Heisenberg finite chain. Nucl. Phys. B **554**, 647 (1999). doi:10.1016/S0550-3213(99)00295-3
162. N. Kitanine, J. Maillet, V. Terras, Correlation functions of the XXZ Heisenberg spin chain in a magnetic field. Nucl. Phys. B **567**, 554 (2000). doi:10.1016/S0550-3213(99)00619-7

163. R.M. Konik, Haldane-gapped spin chains: exact low-temperature expansions of correlation functions. Phys. Rev. B **68**, 104435 (2003). doi:10.1103/PhysRevB.68.104435
164. F.H.L. Essler, R.M. Konik, Finite-temperature lineshapes in gapped quantum spin chains. Phys. Rev. B **78**, 100403 (2008). doi:10.1103/PhysRevB.78.100403
165. A. James, F.H.L. Essler, R.M. Konik, Finite-temperature dynamical structure factor of alternating Heisenberg chains. Phys. Rev. B **78**, 094411 (2008). doi:10.1103/PhysRevB.78.094411
166. F.H.L. Essler, R.M. Konik, Finite-temperature dynamical correlations in massive integrable quantum field theories. J. Stat. Mech. Theory Exp. **2009**, P09018 (2009). doi:10.1088/1742-5468/2009/09/P09018
167. W.D. Goetze, U. Karahasanovic, F.H.L. Essler, Low-temperature dynamical structure factor of the two-leg spin-1/2 Heisenberg ladder. Phys. Rev. B **82**, 104417 (2010). doi:10.1103/PhysRevB.82.104417
168. D.A. Tennant, B. Lake, A.J.A. James, F.H.L. Essler, S. Notbohm, H.-J. Mikeska, J. Fielden, P. Kögerler, P.C. Canfield, M.T.F. Telling, Anomalous dynamical line shapes in a quantum magnet at finite temperature. Phys. Rev. B **85**, 014402 (2012). doi:10.1103/PhysRevB.85.014402
169. M. Rigol, V. Dunjko, V. Yurovsky, M. Olshanii, Relaxation in a completely integrable many-body quantum system: an ab initio study of the dynamics of the highly excited states of 1D lattice hard-core Bosons. Phys. Rev. Lett. **98**, 4 (2007). doi:10.1103/PhysRevLett.98.050405
170. M. Fagotti, F.H.L. Essler, Stationary behaviour of observables after a quantum quench in the spin-1/2 Heisenberg XXZ chain. J. Stat. Mech. Theory Exp. **2013**, P07012 (2013). doi:10.1088/1742-5468/2013/07/P07012
171. M. Fagotti, M. Collura, F.H.L. Essler, P. Calabrese, Relaxation after quantum quenches in the spin-1/2 Heisenberg XXZ chain. Phys. Rev. B **89**, 125101 (2014). doi:10.1103/PhysRevB.89.125101
172. E. Ilievski, M. Medenjak, T. Prosen, Quasilocal conserved operators in isotropic Heisenberg spin 1/2 chain (2015), arXiv:1506.05049
173. E. Ilievski, J. De Nardis, B. Wouters, J.-S. Caux, F.H.L. Essler, T. Prosen, Complete generalized Gibbs ensembles in an interacting theory. Phys. Rev. Lett. **115**, 157201 (2015). doi:10.1103/PhysRevLett.115.157201
174. E.T. Jaynes, Information theory and statistical mechanics. Phys. Rev. **106**, 171 (1957). doi:10.1103/PhysRev.106.620
175. E.T. Jaynes, Information theory and statistical mechanics II. Phys. Rev. **108**, 171 (1957). doi:10.1103/PhysRev.108.171
176. D. Fioretto, G. Mussardo, Quantum quenches in integrable field theories. New J. Phys. **12**, 055015 (2010). doi:10.1088/1367-2630/12/5/055015
177. M. Fagotti, F.H.L. Essler, Reduced density matrix after a quantum quench. Phys. Rev. B **87**, 245107 (2013). doi:10.1103/PhysRevB.87.245107
178. F.H.L. Essler, G. Mussardo, M. Panfil, Generalized Gibbs ensembles for quantum field theories. Phys. Rev. A **91**, 051602 (2015). doi:10.1103/PhysRevA.91.051602
179. F.H.L. Essler, S. Kehrein, S.R. Manmana, N.J. Robinson, Quench dynamics in a model with tuneable integrability breaking. Phys. Rev. B **89**, 165104 (2014). doi:10.1103/PhysRevB.89.165104
180. S. Sotiriadis, Zamolodchikov–Faddeev algebra and quantum quenches in integrable field theories. J. Stat. Mech. Theory Exp. **2012**, P02017 (2012), http://iopscience.iop.org/1742-5468/2012/02/P02017
181. J.-S. Caux, R.M. Konik, Constructing the generalized Gibbs ensemble after a quantum quench. Phys. Rev. Lett. **109**, 175301 (2012). doi:10.1103/PhysRevLett.109.175301
182. J. Mossel, J.-S. Caux, Generalized TBA and generalized Gibbs. J. Phys. A Math. Theor. **45**, 255001 (2012). doi:10.1088/1751-8113/45/25/255001
183. E. Demler, A.M. Tsvelik, Universal features of the excitation spectrum in a generalized Gibbs distribution ensemble. Phys. Rev. B **86**, 115448 (2012). doi:10.1103/PhysRevB.86.115448
184. L. Bonnes, F.H.L. Essler, A.M. Läuchli, Light-cone dynamics after quantum quenches in spin chains. Phys. Rev. Lett. **113**, 187203 (2014). doi:10.1103/PhysRevLett.113.187203

185. M. Kollar, F.A. Wolf, M. Eckstein, Generalized Gibbs ensemble prediction of prethermalization plateaus and their relation to nonthermal steady states in integrable systems. Phys. Rev. B **84**, 054304 (2011). doi:10.1103/PhysRevB.84.054304
186. M. Fagotti, On conservation laws, relaxation and pre-relaxation after a quantum quench. J. Stat. Mech. Theory Exp. **2014**, P03016 (2014). doi:10.1088/1742-5468/2014/03/P03016
187. B. Bertini, F.H.L. Essler, S. Groha, N.J. Robinson, Prethermalization and thermalization in models with weak integrability breaking. Phys. Rev. Lett. **115**, 180601 (2015). doi:10.1103/PhysRevLett.115.180601
188. T. Langen, T. Gasenzer, J. Schmiedmayer, Prethermalization and universal dynamics in near-integrable quantum systems. J. Stat. Mech. Theory Exp. **2016**, 064009 (2016), http://stacks.iop.org/1742-5468/2016/i=6/a=064009
189. F. Verstraete, V. Murg, J.I. Cirac, Matrix product states, projected entangled pair states, and variational renormalization group methods for quantum spin systems. Adv. Phys. **57**, 143 (2008). doi:10.1080/14789940801912366
190. N. Schuch, Condensed Matter Applications of Entanglement Theory (2013), arXiv:1306.5551
191. R. Orús, A practical introduction to tensor networks: matrix product states and projected entangled pair states. Ann. Phys. **349**, 117 (2014). doi:10.1016/j.aop.2014.06.013
192. K.G. Wilson, The renormalization group: critical phenomena and the Kondo problem. Rev. Mod. Phys. **47**, 773 (1975). doi:10.1103/RevModPhys.47.773
193. J. Kondo, Resistance minimum in dilute magnetic alloys. Prog. Theor. Phys. **32**, 37 (1964). doi:10.1143/PTP.32.37
194. P.W. Anderson, Localized magnetic states in metals. Phys. Rev. **124**, 41 (1961). doi:10.1103/PhysRev.124.41
195. A.C. Hewson, *The Kondo Problem to Heavy Fermions* (Cambridge University Press, Cambridge, 1993)
196. R. Bulla, T. Costi, T. Pruschke, Numerical renormalization group method for quantum impurity systems. Rev. Mod. Phys. **80**, 395 (2008). doi:10.1103/RevModPhys.80.395
197. S.R. White, R.M. Noack, Real-space quantum renormalization groups. Phys. Rev. Lett. **68**, 3487 (1992). doi:10.1103/PhysRevLett.68.3487
198. S.R. White, Density matrix formulation for quantum renormalization groups. Phys. Rev. Lett. **69**, 2863 (1992). doi:10.1103/PhysRevLett.69.2863
199. S.R. White, Density-matrix algorithms for quantum renormalization groups. Phys. Rev. B **48**, 10345 (1993). doi:10.1103/PhysRevB.48.10345
200. J.I. Latorre, E. Rico, G. Vidal, Ground state entanglement in quantum spin chains. Quantum Inf. Comput. **4**, 48 (2004), arXiv:0304098
201. U. Schollwöck, The density-matrix renormalization group. Rev. Mod. Phys. **77**, 259 (2005). doi:10.1103/RevModPhys.77.259
202. U. Schollwöck, The density-matrix renormalization group in the age of matrix product states. Ann. Phys. **326**, 96 (2011). doi:10.1016/j.aop.2010.09.012
203. S.R. White, D. Huse, Numerical renormalization-group study of low-lying eigenstates of the antiferromagnetic $S = 1$ Heisenberg chain. Phys. Rev. B **48**, 3844 (1993). doi:10.1103/PhysRevB.48.3844
204. I. Peschel, X. Wang, M. Kaulke, K. Hallberg, Density-Matrix Renormalization - A New Numerical Method in Physics (Springer, Berlin, 1999). doi:10.1007/BFb0106062
205. K. Hallberg, Density Matrix Renormalization: A Review of the Method and its Applications (2003), arXiv:0303557
206. S. Östlund, S. Rommer, Thermodynamic limit of the density matrix renormalization for the spin-1 Heisenberg chain. Phys. Rev. Lett. **75**, 13 (1995). doi:10.1103/PhysRevLett.75.3537
207. S. Rommer, S. Östlund, Class of ansatz wave functions for one-dimensional spin systems and their relation to the density matrix renormalization group. Phys. Rev. B **55**, 2164 (1997). doi:10.1103/PhysRevB.55.2164
208. J. Dukelsky, M.A. Martín-Delgado, T. Nishino, G. Sierra, Equivalence of the variational matrix product method and the density matrix renormalization group applied to spin chains. Europhys. Lett. **43**, 457 (1998). doi:10.1209/epl/i1998-00381-x

209. I.P. McCulloch, From density-matrix renormalization group to matrix product states. J. Stat. Mech. Theory Exp. **2007**, P10014 (2007). doi:10.1088/1742-5468/2007/10/P10014
210. C.K. Majumdar, D.K. Ghosh, On next-nearest-neighbor interaction in linear chain I. J. Math. Phys. **10**, 1388 (1969). doi:10.1063/1.1664978
211. C.K. Majumdar, D.K. Ghosh, On Next-Nearest-Neighbor Interaction in Linear Chain II. J. Math. Phys. **10**, 1388 (1969). doi:10.1063/1.1664978
212. C.K. Majumdar, Antiferromagnetic model with known ground state. J. Phys. C Solid State Phys. **3**, 911 (1970). doi:10.1088/0022-3719/3/4/019
213. M. Fannes, B. Nachtergaele, R.F. Werner, Finitely correlated states on quantum spin chains. Commun. Math. Phys. **144**, 443 (1992). doi:10.1007/BF02099178
214. G. Vidal, Efficient classical simulation of slightly entangled quantum computations. Phys. Rev. Lett. **91**, 4 (2003). doi:10.1103/PhysRevLett.91.147902
215. G. Vidal, Efficient simulation of one-dimensional quantum many-body systems. Phys. Rev. Lett. **93**, 4 (2004). doi:10.1103/PhysRevLett.93.040502
216. F. Verstraete, J.I. Cirac, Valence-bond states for quantum computation. Phys. Rev. A **70**, 060302 (2004). doi:10.1103/PhysRevA.70.060302
217. F. Verstraete, D. Porras, J.I. Cirac, Density matrix renormalization group and periodic boundary conditions: a quantum information perspective. Phys. Rev. Lett. **93**, 227205 (2004). doi:10.1103/PhysRevLett.93.227205
218. C. Schön, E. Solano, F. Verstraete, J.I. Cirac, M.M. Wolf, Sequential generation of entangled multiqubit states. Phys. Rev. Lett. **95**, 4 (2005). doi:10.1103/PhysRevLett.95.110503
219. C. Schön, K. Hammerer, M.M. Wolf, J.I. Cirac, E. Solano, Sequential generation of matrix-product states in cavity QED. Phys. Rev. A **75**, 11 (2007). doi:10.1103/PhysRevA.75.032311
220. T.J. Osborne, J. Eisert, F. Verstraete, Holographic quantum states. Phys. Rev. Lett. **105**, 6 (2010). doi:10.1103/PhysRevLett.105.260401
221. D. Pérez-García, F. Verstraete, M.M. Wolf, J.I. Cirac, Matrix product state representations. Quantum Inf. Comput. **7**, 401 (2007), arXiv:0608197
222. D.E. Evans, R. Hoegh-Krohn, Spectral properties of positive maps on C*-algebras. J. Lond. Math. Soc. **s2-17**, 345 (1978). doi:10.1112/jlms/s2-17.2.345
223. J.I. Cirac, D. Perez-Garcia, N. Schuch, F. Verstraete, Matrix Product Density Operators: Renormalization Fixed Points and Boundary Theories (2016), arXiv:1606.00608
224. N. Schuch, M.M. Wolf, F. Verstraete, J.I. Cirac, Entropy scaling and simulability by matrix product states. Phys. Rev. Lett. **100**, 4 (2008). doi:10.1103/PhysRevLett.100.030504
225. F. Pollmann, S. Mukerjee, A.M. Turner, J.E. Moore, Theory of finite-entanglement scaling at one-dimensional quantum critical points. Phys. Rev. Lett. **102**, 255701 (2009). doi:10.1103/PhysRevLett.102.255701
226. D. Pérez-García, M. Wolf, M. Sanz, F. Verstraete, J.I. Cirac, String order and symmetries in quantum spin lattices. Phys. Rev. Lett. **100**, 167202 (2008). doi:10.1103/PhysRevLett.100.167202
227. N. Schuch, D. Pérez-García, I. Cirac, Classifying quantum phases using matrix product states and projected entangled pair states. Phys. Rev. B **84**, 165139 (2011). doi:10.1103/PhysRevB.84.165139
228. X. Chen, Z.-C. Gu, X.-G. Wen, Classification of gapped symmetric phases in one-dimensional spin systems. Phys. Rev. B **83**, 035107 (2011). doi:10.1103/PhysRevB.83.035107
229. F. Pollmann, E. Berg, A.M. Turner, M. Oshikawa, Symmetry protection of topological phases in one-dimensional quantum spin systems. Phys. Rev. B **85**, 075125 (2012). doi:10.1103/PhysRevB.85.075125
230. G. Vidal, Classical simulation of infinite-size quantum lattice systems in one spatial dimension. Phys. Rev. Lett. **98**, 5 (2007). doi:10.1103/PhysRevLett.98.070201
231. R. Orús, G. Vidal, Infinite time-evolving block decimation algorithm beyond unitary evolution. Phys. Rev. B **78**, 155117 (2008). doi:10.1103/PhysRevB.78.155117
232. J. Haegeman, J.I. Cirac, T.J. Osborne, I. Pizorn, H. Verschelde, F. Verstraete, Time-dependent variational principle for quantum lattices. Phys. Rev. Lett. **107**, 070601 (2011). doi:10.1103/PhysRevLett.107.070601

233. S.R. White, Density matrix renormalization group algorithms with a single center site. Phys. Rev. B **72**, 180403 (2005). doi:10.1103/PhysRevB.72.180403
234. I.P. McCulloch, Infinite size density matrix renormalization group, revisited (2008), arXiv:0804.2509
235. J.A. Kjäll, M.P. Zaletel, R.S.K. Mong, J.H. Bardarson, F. Pollmann, Phase diagram of the anisotropic spin-2 XXZ model: infinite-system density matrix renormalization group study. Phys. Rev. B **87**, 235106 (2013). doi:10.1103/PhysRevB.87.235106
236. J. Haegeman, T.J. Osborne, F. Verstraete, Post-matrix product state methods: to tangent space and beyond. Phys. Rev. B **88**, 075133 (2013). doi:10.1103/PhysRevB.88.075133
237. A. Milsted, J. Haegeman, T.J. Osborne, Matrix product states and variational methods applied to critical quantum field theory. Phys. Rev. D **88**, 085030 (2013). doi:10.1103/PhysRevD.88.085030
238. J. Haegeman, M. Mariën, T.J. Osborne, F. Verstraete, Geometry of matrix product states: metric, parallel transport, and curvature. J. Math. Phys. **55**, 021902 (2014). doi:10.1063/1.4862851
239. P. Pippan, S.R. White, H.G. Evertz, Efficient matrix-product state method for periodic boundary conditions. Phys. Rev. B **81**, 81103 (2010). doi:10.1103/PhysRevB.81.081103
240. B. Pirvu, F. Verstraete, G. Vidal, Exploiting translational invariance in matrix product state simulations of spin chains with periodic boundary conditions. Phys. Rev. B **83**, 125104 (2011). doi:10.1103/PhysRevB.83.125104
241. D. Porras, F. Verstraete, J.I. Cirac, Renormalization algorithm for the calculation of spectra of interacting quantum systems. Phys. Rev. B **73** (2006). doi:10.1103/PhysRevB.73.014410
242. B. Pirvu, V. Murg, J.I. Cirac, F. Verstraete, Matrix product operator representations. New J. Phys. **12**, 025012 (2010). doi:10.1088/1367-2630/12/2/025012
243. A.J. Daley, C. Kollath, U. Schollwöck, G. Vidal, Time-dependent density-matrix renormalization-group using adaptive effective Hilbert spaces. J. Stat. Mech. Theory Exp. **2004**, P04005 (2004). doi:10.1088/1742-5468/2004/04/P04005
244. S.R. White, A.E. Feiguin, Real-time evolution using the density matrix renormalization group. Phys. Rev. Lett. **93**, 076401 (2004). doi:10.1103/PhysRevLett.93.076401
245. F. Verstraete, J.J. García-Ripoll, J.I. Cirac, Matrix product density operators: simulation of finite-temperature and dissipative systems. Phys. Rev. Lett. **93**, 207204 (2004). doi:10.1103/PhysRevLett.93.207204
246. J. Haegeman, C. Lubich, I. Oseledets, B. Vandereycken, F. Verstraete, Unifying time evolution and optimization with matrix product states (2014), arXiv:1408.5056
247. V. Zauner, M. Ganahl, H.G. Evertz, T. Nishino, Time evolution within a comoving window: scaling of signal fronts and magnetization plateaus after a local quench in quantum spin chains. J. Phys. Condens. Matter **27**, 425602 (2012). doi:10.1088/0953-8984/27/42/425602
248. H.N. Phien, G. Vidal, I.P. McCulloch, Infinite boundary conditions for matrix product state calculations. Phys. Rev. B **86**, 245107 (2012). doi:10.1103/PhysRevB.86.245107
249. A. Milsted, J. Haegeman, T.J. Osborne, F. Verstraete, Variational matrix product ansatz for nonuniform dynamics in the thermodynamic limit. Phys. Rev. B **88**, 155116 (2013). doi:10.1103/PhysRevB.88.155116
250. K. Hallberg, Density-matrix algorithm for the calculation of dynamical properties of low-dimensional systems. Phys. Rev. B **52**, R9827 (1995). doi:10.1103/PhysRevB.52.R9827
251. P.E. Dargel, A. Honecker, R. Peters, R.M. Noack, T. Pruschke, Adaptive Lanczos-vector method for dynamic properties within the density matrix renormalization group. Phys. Rev. B **83**, 161104 (2011). doi:10.1103/PhysRevB.83.161104
252. P.E. Dargel, A. Wöllert, A. Honecker, I.P. McCulloch, U. Schollwöck, T. Pruschke, Lanczos algorithm with matrix product states for dynamical correlation functions. Phys. Rev. B **85**, 205119 (2012). doi:10.1103/PhysRevB.85.205119
253. S. Ramasesha, S.K. Pati, H.R. Krishnamurthy, Z. Shuai, J.L. Brédas, Symmetrized density-matrix renormalization-group method for excited states of Hubbard models. Phys. Rev. B **54**, 7598 (1996). doi:10.1103/PhysRevB.54.7598

254. T. Kühner, S. White, Dynamical correlation functions using the density matrix renormalization group. Phys. Rev. B **60**, 335 (1999). doi:10.1103/PhysRevB.60.335
255. E. Jeckelmann, Dynamical density-matrix renormalization-group method. Phys. Rev. B **66**, 045114 (2002). doi:10.1103/PhysRevB.66.045114
256. A. Weichselbaum, F. Verstraete, U. Schollwöck, J.I. Cirac, J. von Delft, Variational matrix-product-state approach to quantum impurity models. Phys. Rev. B **80**, 165117 (2009). doi:10.1103/PhysRevB.80.165117
257. S.R. White, I. Affleck, Spectral function for the S = 1 Heisenberg antiferromagetic chain. Phys. Rev. B **77**, 134437 (2008). doi:10.1103/PhysRevB.77.134437
258. R. Pereira, S. White, I. Affleck, Exact edge singularities and dynamical correlations in spin-1/2 chains. Phys. Rev. Lett. **100**, 4 (2008). doi:10.1103/PhysRevLett.100.027206
259. T. Barthel, U. Schollwöck, S. White, Spectral functions in one-dimensional quantum systems at finite temperature using the density matrix renormalization group. Phys. Rev. B **79**, 245101 (2009). doi:10.1103/PhysRevB.79.245101
260. J. Kjäll, F. Pollmann, J. Moore, Bound states and E_{8} symmetry effects in perturbed quantum Ising chains. Phys. Rev. B **83**, 020407 (2011). doi:10.1103/PhysRevB.83.020407
261. L. Seabra, F. Pollmann, Exotic Ising dynamics in a Bose-Hubbard model. Phys. Rev. B **88**, 5 (2013). doi:10.1103/PhysRevB.88.125103
262. A. Holzner, A. Weichselbaum, I.P. McCulloch, U. Schollwöck, J. von Delft, Chebyshev matrix product state approach for spectral functions. Phys. Rev. B **83**, 195115 (2011). doi:10.1103/PhysRevB.83.195115
263. A. Feiguin, S. White, Finite-temperature density matrix renormalization using an enlarged Hilbert space. Phys. Rev. B **72**, 220401 (2005). doi:10.1103/PhysRevB.72.220401
264. S. White, Minimally entangled typical quantum states at finite temperature. Phys. Rev. Lett. **102**, 190601 (2009). doi:10.1103/PhysRevLett.102.190601
265. C. Karrasch, J.H. Bardarson, J.E. Moore, Finite-temperature dynamical density matrix renormalization group and the drude weight of spin-1/2 chains. Phys. Rev. Lett. **108**, 227206 (2012). doi:10.1103/PhysRevLett.108.227206
266. S.R. White, Spin gaps in a frustrated Heisenberg model for CaV_4O_9. Phys. Rev. Lett. **77**, 3633 (1996). doi:10.1103/PhysRevLett.77.3633
267. E. Stoudenmire, S.R. White, Studying two-dimensional systems with the density matrix renormalization group. Annu. Rev. Condens. Matter Phys. **3**, 111 (2012). doi:10.1146/annurev-conmatphys-020911-125018
268. F. Verstraete, J.I. Cirac, Renormalization algorithms for Quantum-Many Body Systems in two and higher dimensions (2004), arXiv:0407066
269. D. Perez-Garcia, F. Verstraete, J.I. Cirac, M.M. Wolf, PEPS as unique ground states of local Hamiltonians. Quantum Inf. Comput. **8**, 0650 (2007), http://www.rintonpress.com/journals/qiconline.html
270. F. Verstraete, M.M. Wolf, D. Perez-Garcia, J.I. Cirac, Criticality, the area law, and the computational power of projected entangled pair states. Phys. Rev. Lett. **96**, 220601 (2006). doi:10.1103/PhysRevLett.96.220601
271. N. Schuch, D. Poilblanc, J.I. Cirac, D. Pérez-García, Resonating valence bond states in the PEPS formalism. Phys. Rev. B **86**, 115108 (2012). doi:10.1103/PhysRevB.86.115108
272. Z.-C. Gu, M. Levin, B. Swingle, X.-G. Wen, Tensor-product representations for string-net condensed states. Phys. Rev. B **79**, 085118 (2009). doi:10.1103/PhysRevB.79.085118
273. O. Buerschaper, M. Aguado, G. Vidal, Explicit tensor network representation for the ground states of string-net models. Phys. Rev. B **79**, 085119 (2009). doi:10.1103/PhysRevB.79.085119
274. N. Schuch, M. Wolf, F. Verstraete, J.I. Cirac, Computational complexity of projected entangled pair states. Phys. Rev. Lett. **98**, 140506 (2007). doi:10.1103/PhysRevLett.98.140506
275. J.I. Cirac, D. Poilblanc, N. Schuch, F. Verstraete, Entanglement spectrum and boundary theories with projected entangled-pair states. Phys. Rev. B **83**, 245134 (2011). doi:10.1103/PhysRevB.83.245134

276. N. Schuch, D. Poilblanc, J.I. Cirac, D. Pérez-García, Topological order in the projected entangled-pair states formalism: transfer operator and boundary Hamiltonians. Phys. Rev. Lett. **111**, 090501 (2013). doi:10.1103/PhysRevLett.111.090501
277. S. Yang, L. Lehman, D. Poilblanc, K. Van Acoleyen, F. Verstraete, J.I. Cirac, N. Schuch, Edge theories in projected entangled pair state models. Phys. Rev. Lett. **112**, 036402 (2014). doi:10.1103/PhysRevLett.112.036402
278. N. Schuch, J.I. Cirac, D. Pérez-García, PEPS as ground states: degeneracy and topology. Ann. Phys. **325**, 2153 (2010). doi:10.1016/j.aop.2010.05.008
279. O. Buerschaper, Twisted injectivity in projected entangled pair states and the classification of quantum phases. Ann. Phys. **351**, 447 (2014). doi:10.1016/j.aop.2014.09.007
280. M.B. Şahinoğlu, D. Williamson, N. Bultinck, M. Mariën, J. Haegeman, N. Schuch, F. Verstraete, Characterizing Topological Order with Matrix Product Operators (2014), arXiv:1409.2150
281. N. Bultinck, M. Mariën, D.J. Williamson, M.B. Şahinoğlu, J. Haegeman, F. Verstraete, Anyons and matrix product operator algebras (2015), arXiv:1511.08090
282. J. Haegeman, V. Zauner, N. Schuch, F. Verstraete, Shadows of anyons and the entanglement structure of topological phases. Nat. Commun. **6**, 8284 (2015). doi:10.1038/ncomms9284
283. M. Mariën, J. Haegeman, P. Fendley, F. Verstraete, Condensation-Driven Phase Transitions in Perturbed String Nets (2016), arXiv:1607.05296
284. G. Vidal, Entanglement renormalization. Phys. Rev. Lett. **99**, 220405 (2007). doi:10.1103/PhysRevLett.99.220405
285. F. Verstraete, J.I. Cirac, J. Latorre, E. Rico, M. Wolf, Renormalization-group transformations on quantum states. Phys. Rev. Lett. **94**, 5 (2005). doi:10.1103/PhysRevLett.94.140601
286. G. Evenbly, G. Vidal, Algorithms for entanglement renormalization: boundaries, impurities and interfaces. J. Stat. Phys. **157**, 931 (2014). doi:10.1007/s10955-014-0983-1
287. V. Zauner, D. Draxler, L. Vanderstraeten, M. Degroote, J. Haegeman, M.M. Rams, V. Stojevic, N. Schuch, F. Verstraete, Transfer matrices and excitations with matrix product states. New J. Phys. **17**, 053002 (2015). doi:10.1088/1367-2630/17/5/053002
288. M. Bal, M.M. Rams, V. Zauner, J. Haegeman, F. Verstraete, Matrix product state renormalization (2015), arXiv:1509.01522
289. F. Verstraete, J.I. Cirac, Continuous matrix product states for quantum fields. Phys. Rev. Lett. **104** (2010). doi:10.1103/PhysRevLett.104.190405
290. J. Haegeman, J.I. Cirac, T.J. Osborne, H. Verschelde, F. Verstraete, Applying the variational principle to $(1+1)$-dimensional quantum field theories. Phys. Rev. Lett. **105**, 251601 (2010). doi:10.1103/PhysRevLett.105.251601
291. J. Haegeman, J.I. Cirac, T.J. Osborne, F. Verstraete, Calculus of continuous matrix product states. Phys. Rev. B **88**, 085118 (2013). doi:10.1103/PhysRevB.88.085118
292. D. Draxler, J. Haegeman, T.J. Osborne, V. Stojevic, L. Vanderstraeten, F. Verstraete, Particles, holes, and solitons: a matrix product state approach. Phys. Rev. Lett. **111**, 020402 (2013). doi:10.1103/PhysRevLett.111.020402
293. J. Rincon, M. Ganahl, G. Vidal, Lieb-Liniger model with exponentially-decaying interactions: a continuous matrix product state study. Phys. Rev. B **92**, 115107 (2015). doi:10.1103/PhysRevB.92.115107
294. D. Draxler, J. Haegeman, F. Verstraete, M. Rizzi, Atomtronics - a continuous matrix product state approach (2016), arXiv:1609.09704
295. J. Haegeman, D. Draxler, V. Stojevic, J.I. Cirac, T.J. Osborne, F. Verstraete, Quantum Gross–Pitaevskii Equation (2015), arXiv:1501.06575
296. K. Wilson, J.B. Kogut, The renormalization group and the ϵ expansion. Phys. Rep. **12**, 75 (1974). doi:10.1016/0370-1573(74)90023-4
297. M.E. Peskin, D.V. Schroeder, *An Introduction to Quantum Field Theory* (Westview Press, Colorado, 1995)
298. E. Fradkin, *Field Theories of Condensed Matter Physics*, 2nd edn. (Cambridge University Press, Cambridge, 2013)
299. P. Nozières, *Theory Of Interacting Fermi Systems* (W.A. Benjamin Inc., 1964)

300. R. Shankar, Renormalization-group approach to interacting fermions. Rev. Mod. Phys. **66**, 129 (1994). doi:10.1103/RevModPhys.66.129
301. L.D. Landau, Oscillations in a Fermi liquid. JETP **30**, 1058 (1956)
302. L.D. Landau, The theory of a Fermi liquid. JETP **3**, 920 (1957)

Chapter 3
Effective Particles in Quantum Spin Chains: The Framework

In the previous chapter we have already highlighted the importance of emergent particles as the effective degrees of freedom that determine the properties of strongly-correlated quantum matter at low energies. In this chapter we will present a particular realization of this theme for one-dimensional quantum spin systems; the central idea will be to determine the elementary excitations variationally and treat them as particles in a many-particle description. The rationale behind this approach is perfectly explained by P.W. Anderson in his *Concepts in Solids*:

> There are two ideas behind the concept of elementary excitations. First is the idea that the total binding energy of the ground state is not a very important physical quantity, and does not have much to do with the behavior of a physical system. What *is* important physically is the behavior of the lower excited states relative to the ground state: those states, that is, which are likely to be excited at relatively low temperatures or by weak external fields. We think immediately of a metal or a semiconductor, in which all the behavior is determined by the low excited states which we speak of as having a few moving charge carriers, or of the elastic or thermal properties of a solid, determined by the presence of a small number of lattice waves, which we call phonons. Thus our interest is often focused on the set of low-lying excited states of a system as the physically most fundamental property of it.
>
> The second idea is that the low-lying states often – in fact almost always – are of a particularly simple character, and can be treated with much greater mathematical rigor and physical understanding than other states. Let me explain the reason for this [1, p. 99].

Anderson's reasons are in fact quite simple. The operator q_k that connects the ground state ψ_g and an excited state ψ_k with momentum k (such that $\psi_k = q_k^\dagger \psi_g$) must inevitably represent only a very small displacement of the entire system. This means that ψ_k will look very much the same as ψ_g and that wave packets can be built

$$q^\dagger_{\text{packet}}(k_0, r_0) = \int \mathrm{d}k \exp\left(-ikr_0 - \frac{1}{2}\left(\frac{k-k_0}{\Delta k}\right)^2 \right) q_k^\dagger,$$

L. Vanderstraeten, *Tensor Network States and Effective Particles for Low-Dimensional Quantum Spin Systems*, Springer Theses,
DOI 10.1007/978-3-319-64191-1_3

which only disturb the ground state in a local region near r_0 of size $1/\Delta k$. If now a second wave packet is introduced, it is easy to see that in a large system the two packets will not interfere very much. Thus, if the excited state has excitation energy E_k, the state containing two such excitations has an energy equal to the sum of the two energies up to corrections of $\mathcal{O}(1/N)$ in system size. When only a small number of excitations is present, interactions between the excitations will be weak and the system can be treated as a gas of independent particles. Only after the particles have been identified as such, the interactions are taken into account.

We work out this fundamental idea within the framework of spin chains and matrix product states. In contrast to more common approaches based on perturbation theory, our approach is variational: we start from the quasi-exact ground state in the form of a matrix product state, on which we build momentum superpositions of localized perturbations that form an extremely good approximation to the excitations of the system. We will interpret these excitations as independent particles and characterize their interactions through the two-particle S matrix and the spectral function. This will give us the key for doing many-particle physics above a strongly-correlated ground state that captures a system's properties at low energies.

3.1 The Variational Approach

The history of this variational approach for describing effective particles traces back to the works of R.P. Feynman on liquid helium [2–6]. Because superfluid ^{4}He is a Bose liquid, the standard Fermi-liquid theory could not be applied to understand the nature of the low-lying excitations [7]. Still Feynman wanted to understand the spectrum in terms of elementary excitations that have a simple physical interpretation. He built on the work of L. Landau who first stated that "every weakly excited state must be considered as a combination of simple "elementary excitations" " [8]. Based on quantum hydrodynamics, Landau showed that the spectrum should consist of gapless phonons with linear dispersion $\epsilon = cp$ and gapped rotons dispersing as $\epsilon = \Delta + p^2/2\mu$. Feynman presented a physical picture for these elementary excitations that fully incorporated the Bose statistics of liquid ^{4}He. He developed a variational wavefunction for the excitations,

$$\phi_k(\vec{r}_1, \ldots, \vec{r}_N) = \rho_k \, \psi(\vec{r}_1, \ldots, \vec{r}_N)$$

where ψ is the ground-state wave function for a collection of N particles and ρ_k the momentum representation of the density operator

$$\rho_k = \frac{1}{\sqrt{N}} \sum_{j=1}^{N} \mathrm{e}^{-i\vec{k}\cdot\vec{r}_j} \rho(\vec{r}_j).$$

The essential idea of this ansatz wave function is that it has all the right correlations automatically built in by using the ground-state wave function as a factor. Moreover, one can actually compute the excitation energy as

$$\Delta(k) = \frac{\langle\phi_k|H - E_0|\phi_k\rangle}{\langle\phi_k|\phi_k\rangle} = \frac{\langle\psi|\left[\rho_k^\dagger, [H, \rho_k]\right]|\psi\rangle}{\langle\psi|\rho_k^\dagger\rho_k|\psi\rangle} = \frac{f(k)}{s(k)}.$$

Because the particles' interactions are simply function of the position, the commutator vanishes for the interaction part of the Hamiltonian and only the kinetic energy contributes to $f(k)$. The denominator $s(k)$ is the static structure factor, which can be directly measured with neutron scattering. Feynman obtained a linear dispersion of the phonon modes and a "roton minimum" at higher momentum, confirming the earlier results by Landau. In this analysis, Feynman could show that this second minimum is a consequence of a peak in the structure factor $s(k)$, which is a static property of the ground state. Later on, Feynman and Cohen showed how they could improve the roton gap by including a "backflow" factor in the ansatz wave function. This factor contained a free variational parameter, which was optimized numerically.

In later years the Feynman–Bijl ansatz[1] or "single-mode approximation" was used for describing the collective modes in strongly-correlated systems such as the plasmon modes in three-dimensional electron systems [10, 11] and the collective excitation spectrum in the fractional quantum Hall effect [12, 13]. In these works, it is always the case that the characteristic dispersion of these collective modes is understood from a kinetic energy term renormalized by the factor $1/s(k)$ which represents the effect of the correlated motion of the particles in the ground state.

In one-dimensional quantum spin systems, where quantum fluctuations are especially strong, the elementary excitations have a collective nature so that the single-mode approximation is expected to work quite well. The first application to spin chains was done by Arovas, Auerbach, and Haldane in [14] on the excitation spectrum of the AKLT chain [15, 16]. Starting from the AKLT ground state $|\Psi\rangle_{\mathrm{AKLT}}$, they proposed the following ansatz for capturing the elementary threefold-degenerate magnon excitation

$$|\Phi(\kappa)\rangle_{\mathrm{SMA}} = \sum_n \mathrm{e}^{i\kappa n} S_n^\alpha |\Psi\rangle_{\mathrm{AKLT}}, \qquad \alpha = x, y, z. \tag{3.1}$$

Because the AKLT ground state has a simple form, the single-mode energy can be easily computed (in contrast to the case of liquid helium, where $s(k)$ had to be measured experimentally); the result for the magnon dispersion relation was in qualitative agreement with results from exact diagonalization. With the advent of numerical techniques for simulating the ground state properties of spin chains, the single-mode approximation was also tested on e.g. the magnon spectrum of the spin-1 Heisenberg chain [17–19].

[1] The ansatz was named after Feynman – for obvious reasons – and Bijl, who supposedly wrote down a similar wavefunction in Ref. [9].

In order to improve the single-mode approximation and to obtain quantitatively more accurate excitation spectra, it is paramount to allow for more variational freedom in the ansatz wave function – similar to the backflow factor that Feynman introduced in his wave function. For one-dimensional spin systems, the MPS framework makes this possible: MPS provide a quasi-exact description of ground states and allow to modify this ground state locally in the spirit of Feynman. In Refs. [20, 21] it was shown how, instead of acting with a physical operator as in (3.1), one can act on the virtual level of the MPS to construct a variational ansatz for elementary excitations. Recently this ansatz was greatly improved, as it was realized that acting both on the physical *and* virtual level of the MPS [22, 23] gives rise to extremely accurate results on the spectra of gapped spin chains.[2]

In the spirit of Anderson, the next question that opens up is whether these elementary excitations can be treated as particles on a strongly-correlated background, and whether a many-particle theory can be written down. Feynman already argued in favour of such a picture: By constructing wave packets that are confined to a local region, one can think of adding multiple excitations in distinct regions such that the interactions can be neglected. The wave function for two excitations would correspond to

$$\phi_{k_1,k_2}(\vec{r}_1,\ldots,\vec{r}_N) \propto \left[\sum_{j=1}^{N} \mathrm{e}^{(-i\vec{k}_1\cdot\vec{r}_j)}\rho(\vec{r}_j)\right]\left[\sum_{j=1}^{N} \mathrm{e}^{(-i\vec{k}_2\cdot\vec{r}_j)}\rho(\vec{r}_j)\right]\psi(\vec{r}_1,\ldots,\vec{r}_N)$$

with an energy of $E(k_1) + E(k_2)$. Feynman concludes that "the excitations should behave much like interacting Bose particles which may be created and destroyed, and whose energy as a function of momentum is given by $E(k) = \hbar^2 k^2/2ms(k)$" [4].

Because the particles interact, Feynman's wave function for two excitations cannot be correct: the two excitations will scatter off each other and the elements of the S matrix will show up in the stationary scattering states. A priori it is not clear how a treatment of the particles' interactions fits in the variational picture, since scattering theory is typically derived in a more dynamical setting. In this chapter we will show how stationary scattering theory allows to do just that: We will characterize the particle interactions by computing the two-particle wavefunctions in a variational way. Based on this characterization, we will develop a many-particle description that takes the particle interactions fully into account.

One could ask, however, whether this approach for tackling the many-body problem can be justified in a rigorous fashion and under general conditions. In the context of relativistic quantum field theory [26], this has been realized by the advent of axiomatic quantum field theory [27, 28] and, more specifically, the formulation of Haag–Ruelle scattering theory [29, 30]. At first it seems impossible to develop a similar picture for lattice systems, because there is no strict causality or Lorentz invariance. Indeed, the spectrum cannot be built up by simply boosting the different

[2]Other MPS implementations of the single-mode approximation include Refs. [24, 25].

particles and, in practice, the spectrum is a lot more complicated (see Fig. 3.1). A priori, it is not even clear that elementary excitations have a local particle-like nature at all.

Recently though, using Lieb–Robinson bounds [31] it has been proven [41] that the particle picture is also applicable to lattice systems. More specifically, it has been shown that excited states living on an isolated branch can be approximated by acting with a momentum superposition of a localized operator with an error that is exponentially small in the linear size of the support of the operator. The proviso is that the excited state should have a finite overlap with a local operator, i.e. it should contribute to *some* spectral function (this overlap can in principle be arbitrarily small). The rate at which the approximation converges to the exact excitation as a function of the size of the operator l depends on the energy gap δE as

$$\text{error} \leq p(l) \exp\left(-\frac{\delta E}{2v} l\right)$$

with $p(l)$ a polynomial in l and v a characteristic velocity; the energy gap is the minimal energy difference to another eigenstate within the same momentum sector. This was subsequently extended in formulating a consistent lattice version of Haag–Ruelle scattering theory [42].

A picture of the elementary particle-like excitation spectrum of a quantum spin system in the thermodynamic limit emerges that looks much like the one in Fig. 3.1. The isolated branches in these momentum-energy plots correspond to dispersive localized particle-like excitations that can be targeted by some version of the single-mode approximation. Crucially, these particles correspond to the exact eigenstates of the fully-interacting Hamiltonian and, consequently, have an infinite lifetime. In this sense, the variational approach is orthogonal to perturbative approaches where the particles are defined from a non-interacting limit and acquire a finite lifetime when the interactions are turned on. Put differently, the variational particles show up as δ functions in spectral functions, not as broadened resonances inside a continuous spectrum.

Higher up in the spectrum multi-particle scattering states appear as continuous regions. In the absence of Lorentz symmetry, particles can become unstable or have multiple minima in their dispersion relations. In a low-energy regime, these different minima might even be identified as different "particles" – think about the roton minimum in ^{4}He.[3]

A last remark is in order that situates our work within a broader context. In standard quantum many-body physics, a difference is made between "elementary excitations" and "quasiparticles" [44, 79]. The former are the exact eigenfunctions of the fully-interacting Hamiltonian and are more of a "bookkeeping arrangement" because there is no simple operator that can create them, whereas the latter correspond to poles in

[3]Importantly, every isolated branch is interpreted as a new "elementary particle", so, in the absence of some particle number, there is no way to differentiate a bound state that is supposedly composed of two elementary particles in the system (see Ref. [43] for the analogous result in QFT).

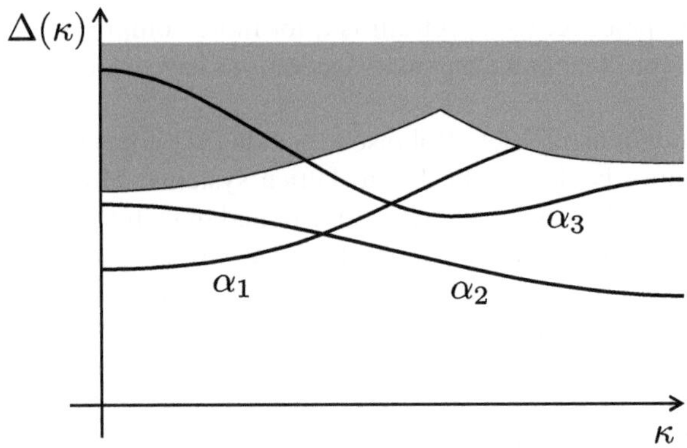

Fig. 3.1 A typical momentum-energy excitation spectrum of a one-dimensional lattice system. We have depicted three elementary (one-particle) excitations (*full lines*) and the many-particle continuum (*grey*). Both α_2 and α_3 are stable in the whole Brillouin zone; the latter remains stable even inside the continuum, possibly because it cannot decay in a two-particle state through symmetry constraints. Particle α_1 becomes unstable upon entering the continuum, so that it ceases to be a one-particle excitation (i.e. cannot be created by a local operator)

the Fourier transform of dynamical correlation functions and have a complex energy. The relation between the two is not straightforward:

> Nevertheless, there is a reputed connection, which has never been made very precise between quasiparticles and elementary excitations in the sense we have used them above. The difference between the two, and the reason one decays and the other does not, lies in the form of the wave functions. The elementary excitations refer to exact eigenfunctions which unfortunately, for an interacting system, do not have the plane wave character of the excitations of a noninteracting gas. The quasiparticle, on the other hand, may be thought of as an attempt to find (inexact) wave functions which do have a plane wave character [79].

From this perspective, our variational approach identifies the two concepts: we target exact eigenfunctions of the interacting Hamiltonian, we describe them as plane waves, and they contribute as singularities to dynamical correlation functions. Let us just call them *particles* for the remainder of this chapter.

3.2 The Ground State and Its Correlations

The central idea of the single-mode approximation was the use of the correlations in the ground state to build a good variational ansatz for the low-lying excitations. For gapped one-dimensional systems, matrix product states provide an excellent parametrization of ground states and carry the right correlations to target the low-lying excited states. Moreover, because they provide a local description of ground state wavefunctions, they can also be perturbed locally in a systematic way. For that reason, they provide the ideal background to build elementary excitations in the spirit of the single-mode approximation. In this section, we will elaborate on the form of uniform matrix product states (uMPS) in the thermodynamic limit, show how to compute observables and focus on the ground-state correlations through the introduction of the transfer matrix.

Let us consider a one-dimensional translation invariant lattice system in the thermodynamic limit, where every site hosts a quantum spin of dimension d. The system is specified by a local nearest-neighbour Hamiltonian[4]

$$H = \sum_n h_{n,n+1},$$

a sum of hermitian operators acting on sites n and $n+1$. A uMPS can be defined as

$$|\Psi(A)\rangle = \sum_{\{s\}} \boldsymbol{v}_L^\dagger \left[\prod_{m\in\mathbb{Z}} A^{s_m} \right] \boldsymbol{v}_R |\{s\}\rangle \tag{3.2}$$

where the sum runs over all possible spin configurations. In this expression, A^s is a set of d matrices or, if we interpret the label s as another index, a three-dimensional tensor. The size of the matrix dimensions of A is called the bond dimension D and can be arbitrary large. The vectors $\boldsymbol{v}_L^\dagger$ and $\boldsymbol{v}_R$ are boundary vectors and, as we will see, do not determine the state $|\Psi(A)\rangle$ in any way; the state depends solely on the tensor A. Graphically, a uMPS can be represented as

$|\Psi(A)\rangle = \Big|$ 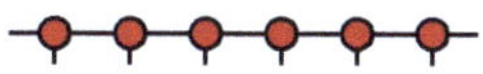 $\Big\rangle$.

This graphical representation shows indeed that one tensor A parametrizes a translation-invariant quantum state in the bulk of the system, where the virtual connections carry the correlations between the different sites.

For computing the norm of the uMPS, we first define the *transfer matrix* E associated to the tensor A as[5]

$$E = \sum_{s=1}^{d} A^s \otimes \bar{A}^s,$$

[4]In this chapter we will only consider nearest-neighbour Hamiltonians, but everything can be straightforwardly extended to larger, yet local interactions – the case of long-range interactions is not included.

[5]In this chapter all computations will involve this transfer matrix or variations of it. Let us therefore introduce the notations

$$E = E_A^A = \sum_{s=1}^{d} A^s \otimes \bar{A}^s,$$

for the transfer matrix,

$$O_A^A = \sum_{s,t} A^s \otimes \bar{A}^t \langle s|O|t\rangle.$$

for the transfer matrix with a one-site operator, and

where a bar indicates complex conjugation of the tensor. The transfer matrix is a completely positive map, so that its largest eigenvalue[6] is positive, and can be put to one by rescaling the tensor A. All the other eigenvalues lie within the unit circle and can, in general, be complex. The left and right eigenvectors corresponding to the largest eigenvalue are denoted as $(l|$ and $|r)$ and, after having rescaled the tensor A, obey the relations[7]

$$(l|E = (l| \quad \text{and} \quad E|r) = |r).$$

These vectors will be normalized such that the operator $|l)(r|$ is a projector on the subspace corresponding to the largest eigenvalue of E. The norm of the MPS is then given by

$$\langle\Psi(A)|\Psi(A)\rangle = \left(\boldsymbol{v_L}\boldsymbol{v_L}^\dagger\right)\left(\prod_{m\in\mathbb{Z}} E\right)\left(\boldsymbol{v_R}\boldsymbol{v_R}^\dagger\right).$$

The infinite product reduces to the projector $|r)(l|$, such that the norm is determined by the overlap between the boundary vectors and the left and right fixed points. As the fixed points are positive matrices, this overlap is non-zero, so that the norm reduces to a finite value. In all computations of bulk properties, this value will drop out, so we can safely put it to one; in the following we will always assume that $\langle\Psi(A)|\Psi(A)\rangle = 1$.

A local expectation value can be computed in a similar way. Indeed, left and right of the operator an infinite product of transfer matrices appear, which can be replaced by the left and right fixed points

(Footnote 5 continued)

$$O_{AA}^{AA} = \sum_{s,t,s',t'} A^s A^{s'} \otimes \bar{A}^t \bar{A}^{t'} \langle ss'|O|tt'\rangle.$$

for a two-site operator. We will also need "transfer matrices" with different tensors; notations:

$$E_B^A = \sum_{s=1}^{d} A^s \otimes \bar{B}^s$$

$$O_B^A = \sum_{s,t} A^s \otimes \bar{B}^t \langle s|O|t\rangle$$

$$O_{CD}^{AB} = \sum_{s,t,s',t'} A^s B^{s'} \otimes \bar{C}^t \bar{D}^{t'} \langle ss'|O|tt'\rangle.$$

[6]It is assumed that the largest eigenvalue is unique, which is only true for so-called injective MPS. In the following, we will always assume this is the case – for all practical applications this assumption is true – and we refer to e.g. Ref. [85] for more details.

[7]One can associate a completely positive map to the transfer matrix, for which the left and right fixed points are guaranteed to be positive definite matrices when reshaped to $D \times D$ matrices l and r.

$$\langle\Psi(A)|O|\Psi(A)\rangle = \boldsymbol{v_L}\boldsymbol{v_L}^{\dagger}\left(\prod_{m<0} E\right) O_A^A \left(\prod_{m>0} E\right)\boldsymbol{v_R}\boldsymbol{v_R}^{\dagger} = (l|O_A^A|r).$$

In all expectation values, the fixed points $(l|$ and $|r)$ determine the environment of a local region due to the infinite product of transfer matrices. In this sense, they allow to only focus on bulk observables.

The other eigenvalues of the transfer matrix determine the correlations in the ground state. Indeed, static correlation functions are computed as

$$\langle\Psi(A)|Q_m P_{m+n}|\Psi(A)\rangle = (l|Q_A^A E^{n-1} P_A^A|r)$$

and, by decomposing the transfer matrix in its spectral representation, we obtain

$$\langle\Psi(A)|Q_m P_{m+n}|\Psi(A)\rangle = (l|Q_A^A|r)(l|P_A^A|r) + \sum_{\lambda} \lambda^n (l|Q_A^A|r_\lambda)(l_\lambda|P_A^A|r),$$

where λ runs over all eigenvalues of E with left and right eigenvectors $(l_\lambda|$ and $|r_\lambda)$. The first part is the product of the ground state expectation values, while the second part is the connected correlation function. Because all λ's lie within the unit circle, the correlations of an MPS always reduce to a sum of decaying exponentials. The largest λ, i.e. the second largest eigenvalue of the transfer matrix, determines the correlation length ξ and the pitch vector of the correlations Q

$$\xi = -\frac{1}{\log|\lambda_{\max}|} \quad \text{and} \quad Q = \arg(\lambda_{\max}). \tag{3.3}$$

In the previous section it was argued that the correlations in the ground state determine the dispersion of the collective excitation. More specifically, strong maxima in the structure factor will show up as minima in the dispersion relation. From the spectral representation of the correlation functions, it is clear that the phase of the largest eigenvalues of the transfer matrix will determine the momenta where we can expect minima in the dispersion relation of elementary excitations. This relation between the transfer matrix of a ground state and excitation spectrum seems to be valid generally [32] and confirms the validity of the single-mode approximation.

A uniform matrix product state is uniquely defined by the tensor A, but the converse is not true; there is a redundancy in the parametrization because different A tensors can give rise to the same physical state. Indeed, one can easily seen that the replacement

$$A^s \to X^{-1} A^s X \qquad \forall\, s$$

with X an invertible matrix, leaves the state (3.2) invariant. This gauge invariance can be turned into an advantage, because it allows for the determination of a so-called canonical form [85] of the tensor A. Indeed, a transformation of the above form can always be found that makes sure that e.g. the left fixed point of the new tensor $\tilde{A}$ reduces to the unit matrix. This implies that $\tilde{A}$ is an isometry if it is interpreted as a a matrix from the left virtual and physical index to the right virtual index. This or similar choices simplify ground state calculations a lot, but it doesn't matter very much for the purposes in this chapter, so we leave the gauge choice of the ground state undecided.

The question remains of how to find the tensor A that provides the best approximation of the ground state for a given Hamiltonian in the thermodynamic limit. According to the variational principle, this question reduces to the minimization of the energy density expectation value

$$e = \langle\Psi(A)|H|\Psi(A)\rangle = (l|H_{AA}^{AA}|r)$$

for normalized states. This highly non-linear optimization problem can be solved with a number of methods, such as infinite DMRG [33], imaginary time-evolution with the time-evolving block decimation [34, 35] or the time-dependent variational principle [36], and hybrid schemes [38]. Again, for the purposes of this chapter, which focuses solely on excitations on top of the ground state, we leave in the middle which algorithm to use to find this ground state.

What is important, though, is that the uMPS indeed provides a good approximation to the exact ground state of the Hamiltonian; because a matrix product state is variational, this is not always guaranteed to be the case. The quantity that determines how well a variational state approximates a true eigenstate is the variance, defined as

$$v = \langle\Psi(A)|(H-e)^2|\Psi(A)\rangle.$$

In Appendix A.1 we show how to compute this quantity.

3.3 One-Particle Excitations

The ground state, in the form of a uMPS in the thermodynamic limit, now serves as a reference state on which to build elementary excitations that are, in the spirit of the single-mode approximation, local perturbations in a momentum superposition. In this section, we will introduce the variational excitation ansatz, show how to compute expectation values and how to solve the variational optimization problem in order to find the low-lying excitations.[8]

3.3.1 The Variational Ansatz

The variational ansatz for an elementary excitation with momentum[9] κ is given by [23, 40]

[8] Additional details can be found in Ref. [40].

[9] Because we work in the thermodynamic limit, the momentum can take on every value between 0 and 2π.

$$|\Phi_\kappa(B)\rangle = \sum_n e^{i\kappa n} \sum_{\{s\}} v_L^\dagger \left[\prod_{m<n} A^{s_m}\right] B^{s_n} \left[\prod_{m>n} A^{s_m}\right] v_R |\{s\}\rangle. \quad (3.4)$$

This ansatz captures the effect of a local one-site operator, but has the ability to modify the ground state over a much larger region, because it also has access to the virtual degrees of freedom of the MPS. In fact, an operation on the virtual level can be lifted to the effect of a physical operator on $2 \log D$ sites, which can serve as an estimate of the number of sites that are directly perturbed by the insertion of the B tensor in the above ansatz.[10] The ansatz can be written down graphically as

$$|\Phi_\kappa(B)\rangle = \sum_n e^{i\kappa n} \Big|$$ 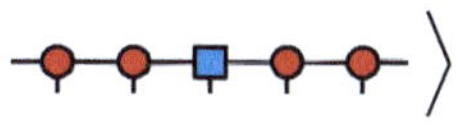 $$\Big\rangle,$$

which shows that the ansatz is creating a particle with momentum κ as a local perturbation on the MPS background state, carrying a small excess of energy on top of the energy density of the ground state.

All variational freedom is contained within the tensor B. The multiplicative gauge invariance of the ground state tensor A, translates into an additive gauge invariance of the excitation tensor B.[11] Indeed, by substituting

$$B^s \to B + XA^s - e^{i\kappa} A^s X \quad (3.5)$$

exactly the same state is obtained. This implies that for tensors $B_0 = XA^s - e^{i\kappa} A^s X$ the state $|\Phi_\kappa[B_0]\rangle$ has zero norm corresponding to so-called null modes in the variational space. For the variational subspace to be well-defined, gauge fixing conditions are imposed and a corresponding parametrization is written down that rule out all null modes automatically. Two choices for effective parametrizations are especially convenient:

1. *Left gauge*. We construct the $(qD \times D)$-matrix $L_{a,(b,s)} = ((A^\dagger)^s l^{1/2})_{a,b}$ and find the right null space V_L of L, so that $LV_L = 0$. This matrix V_L has dimensions $qD \times (q-1)D$ and is orthonormalized as $V_L^\dagger V_L = 1$. The left gauge fixing condition and its reduced parametrization in terms of the $(D(d-1) \times D)$ matrix X are then given by

$$(l|E_A^B = 0 \quad \to \quad B_L[X] = l^{-1/2} V_L^s X r^{-1/2}.$$

[10]Note that this does not mean that, by systematically growing the bond dimension, the ansatz can reproduce the effect of ever larger operators. Every B-tensor can be written in terms of operators of size larger than $2 \log D$, but not vice versa. There is no a priori no reason to expect that variational excitation energies will converge to the exact value for large enough D.

[11]At momentum zero, the gauge transformations of the B tensor are related to linearized infinitesimal gauge transformations of the A tensors. We refer to the tangent space interpretation of the excitation ansatz in Sect. 3.5 for more details.

2. *Right gauge*. We construct the $(qD \times D)$-matrix $R_{(a,s),b} = (r^{1/2}(A^\dagger)^s)_{a,b}$ and find the left null space V_R of R, so that $V_R R = 0$. This matrix V_R has dimensions $(q-1)D \times qD$ and is orthonormalized as $V_R V_r^\dagger = 1$. The right gauge fixing condition and its reduced parametrization in terms of the $(D \times D(d-1))$ matrix X are then given by

$$E_A^B|r) = 0 \quad \rightarrow \quad B_R[X] = l^{-1/2} X V_R^s r^{-1/2}.$$

We see that the overlap with the ground state, given by

$$\langle \Psi(A)|\Phi_\kappa(B)\rangle = \sum_n \mathrm{e}^{i\kappa n}(l|E_A^B|r),$$

is zero by construction under both gauges. By restricting to variational states that are orthogonal to the ground state, we make sure that we are targeting only excited states. If orthogonality to the ground state is indeed obeyed, it can be easily shown that an arbitrary B tensor can be brought into one of the two gauges.

The structure of the variational subspace is further illuminated by computing the overlap of two excited states. This reduces to a double infinite sum as

$$\begin{aligned}\langle \Phi_{\kappa'}(B')|\Phi_\kappa(B)\rangle &= \sum_{n,n'} \mathrm{e}^{i\kappa n}\mathrm{e}^{-i\kappa' n'}\big[B \text{ at site } n,\ B' \text{ at site } n'\big] \\ &= \sum_n \mathrm{e}^{i(\kappa-\kappa')n} \sum_m \mathrm{e}^{-i\kappa' m}\big[B \text{ at site } n,\ B' \text{ at site } n+m\big].\end{aligned}$$

Because of translation invariance, the sum over m does not depend on n so that we can transform the infinite momentum sum to a $\delta(\kappa-\kappa')$ function (this δ-function normalization is expected for momentum superpositions in the thermodynamic limit). The sum over m can be explicitly written out as

$$\begin{aligned}\langle \Phi_{\kappa'}(B')|\Phi_\kappa(B)\rangle = 2\pi\delta(\kappa-\kappa')\Bigg(&\sum_{n=-\infty}^{-1} \mathrm{e}^{-i\kappa n}(l|E_A^B E^{(|n|-1)})E_{B'}^A|r) \\ &+ (l|E_{B'}^B|r) + \sum_{n=1}^{+\infty} \mathrm{e}^{-i\kappa n}(l|E_{B'}^A E^{(m-1)} E_A^B|r)\Bigg)\end{aligned}$$

The geometric series appearing in this expression do not always converge, because the transfer matrix E has an eigenvalue equal to 1. By splitting of the component along the corresponding eigenvectors, the geometric series does converge and can be explicitly computed by taking the inverse of the "regularized" transfer matrix $\tilde{E}$

$$E = |r)(l| + \tilde{E} \rightarrow \sum_{n=0}^{+\infty} \mathrm{e}^{i\kappa n}\tilde{E}^n = (1-\mathrm{e}^{i\kappa}\tilde{E})^{-1} = (1-\mathrm{e}^{i\kappa}E)^P. \tag{3.6}$$

The expression $(\dots)^P$ denotes the pseudo-inverse of an operator, where the components along the zero eigenvector are projected away. The part along the fixed points $(l|$ and $|r)$ give rise to a δ function, because

$$\sum_{n=-\infty}^{\infty} \mathrm{e}^{i\kappa n} \to 2\pi\delta(\kappa),$$

so that the final result is obtained

$$\begin{aligned} \langle \Phi_{\kappa'}(B')|\Phi_\kappa(B)\rangle = 2\pi\delta(\kappa-\kappa')\Big(& (l|E^B_{B'}|r) + \mathrm{e}^{-i\kappa}(l|E^B_A(1-\mathrm{e}^{-i\kappa}E^A_{B'})^P|r) \\ & + \mathrm{e}^{i\kappa}(l|E^A_{B'}(1-\mathrm{e}^{i\kappa}E)^P E^B_A|r) \\ & + (2\pi\delta(\kappa)-1)\,(l|E^B_A|r)(l|E^A_{B'}|r)\Big). \end{aligned} \tag{3.7}$$

The first three terms are all finite, but the last term represents a possible divergence at momentum zero. We quickly see, however, that this last term represents the overlap of the excitation with the ground state, which we have put to zero by going over to one of the two gauge fixing conditions. Moreover, under both gauge conditions the second and third term also vanish and in the effective parametrization the norm of the excitations in the physical Hilbert space reduces to the Euclidean norm on the tensors X and X'[12]

$$\begin{aligned} \langle \Phi_{\kappa'}(B_{L/R}[X'])|\Phi_\kappa(B_{L/R}[X])\rangle &= 2\pi\delta(\kappa-\kappa')(l|E^B_{B'}|r) \\ &= 2\pi\delta(\kappa-\kappa')\mathrm{Tr}((X')^\dagger X). \end{aligned}$$

This will prove especially convenient for variationally optimizing the energy of the excitation: generalized eigenvalue problems become ordinary ones.

3.3.2 *Computing Expectation Values*

But let us first calculate the overlap of two excited states with a one-site operator $O = \sum_n o_n$. This operator is assumed to be translation invariant, $o_n = o$. If the ground state has a non-zero expectation value (density) for this operator, i.e. $\langle \Psi[A]|o|\Psi[A]\rangle \neq 0$, then we redefine the operator as $o \to o - \langle \Psi[A]|o|\Psi[A]\rangle$, so that we calculate only the net effect of the excitation.

[12]This "gauging away" of the momentum dependence of the norm seems in contradiction with our remarks that it is the norm that determines the dispersion in the single-mode approximation. But note that the gauge transformation that brings the tensor B in the right gauge *is* momentum dependent. It is due to the special structure of an MPS that this particular gauge choice is possible: the momentum information in the norm is gauged away through the virtual dimension of the MPS, such that only the numerator contains all momentum dispersion.

We get an expression of the form

$$\langle\Phi_{\kappa'}(B')|O|\Phi_\kappa(B)\rangle = \sum_n \sum_{n'} \sum_{n_0} e^{i\kappa n} e^{-i\kappa' n'} \times \Big[B \text{ at position } n,\ B' \text{ at position } n',\ o \text{ at position } n_0 \Big].$$

Global translation invariance gives a $2\pi\delta(\kappa - \kappa')$-factor and we redefine n, resp. n_0, as being the position of B', resp. o, relative to B:

$$\langle\Phi_{\kappa'}(B')|O|\Phi_\kappa(B)\rangle = 2\pi\delta(\kappa - \kappa') \sum_n \sum_{n_0} e^{-i\kappa n} \times \Big[B' \text{ at position } n \text{ from } B,\ o \text{ at position } n_0 \text{ from } B \Big].$$

The total expression is

$$\begin{aligned}
\langle\Phi_{\kappa'}(B')|O|\Phi_\kappa(B)\rangle &= 2\pi\delta(\kappa - \kappa')\Big[\\
&(l|O^B_{B'}|r) + \sum_{n_0=1}^{+\infty} (l|E^B_{B'} E^{n_0-1} O^A_A|r) + \sum_{n_0=-\infty}^{-1} (l|O^A_A E^{|n_0|-1} E^B_{B'}|r) \\
&+ \sum_{n=1}^{+\infty} e^{-i\kappa n} \Bigg((l|E^B_A E^{n-1} O^{B'}_A|r) + (l|O^B_A E^{n-1} E^A_{B'}|r) \\
&\quad + \sum_{n_0=-\infty}^{-1} (l|O^A_A E^{|n_0|-1} E^B_A E^{n-1} E^A_{B'}|r) \\
&\quad + \sum_{n_0=1}^{n-1} (l|E^B_A E^{n_0-1} O^A_A E^{n-n_0-1} E^A_{B'}|r) \\
&\quad + \sum_{n_0=n+1}^{+\infty} (l|E^B_A E^{n-1} E^A_{B'} E^{|n-n_0|-1} O^A_A|r) \Bigg) \\
&+ \sum_{n=-\infty}^{-1} e^{-i\kappa n} \Bigg((l|E^A_{B'} E^{|n|-1} O^B_A|r) + (l|O^A_{B'} E^{|n|-1} E^B_A|r) \\
&\quad + \sum_{n_0=-\infty}^{n-1} (l|O^A_A E^{|n_0-n|-1} E^A_{B'} E^{|n|-1} E^B_A|r) \\
&\quad + \sum_{n_0=n+1}^{-1} (l|E^A_{B'} E^{|n-n_0|-1} O^A_A E^{|n_0|-1} E^B_A|r) \\
&\quad + \sum_{n_0=1}^{+\infty} (l|E^A_{B'} E^{|n|-1} E^B_A E^{n_0-1} O^A_A|r) \Bigg) \Big].
\end{aligned}$$

The simple sums can be computed using a similar strategy as for the computation of the norm [Eq. (3.7)]

$$\begin{aligned}
&\sum_{n=1}^{+\infty} \mathrm{e}^{-i\kappa n}(l|E_A^B E^{|n|-1} O_{B'}^A|r) + \sum_{n=-\infty}^{-1} \mathrm{e}^{-i\kappa n}(l|O_{B'}^A E^{|n|-1} E_A^B|r) \\
&\quad = (l|E_A^B|r)(l|O_{B'}^A|r)\left(\sum_{n=1}^{+\infty} \mathrm{e}^{-i\kappa n} + \sum_{n=-\infty}^{-1} \mathrm{e}^{-i\kappa n}\right) \\
&\qquad + \mathrm{e}^{-i\kappa}(l|E_A^B(1-\mathrm{e}^{-i\kappa}E)^P O_{B'}^A|r) + \mathrm{e}^{i\kappa}(l|O_{B'}^A(1-\mathrm{e}^{i\kappa})^P E_A^B|r) \\
&\quad = (l|E_A^B|r)(l|O_{B'}^A|r)\,(2\pi\delta(\kappa)-1) \\
&\qquad + \mathrm{e}^{-i\kappa}(l|E_A^B(1-\mathrm{e}^{-i\kappa}E)^P O_{B'}^A|r) + \mathrm{e}^{i\kappa}(l|O_{B'}^A(1-\mathrm{e}^{i\kappa})^P E_A^B|r).
\end{aligned}$$

Again we have a divergence $2\pi\delta(\kappa)$ that vanishes if the excitation is orthogonal to the ground state, i.e. $(l|E_A^B|r)$. As we have said, this is always the case by construction, so that the double sums can be computed as, e.g.

$$\begin{aligned}
&\sum_{n=1}^{+\infty} \mathrm{e}^{-i\kappa n} \sum_{n_0=-\infty}^{-1} (l|O_A^A E^{|n_0|-1} E_A^B E^{n-1} E_{B'}^A|r) \\
&\qquad\qquad = \mathrm{e}^{i\kappa}(l|O_A^A(1-E)^P E_A^B(1-\mathrm{e}^{-i\kappa}E)^P E_{B'}^A|r).
\end{aligned}$$

Lastly we can treat the intermediate term by rearranging the different sums

$$\begin{aligned}
&\sum_{n=1}^{+\infty} \mathrm{e}^{-i\kappa n} \sum_{n_0=1}^{n-1} (l|E_A^B E^{n_0-1} O_A^A E^{n-n_0-1} E_{B'}^A|r) \\
&\qquad = \sum_{n_0=1}^{+\infty} \mathrm{e}^{-i\kappa n_0} \sum_{n=1}^{+\infty} \mathrm{e}^{-i\kappa n}(l|E_A^B E^{n_0-1} O_A^A E^{n-1} E_{B'}^A|r) \\
&\qquad = \mathrm{e}^{-2i\kappa}(l|E_A^B(1-\mathrm{e}^{-i\kappa}E)^P O_A^A(1-\mathrm{e}^{-i\kappa}E)^P E_{B'}^A|r).
\end{aligned}$$

If we recollect all different terms, we obtain the final result

$$\begin{aligned}
\langle \Phi_{\kappa'}(B')|O|\Phi_\kappa(B)\rangle = 2\pi\delta(\kappa-\kappa')\Big[\\
&(l|O_{B'}^B|r) + (l|E_{B'}^B(1-E)^P O_A^A|r) + (l|O_A^A(1-E)^P E_{B'}^B|r) \\
&+ \mathrm{e}^{-i\kappa}(l|E_A^B(1-\mathrm{e}^{-i\kappa}E)^P O_A^{B'}|r) + \mathrm{e}^{-i\kappa}(l|O_A^B(1-\mathrm{e}^{-i\kappa}E)^P E_{B'}^A|r) \\
&+ \mathrm{e}^{i\kappa}(l|E_{B'}^A(1-\mathrm{e}^{i\kappa}E)^P O_A^B|r) + \mathrm{e}^{i\kappa}(l|O_{B'}^A(1-\mathrm{e}^{i\kappa}E)^P E_A^B|r) \\
&+ \mathrm{e}^{-i\kappa}(l|O_A^A(1-E)^P E_A^B(1-\mathrm{e}^{-i\kappa}E)^P E_{B'}^A|r) \\
&+ \mathrm{e}^{-i\kappa}(l|E_A^B(1-\mathrm{e}^{-i\kappa}E)^P E_{B'}^A(1-E)^P O_A^A|r) \\
&+ \mathrm{e}^{i\kappa}(l|O_A^A(1-E)^P E_{B'}^A(1-\mathrm{e}^{i\kappa}E)^P E_A^B|r) \\
&+ \mathrm{e}^{i\kappa}(l|E_{B'}^A(1-\mathrm{e}^{i\kappa})^P E_A^B(1-E)^P O_A^A|r)
\end{aligned}$$

$$+ \mathrm{e}^{2i\kappa}(l|E^A_{B'}(1-\mathrm{e}^{i\kappa}E)^P O^A_A(1-\mathrm{e}^{i\kappa}E)^P E^B_A|r)$$
$$+ \mathrm{e}^{-2i\kappa}(l|E^B_A(1-\mathrm{e}^{-i\kappa})^P O^A_A(1-\mathrm{e}^{-i\kappa})^P E^A_{B'}|r)\Big].$$

This expression is further simplified by imposing a gauge fixing condition. For left-gauge fixed excitations, the overlap is given by

$$\begin{aligned}\langle\Phi_{\kappa'}(B')|O|\Phi_\kappa(B)\rangle = 2\pi\delta(\kappa-\kappa')\Big[\\ (l|O^B_{B'}|r) + (l|O^A_A(1-E)^P E^B_{B'}|r) + (l|E^B_{B'}(1-E)^P O^A_A|r) \\ + \mathrm{e}^{i\kappa}(l|O^A_{B'}(1-\mathrm{e}^{i\kappa}E)^P E^B_A|r) + \mathrm{e}^{-i\kappa}(l|O^B_A(1-\mathrm{e}^{-i\kappa}E)^P E^A_{B'}|r) \\ + \mathrm{e}^{-i\kappa}(l|O^A_A(1-E)^P E^B_A(1-\mathrm{e}^{-i\kappa}E)^P E^A_{B'}|r) \\ + \mathrm{e}^{i\kappa}(l|O^A_A(1-E)^P E^A_{B'}(1-\mathrm{e}^{i\kappa}E^A_A)^P E^B_A|r)\Big].\end{aligned}$$

A similar expression is obtained for right-gauge fixed excitations.

3.3.3 *Solving the Eigenvalue Problem*

The excitation ansatz (3.4) provides a variational subspace of size $D^2(d-1)$ for every value of momentum in $[0, 2\pi)$. In order to find the physical excitations one has to find the tensors B that minimize the energy within this subspace. From its form, it is obvious that the subspace is linear in the elements of the B tensor so that the optimization

$$\min_X \frac{\langle\Phi_\kappa[B_{L/R}(X)]|H|\Phi_\kappa[B_{L/R}(X)]\rangle}{\langle\Phi_\kappa[B_{L/R}(X)]|\Phi_\kappa[B_{L/R}(X)]\rangle}$$

reduces to finding the eigenspectrum of the *effective one-particle Hamiltonian matrix*

$$\mathsf{H}_{\mathrm{eff,1p}}(\kappa)\boldsymbol{x} = \lambda\boldsymbol{x}, \tag{3.8}$$

where $\boldsymbol{x}$ are the vectorized (reshaped) versions of the reduced parametrization X.[13] The effective Hamiltonian matrix is obtained by projecting the full Hamiltonian (with the ground state energy E_0 subtracted) on the one-particle subspace

$$\boldsymbol{y}^\dagger\mathsf{H}_{\mathrm{eff,1p}}(\kappa)\boldsymbol{x} = \langle\Phi_\kappa[B_{L/R}(Y)]|H - E_0|\Phi_\kappa[B_{L/R}(X)]\rangle. \tag{3.9}$$

[13]Again, the norm of the excitations in terms of the vector $\boldsymbol{x}$ is just the Euclidean inner product, $\langle\Phi_\kappa[B_{L/R}(Y)]|\Phi_\kappa[B_{L/R}(X)]\rangle = \boldsymbol{y}^\dagger\boldsymbol{x}$, so that the normalization of the states does not show up in the eigenvalue problem.

The explicit expression is obtained analogously to the computation of the one-site operator in the previous section. The final result is[14]

$$
\begin{aligned}
\langle\Phi_{\kappa'}[B']|H|\Phi_\kappa[B]\rangle = 2\pi\delta(\kappa-\kappa')\Big[\\
&(l|H^{BA}_{B'A}|r) + (l|H^{AB}_{AB'}|r) + e^{-i\kappa}(l|H^{BA}_{AB'}|r) + e^{+i\kappa}(l|H^{AB}_{B'A}|r) \\
&+ (l|E^{B}_{B'}(1-E)^P H^{AA}_{AA}|r) + (l|H^{AA}_{AA}(1-E)^P E^{B}_{B'}|r) \\
&+ e^{-i\kappa}(l|E^{B}_{A}(1-e^{-i\kappa}E)^P H^{AA}_{B'A}|r) + e^{-2i\kappa}(l|E^{B}_{A}(1-e^{-i\kappa}E)^P H^{AA}_{AB'}|r) \\
&+ e^{-i\kappa}(l|H^{AB}_{AA}(1-e^{-i\kappa}E)^P E^{A}_{B'}|r) + e^{-2i\kappa}(l|H^{BA}_{AA}(1-e^{-i\kappa}E)^P E^{A}_{B'}|r) \\
&+ e^{+i\kappa}(l|H^{AA}_{AB'}(1-e^{+i\kappa}E)^P E^{B}_{A}|r) + e^{2i\kappa}(l|H^{AA}_{B'A}(1-e^{+i\kappa}E)^P E^{B}_{A}|r) \\
&+ e^{+i\kappa}(l|E^{A}_{B'}(1-e^{+i\kappa}E)^P H^{BA}_{AA}|r) + e^{+2i\kappa}(l|E^{A}_{B'}(1-e^{+i\kappa}E)^P H^{AB}_{AA}|r) \\
&+ e^{-i\kappa}(l|H^{AA}_{AA}(1-E)^P E^{B}_{A}(1-e^{-i\kappa}E)^P E^{A}_{B'}|r) \\
&+ e^{-i\kappa}(l|E^{B}_{A}(1-e^{-i\kappa}E)^P E^{A}_{B'}(1-E)^P H^{AA}_{AA}|r) \\
&+ e^{+i\kappa}(l|H^{AA}_{AA}(1-E)^P E^{A}_{B'}(1-e^{+i\kappa}E)^P E^{B}_{A}|r) \\
&+ e^{+i\kappa}(l|E^{A}_{B'}(1-e^{+i\kappa})^P E^{B}_{A}(1-E)^P H^{AA}_{AA}|r) \\
&+ e^{-3i\kappa}(l|E^{B}_{A}(1-e^{-i\kappa}E)^P H^{AA}_{AA}(1-e^{-i\kappa}E)^P E^{A}_{B'}|r) \\
&+ e^{+3i\kappa}(l|E^{A}_{B'}(1-e^{+i\kappa}E)^P H^{AA}_{AA}(1-e^{+i\kappa}E)^P E^{B}_{A}|r)\Big].
\end{aligned}
$$

For left gauge fixed excitations, the Hamiltonian overlap is simplified to

$$
\begin{aligned}
\langle\Phi_{\kappa'}[B']|H|\Phi_\kappa[B]\rangle = 2\pi\delta(\kappa-\kappa')\Big[\\
&(l|H^{BA}_{B'A}|r) + (l|H^{AB}_{AB'}|r) + e^{-i\kappa}(l|H^{BA}_{AB'}|r) + e^{+i\kappa}(l|H^{AB}_{B'A}|r) \\
&+ (l|E^{B}_{B'}(1-E)^P H^{AA}_{AA}|r) + (l|H^{AA}_{AA}(1-E)^P E^{B}_{B'}|r) \\
&+ e^{-i\kappa}(l|H^{AB}_{AA}(1-e^{-i\kappa}E)^P E^{A}_{B'}|r) + e^{-2i\kappa}(l|H^{BA}_{AA}(1-e^{-i\kappa}E)^P E^{A}_{B'}|r) \\
&+ e^{+i\kappa}(l|H^{AA}_{AB'}(1-e^{+i\kappa}E)^P E^{B}_{A}|r) + e^{+2i\kappa}(l|H^{AA}_{B'A}(1-e^{+i\kappa}E)^P E^{B}_{A}|r) \\
&+ e^{-i\kappa}(l|H^{AA}_{AA}(1-E)^P E^{B}_{A}(1-e^{-i\kappa}E)^P E^{A}_{B'}|r) \\
&+ e^{+i\kappa}(l|H^{AA}_{AA}(1-E)^P E^{A}_{B'}(1-e^{+i\kappa}E)^P E^{B}_{A}|r)\Big].
\end{aligned}
$$

The full $D^2(d-1)\times D^2(d-1)$ matrix $\mathsf{H}_{\text{eff,1p}}$ can be computed with a scaling of $\mathcal{O}(D^6)$ in the bond dimension – the hardest step is computing the pseudo-inverse of the transfer matrix $(1-E)^P$. If only a few smallest eigenvalues are wanted, corresponding to the lowest-lying excitations, the eigenvalue equation can be implemented using an iterative eigensolver and the pseudo-inverses can be computed with conjugate-gradient methods; the result is only a $\mathcal{O}(D^3)$ scaling.

If the full the eigenvalue problem is solved, a set of $(d-1)D^2$ eigenvalues for every momentum κ is obtained. Some of the lowest eigenvalues will correspond to

[14]Since we have subtracted the ground-state energy density, the expectation value $(l|H^{AA}_{AA}|r)$ is zero; this implies that we can safely put in the regularized transfer matrix $\tilde{E}$ instead of the full one wherever needed to define the pseudo-inverse.

the elementary one-particle excitations of the Hamiltonian, so that a few continuous dispersion relations $\Delta_\alpha(\kappa)$ can be traced in the spectrum.

Other eigenvalues obtained from Eq. (3.8) will fall in the continuous part of the spectrum of the Hamiltonian, i.e. in the set of scattering states. Scattering states cannot be described by a single local perturbation, so we expect the ansatz to fail. In fact, instead of a scattering state, the variational optimization will create a localized wave packet of two-particle states within some energy range. Obviously, the variational eigenstates of the form in Eq. (3.4) will not provide a good approximation to the exact scattering eigenstates of the full Hamiltonian.

A more quantitative way to assess how well an exact eigenstate is approximated consists of calculating the variance of the Hamiltonian [45], i.e. $\langle\Phi_\kappa[B]|(H-\Delta(\kappa))^2|\Phi_\kappa[B]\rangle$. For elementary excitations, these variances should be small. For the other solutions of the one-particle problem (3.8), which correspond to scattering states, the variance should be larger; for a typical gapped system, the difference will be some orders of magnitude. Consequently, this quantity allows for the identification of one-particle states, even within higher-particle bands and without exploiting symmetries.

The actual computation of this variance requires some work, so we refer to Appendix A.2 for the explicit expressions. What is interesting, though, is the fact that we have to subtract the ground state variance in order to arrive at a finite value for the variance of the excitation. This implies that the excitation's variance is not guaranteed to be a positive quantity, and, in fact, becomes negative if the excitation is locally a better eigenstate than the ground state. In regions where the excitation is a well localized particle – for excitation branches that are strongly isolated from all continuous bands in the spectrum this is expected to be the case – this does indeed happen; see Sect. 4.1 for a numerical example.

Note finally that, without Galilean invariance on the lattice, the tensor $B_\alpha(\kappa)$, which describes the particle α on a dispersion branch $\Delta_\alpha(\kappa)$, is momentum dependent. On the other hand, we expect that for a well-defined particle in a certain momentum range this momentum dependence is not too strong. Indeed, it turns out that by a suitable choice of the basis tensors $\{B_{(i)}, i=1,\dots,(d-1)D^2\}$, we can fully capture $B_\alpha(\kappa)$ for all elementary excitations α and for all momenta κ in the span of just a small number $\ell \ll (d-1)D^2$ basis vectors $\{B_{(i)}, i=1,\dots,\ell\}$. Although more sophisticated optimization strategies should be possible, we construct this reduced basis from a number of B's at different momenta. This reduced basis will be important for solving the scattering problem in the next sections.

3.3.4 One-Particle Form Factors

The excitation ansatz provides a variational approximation for the true low-lying excitations of the full Hamiltonian. The overlap with a local operator acting on the ground state (their spectral weights) contain an important contribution to the spectral function. For further use (see Sect. 3.8), we give the expression for the one-particle form factor (with B_α in the left gauge)

$$\langle\Phi_\kappa(B)|O_0|\Psi[A]\rangle = (l|O_B^A|r) + \sum_{n=1}^{\infty} e^{i\kappa n}(l|O_A^A E^{n-1} E_B^A|r)$$
$$= (l|O_B^A|r) + e^{i\kappa}(l|O_A^A \left(1 - e^{i\kappa}E\right)^P E_{B_\alpha}^A|r)$$

where O_0 is an operator at site 0 and we have assumed that the excitation is in the left gauge.

3.3.5 *Topological Particle Excitations*

The elementary excitations in one-dimensional spin systems are not always of the simple form that we have introduced earlier. In the case of symmetry breaking, where the ground state is degenerate, the elementary excitations are typically kinks or domain walls, i.e. particles that interpolate between the different ground states. These excitations are topological, because (i) they cannot be created by a local operator acting on one of the ground states, and (ii) it would take an infinite amount of energy to connect them to a state in the ground state sector [46]. It is not clear that these topological excitations are local (i.e. that the interpolation between the two ground states is happening over a small region), and, in fact, this is not at all obvious from other approaches such as the Bethe ansatz [47–49]. One expects, however, that the proof for excitations in the trivial sector [41] can be extended to topological excitations as well.[15]

Because it is formulated in the thermodynamic limit directly, our framework can be easily extended to target these topological sectors as well.[16] Suppose we have a twofold-degenerate ground state, approximated by two uMPS $|\Psi[A_1]\rangle$ and $|\Psi[A_2]\rangle$. The obvious ansatz for a domain wall excitation is[17]

$$|\Phi_\kappa[B]\rangle = \sum_n e^{i\kappa n} \sum_{\{s\}} \boldsymbol{v}_L^\dagger \left[\prod_{m<n} A_1^{s_m}\right] B^{s_n} \left[\prod_{m>n} A_2^{s_m}\right] \boldsymbol{v}_R |\{s\}\rangle, \tag{3.10}$$

i.e. the domain wall interpolates between the two ground states [23]. All the calculations of the previous sections can be repeated in order to determine gauge-fixing conditions, compute expectation values and solve the eigenvalue problem. The only difference concerns the appearance of the "mixed transfer matrix" $E_{\text{mix}} = E_{A_2}^{A_1}$,

[15] See also Refs. [50, 51] for a Monte Carlo study on the local nature of spinons.

[16] In finite systems with periodic boundary conditions, topological excitations always have to be described in pairs. In order to capture them in finite systems, non-trivial boundary conditions have to be applied, see e.g. [52].

[17] In quantum field theory, this ansatz has been proposed earlier [53] to study the kink excitations in the sine-Gordon model.

which determines the correlation functions corresponding to string-like operators that interpolate to between the two ground states. This matrix has spectral radius smaller than one – otherwise the two ground states would not be orthogonal – such that the geometric sums involving E_{mix} should be computed with the full inverse $(E_{\text{mix}})^{-1}$.

Yet there is one problem with considering topological excitations. Strictly speaking the momentum of the ansatz [Eq. (3.10)] is not well defined: multiplying the tensor A_2 with an arbitrary phase factor $A_2 \leftarrow A_2 \mathrm{e}^{i\phi}$ shifts the momentum with $\kappa \leftarrow \kappa + \phi$. The origin of this ambiguity is the fact that one domain wall cannot be properly defined when using periodic boundary conditions.

Physically, however, domain walls should come in pairs. For these states the total momentum is well-defined, although the individual momenta can be arbitrarily transferred between the two domain walls. A heuristic way to fix the kink momentum unambiguously is related to "mixed transfer matrix" E_{mix}; it can then be imposed that its spectrum be symmetric with respect to the real axis. This will give rise to a kink spectrum that is symmetric in the momentum κ. This problem disappears, as we will see, when considering excitations with two topological particles.

3.3.6 *Larger Blocks*

As we have mentioned earlier, there is no guarantee that the variational energies converge to the exact excitation energy of the full Hamiltonian, even for a clearly isolated excitation branch. The reason is that the effect of physical operators of growing size cannot be reproduced by the excitation ansatz, even by growing the bond dimension. This can pose a problem for one-particle excitations that are very wide, because e.g. they are very close to a scattering continuum in the spectrum.

The excitation ansatz can be systematically extended, however, in order to capture larger and larger regions. Instead of inserting a one-site tensor, one can introduce larger blocks, which leads to the ansatz [40, 41]

$$|\Phi_\kappa[B]\rangle = \sum_n \mathrm{e}^{i\kappa n} \sum_{\{s\}} \boldsymbol{v}_L^\dagger \left[\prod_{m<n} A^{s_m} \right] B^{s_n, s_{n+1}, \ldots, s_{n+M-1}} \left[\prod_{m>M+n-1} A^{s_m} \right] \boldsymbol{v}_R |\{s\}\rangle,$$

or, pictorially,

$$|\phi_\kappa(B)\rangle = \sum_n \mathrm{e}^{i\kappa n} \Big|$$  $$\Big\rangle$$

In principle this approach is guaranteed to converge to the exact excitation energy – assuming the ground state energy is converged – but the number of the variational parameters in the big B tensor grows exponentially in the number of sites M, so that, practically, this becomes infeasible quickly. One could think of matrix decomposition of the block to make this approach more scalable, but this would fall outside the scope of this chapter.

The same gauge freedom is present for these larger blocks, and the same gauge conditions can be imposed. The left gauge condition reads

$$\sum_s (A^s)^\dagger l B^{s,t_1,\dots,t_{M-1}} = 0 \qquad \forall\, t_1, \dots, t_{M-1}$$

and can be enforced by going to the effective parametrization of the B tensor

$$B_L^{s_1,\dots,s_M}[X] = l^{-1/2} V_L^{s_1} X^{s_2,\dots,s_M} r^{-1/2} \tag{3.11}$$

where $X^{s_2,\dots,s_M}$ is a $(D(d-1) \times d \times \cdots \times d \times D)$ tensor containing all variational parameters. With this effective parametrization, the overlap of states again reduces to the Euclidean norm on the tensor X, while the effective Hamiltonian matrix is given by

$$\begin{aligned}
\langle \Phi_{\kappa'}(B')|H|\Phi_\kappa(B)\rangle = 2\pi\delta(\kappa-\kappa')\Big[\\
& (l|\Big(H_{AA}^{AA}(1-E)^P E_{B_1'\dots B_M'}^{B_1\dots B_M} + H_{AB_1'}^{AB_1} E_{B_2'\dots B_M'}^{B_2\dots B_M} \\
&\qquad + \sum_{i=1}^{M-1} E_{B_1'\dots B_{i-1}'}^{B_1\dots B_{i-1}} H_{B_i' B_{i+1}'}^{B_i B_{i+1}} E_{B_{i+2}'\dots B_M'}^{B_{i+2}\dots B_M} \\
&\qquad + E_{B_1'\dots B_{M-1}'}^{B_1\dots B_{M-1}} H_{B_M' A}^{B_M A} + E_{B_1'\dots B_M'}^{B_1\dots B_M}(1-E)^P H_{AA}^{AA}\Big)|r) \\
&+ \mathrm{e}^{-i\kappa}(l|\Big(H_{AA}^{AA}(1-E)^P E_A^{B_1} E_{B_1'}^{B_2} + H_{AA}^{AB_1} E_{B_1'}^{B_2} + H_{AB_2'}^{B_1 B_2}\Big) s E_{B_2'\dots B_{M-1}' B_M'}^{B_3\dots B_M A}|r) \\
&+ \mathrm{e}^{i\kappa}(l|\Big(H_{AA}^{AA}(1-E)^P E_{B_1'}^{A} E_{B_2'}^{B_1} + H_{AB_1'}^{AA} E_{B_2'}^{B_1} + H_{B_1' B_2'}^{AB_1}\Big) E_{B_3'\dots B_M' A}^{B_2\dots B_{M-1} B_M}|r) \\
&+ \sum_{j=2}^{M-1} \mathrm{e}^{-ij\kappa}(l|\Big(H_{AA}^{AA}(1-E)^P E_A^{B_1} E_A^{B_2} + H_{AA}^{AB_1} E_A^{B_2} + H_{AA}^{B_1 B_2}\Big) \\
&\qquad \times E_{A\dots A}^{B_3\dots B_j} E_{B_1'\dots B_{M-j}'}^{B_{j+1}\dots B_M} E_{B_{M-j+1}'\dots B_M'}^{A\dots A}|r) \\
&+ \sum_{j=2}^{M-1} \mathrm{e}^{ij\kappa}(l|\Big(H_{AA}^{AA}(1-E)^P E_{B_1'}^{A} E_{B_2'}^{A} + H_{AB_1'}^{AA} E_{B_2'}^{A} + H_{B_1' B_2'}^{AA}\Big) \\
&\qquad \times E_{B_3'\dots B_j'}^{A\dots A} E_{B_{j+1}'\dots B_M'}^{B_1\dots B_{M-j}} E_{A\dots A}^{B_{M-j+1}\dots B_M}|r) \\
&+ \mathrm{e}^{-iM\kappa}(l|\Big(H_{AA}^{AA}(1-E)^P E_A^{B_1} E_A^{B_2} + H_{AA}^{AB_1} E_A^{B_2} + H_{AA}^{B_1 B_2}\Big) \\
&\qquad \times E_{A\dots A}^{B_3\dots B_M}(1-\mathrm{e}^{-i\kappa}E)^P E_{B_1'\dots B_M'}^{A\dots A}\Big)|r) \\
&+ \mathrm{e}^{iM\kappa}(l|\Big(H_{AA}^{AA}(1-E)^P E_{B_1'}^{A} E_{B_2'}^{A} + H_{AB_1'}^{AA} E_{B_2'}^{A} + H_{B_1' B_2'}^{AA}\Big)
\end{aligned}$$

$$\times E^{A\ldots A}_{B_3'\ldots B_M'}(1-\mathrm{e}^{i\kappa}E)^P E^{B_1\ldots B_M}_{A\ldots A}\Big)|r)\Big],$$

where the dubious notation

$$B^{s_1,\ldots,s_M} = B_1^{s_1} \times \cdots \times B_M^{s_m}$$

was introduced for making clear which s_i indices have to be contracted (in the contraction the tensor B is not actually factorized).

3.4 Two-Particle States

In the previous section it became clear that we need another ansatz to capture the delocalized nature of a two-particle state. We will start from a one-particle spectrum consisting of a number of different types of particles, labelled by α, with dispersion relations $\Delta_\alpha(\kappa)$. In the thermodynamic limit, constructing the two-particle spectrum is trivial: the momentum and energy are the sum of the individual momenta and energies of the two particles. The two-particle wave function, however, depends on the particle interactions. The interactions, which depend on both the Hamiltonian and the ground state correlations, are reflected in the wave function in two ways: (i) the asymptotic wave function has different terms, with the S matrix elements as the relative coefficients, and (ii) the local part of the wave function.[18]

3.4.1 *Variational Ansatz*

In order to capture both effects of the interactions on the wave function, we introduce the following ansatz for describing states with two localized, particle-like excitations with total momentum K

$$|\Upsilon(K)\rangle = \sum_{n=0}^{+\infty}\sum_{j=1}^{L_n} c^j(n)|\chi_{K,j}(n)\rangle \tag{3.12}$$

where the basis states are

[18]The form of the two-particle wave function that we propose, is inspired by the similar construction that the Bethe ansatz uses to solve the two-magnon problem for e.g. the Heisenberg model – see Sect. 2.5. In addition, our approach is greatly inspired by Feynman's lectures on statistical mechanics [54] and Kohn's variational approach to solve the scattering problem [55].

$$|\chi_{K,j}(n=0)\rangle = \sum_{n_1=-\infty}^{+\infty} e^{iKn_1} \sum_{\{s\}=1}^{d} \boldsymbol{v}_L^\dagger \left[\prod_{m<n_1} A^{s_m}\right] \times B_{(j)}^{s_{n_1}} \left[\prod_{m>n_1} A^{s_m}\right] \boldsymbol{v}_R |\{s\}\rangle \tag{3.13}$$

$$|\chi_{K,(j_1,j_2)}(n>0)\rangle = \sum_{n_1=-\infty}^{+\infty} e^{iKn_1} \sum_{\{s\}=1}^{d} \boldsymbol{v}_L^\dagger \left[\prod_{m<n_1} A^{s_m}\right] B_{(j_1)}^{s_{n_1}} \times \left[\prod_{n_1<m<n_1+n} A^{s_m}\right] B_{(j_2)}^{s_{n_1+n}} \left[\prod_{m>n_1+n} A^{s_m}\right] \boldsymbol{v}_R |\{s\}\rangle. \tag{3.14}$$

We collect the variational coefficients either in one half-infinite vector $\boldsymbol{C}$ with $C^{j,n} = c^j(n)$ or using the finite vectors $\boldsymbol{c}(n)$ with entries $\{c^j(n), j = 1, \ldots, L_n\}$ for every $n = 0, 1, \ldots$. Here, we have $L_0 = (d-1)D^2$ and $L_{n>0} = [(d-1)D^2]^2$. Note that the sum in Eq. (3.12) only runs over values $n \geq 0$, because a sum over all integers would result in an overcomplete basis. The basis states are graphically represented as

$$|\chi_{K,j}(n=0)\rangle = \sum_{n_1=-\infty}^{+\infty} e^{iKn_1} \Big|$$  $$\Big\rangle$$

$$|\chi_{K,(j_1,j_2)}(n>0)\rangle = \sum_{n_1=-\infty}^{+\infty} e^{iKn_1} \Big|$$  $$\Big\rangle.$$

Already at this point, we will reduce the number of variational parameters to keep the problem tractable. The terms with $n = 0$ (corresponding to the basis vectors in Eq. (3.13)) are designed to capture the situation where the two particles are close together. No information on how this part should look like is a priori available, so we keep all variational parameters $c^j(0)$, $j = 1, \ldots, L_0 = D^2(d-1)$. The terms with $n > 0$ corresponding to the basis vectors in Eq. (3.14) represent the situation where the particles are separated. We know that, as $n \to \infty$, the particles decouple and we should obtain a combination of one-particle solutions. With this in mind, we restrict the range of j_1 and j_2 to the first ℓ basis tensors $\{B_{(i)}, i = 1, \ldots, \ell\}$, which were chosen so as to capture the momentum dependent solutions of the one-particle problem. Consequently, the number of basis states of Eq. (3.14) for $n > 0$ satisfies $L_n = \ell^2$, which we will henceforth denote as just L.

This might seem like a big approximation for small n: when the two particles approach, the wave functions might begin to deform, so that the B tensors that were obtained as solutions for the one-particle problem, no longer apply. Note, however, that the local ($n = 0$) and non-local ($n > 0$) part are not orthogonal, so that the local part is able to correct for the part of the non-local wave function where the one-particle description is no longer valid.

As the state (3.12) is again linear in its variational parameters $\boldsymbol{C}$, optimizing the energy amounts to solving a generalized eigenvalue problem

$$\mathsf{H}_{\text{eff}}\boldsymbol{C} = \omega \mathsf{N}_{\text{eff}}\boldsymbol{C} \tag{3.15}$$

with ω the total energy of the state and

$$(\mathsf{H}_{\text{eff}})_{n'j',nj} = \langle \chi_{K,j'}(n') | H - E_0 | \chi_{K,j}(n) \rangle \tag{3.16}$$

$$(\mathsf{N}_{\text{eff}})_{n'j',nj} = \langle \chi_{K,j'}(n') | \chi_{K,j}(n) \rangle \tag{3.17}$$

two half-infinite matrices. They have a block matrix structure, where the submatrices are labelled by (n', n) and are of size $L_{n'} \times L_n$. The computation of the matrix elements is quite involved and technical, so we refer to Appendix A.3 for the explicit formulas.

Since the eigenvalue problem is still infinite, it cannot be diagonalized directly. Since we actually know the possible energies ω for a scattering state with total momentum K (it follows from the one-particle energies), we can also interpret Eq. (3.15) as an overdetermined system of linear equations for the coefficients $C^{j,n} = c^j(n)$. In the next two sections we will show how to reduce this problem to a finite linear equation.

3.4.2 Asymptotic Regime

First we solve the problem in the asymptotic regime, where the two particles are completely decoupled. This regime corresponds to the limit $n', n \to \infty$, where the effective norm and Hamiltonian matrices, consisting of blocks of size $L \times L$, take on a simple form. Indeed, if we properly normalize the basis states, the asymptotic form of the effective norm matrix reduces to the identity, while the effective Hamiltonian matrix is a repeating row of block matrices centred around the diagonal

$$(\mathsf{H}_{\text{eff}})_{n',n} \to \mathsf{A}_{n-n'}, \qquad n, n' \to \infty. \tag{3.18}$$

The blocks decrease exponentially as we go further from the diagonal, so we can, in order to solve the problem, consider them to be zero if $|n - n'| > M$ for a sufficiently large M. In this approximation, the coefficients $\boldsymbol{c}(n)$ obey

$$\sum_{m=-M}^{M} \mathsf{A}_m \boldsymbol{c}(n+m) = \omega \boldsymbol{c}(n), \qquad n \to \infty. \tag{3.19}$$

We can reformulate this as a recurrence relation for the $\boldsymbol{c}(n)$ vectors and therefore look for elementary solutions of the form $\boldsymbol{c}(n) = \mu^n \boldsymbol{v}$. For fixed ω, the solutions μ and $\boldsymbol{v}$ are now determined by the polynomial eigenvalue equation

$$\sum_{m=-M}^{M} \mathsf{A}_m \mu^m \boldsymbol{v} = \omega \boldsymbol{v}. \tag{3.20}$$

From the special structure of the blocks A_m (see Appendix A.3.3) and their relation to the effective one-particle Hamiltonian $\mathsf{H}_{\text{eff,1p}}$, we already know a number of solutions to Eq. (3.20). Indeed, if we can find Γ combinations of two types of particles (α, β) with individual momenta (κ_1, κ_2) such that $K = \kappa_1 + \kappa_2$ and $\omega = \Delta_\alpha(\kappa_1) + \Delta_\beta(\kappa_2)$, then the polynomial eigenvalue problem will have 2Γ solutions μ on the unit circle. These solutions take the form $\mu = \mathrm{e}^{i\kappa_2}$ and the corresponding eigenvectors are given by

$$\boldsymbol{v} = \boldsymbol{u}_\alpha(\kappa_1) \otimes \boldsymbol{u}_\beta(\kappa_2), \tag{3.21}$$

where $\boldsymbol{u}_\alpha(\kappa)$ is a vector corresponding to the one-particle solution of type α with momentum κ with respect to the reduced basis $\{B_{(i)}, i = 1, \ldots, \ell\}$ (in the case of degenerate eigenvalues we can take linear combinations of these eigenvectors that no longer have this product structure). Every combination is counted twice, because we can have particle α on the left and particle β on the right, and vice versa.

Moreover, since $\mathsf{A}_m^\dagger = \mathsf{A}_{-m}$, the number of eigenvalues within and outside the unit circle should be equal. This allows for a classification of the eigenvalues μ as

$$\begin{array}{lll} |\mu_i| < 1 & \text{for} & i = 1, \ldots, LM - \Gamma \\ |\mu_i| = 1 & \text{for} & i = LM - \Gamma + 1, \ldots, LM + \Gamma \\ |\mu_i| > 1 & \text{for} & i = LM + \Gamma + 1, \ldots, 2LM. \end{array}$$

The last eigenvalues with modulus bigger than one are not physical (because the corresponding $\boldsymbol{c}(n) \sim \mu_i^n \boldsymbol{v}_i$ yields a non-normalizable state) and should be discarded. The 2Γ eigenvalues with modulus 1 are the oscillating modes discussed above; we will henceforth label them with $\gamma = 1, \ldots, 2\Gamma$ such that $\mu = \mathrm{e}^{i\kappa_\gamma}$ (κ_γ being the momentum of the particle of the right) and the corresponding eigenvector is given by

$$\boldsymbol{v}_\gamma = \boldsymbol{u}_{\alpha_\gamma}(K - \kappa_\gamma) \otimes \boldsymbol{u}_{\beta_\gamma}(\kappa_\gamma).$$

Finally, the first eigenvalues are exponentially decreasing and represent corrections when the excitations are close to each other. We henceforth denote them as $e^{-\lambda_i}$ with $\mathrm{Re}(\lambda_i) > 0$ for $i = 1, \ldots, LM - \Gamma$ and denote the corresponding eigenvectors as $\boldsymbol{w}_i$.

With these solutions, we can represent the general asymptotic solution as

$$\boldsymbol{c}(n) \to \sum_{i=1}^{LM-\Gamma} p^i \mathrm{e}^{-\lambda_i n} \boldsymbol{w}_i + \sum_{\gamma=1}^{2\Gamma} q^\gamma \mathrm{e}^{i\kappa_\gamma n} \boldsymbol{v}_\gamma. \tag{3.22}$$

Of course, we still have to determine the coefficients $\{p^i, q^\gamma\}$ by solving the local problem.

3.4.3 Solving the Full Eigenvalue Equation

Since the energy ω was fixed when constructing the asymptotic solution, the generalized eigenvalue equation is reduced to the linear equation

$$(\mathsf{H}_{\text{eff}} - \omega \mathsf{N}_{\text{eff}})\boldsymbol{C} = 0.$$

We know that in the asymptotic regime this equation is fulfilled if and only if $\boldsymbol{c}(n)$ is of the form of Eq. (3.22). We will introduce the approximation that the elements for the effective Hamiltonian matrix [Eq. (3.16)] and norm matrix [Eq. (3.17)] have reached their asymptotic values when either $n > M + N$ or $n' > M + N$, where N is a finite value and should be chosen sufficiently large. This implies that we can safely insert the asymptotic form for $n > N$ in the wave function, which we can implement by rewriting the wave function as

$$\boldsymbol{C} = \mathsf{Z} \cdot \boldsymbol{x} \tag{3.23}$$

where

$$\mathsf{Z} = \begin{pmatrix} 1_{\text{local}} & & \\ & \{\mathrm{e}^{-\lambda_i n} \boldsymbol{w}_i\} & \{\mathrm{e}^{-i\kappa_\gamma n} \boldsymbol{v}_\gamma\} \end{pmatrix}.$$

The $\{\mathrm{e}^{-\lambda_i n} \boldsymbol{w}_i\}$ and $\{\mathrm{e}^{-i\kappa_\gamma n} \boldsymbol{v}_\gamma\}$ are the vectors corresponding to the damped, resp. oscillating modes, while the identity matrix is inserted to leave open all parameters in $\boldsymbol{c}(n)$ for $n \leq N$. The number of parameters in x is reduced to the finite value of $D^2(d-1) + NL + LM + \Gamma$.

Since the equation is automatically fulfilled after $M + N$ rows, we can reduce H_{eff} and N_{eff} to the first rows, so we end up with the following linear equation

$$[\mathsf{H} - \omega \mathsf{N}]_{\text{red}} \cdot \mathsf{Z} \cdot \boldsymbol{x} = 0 \tag{3.24}$$

with

$$
\left[\mathsf{H}-\omega\mathsf{N}\right]_{\text{red}} = \left(\begin{array}{c|cccc} & 0 & 0 & \ldots & 0 \\ & \vdots & \vdots & \ddots & \vdots \\ & 0 & 0 & \ldots & 0 \\ (\mathsf{H}-\omega\mathsf{N})_{\text{ex}} & \mathsf{A}_M & 0 & \ldots & 0 \\ & \mathsf{A}_{M-1} & \mathsf{A}_M & \ldots & 0 \\ & \vdots & \vdots & \ddots & \vdots \\ & \mathsf{A}_1 & \mathsf{A}_2 & \ldots & \mathsf{A}_M \end{array}\right).
$$

This "effective scattering matrix" consists of the first $(M+N)\times(M+N)$ blocks of the exact effective Hamiltonian and norm matrix and the A matrices of the asymptotic part [Eq. (3.18)] to make sure that these matrices remain the truncated versions of a hermitian problem. This matrix has $D^2(d-1)+(N+M)L$ rows, which implies that the linear equation (3.24) has Γ exact solutions, which is precisely the number of scattering states we expect to find. Every solution consists of a local part ($D^2(d-1)+NL$ elements), the $LM-\Gamma$ coefficients $\boldsymbol{p}$ of the decaying modes and the 2Γ coefficients $\boldsymbol{q}$ of the asymptotic modes.

3.4.4 S Matrix and Normalization

After having shown how to find the solutions of the scattering problem, we can now elaborate on the structure of the asymptotic wave function and define the S matrix.

We start from Γ linearly independent scattering eigenstates $|\Upsilon_i(K,\omega)\rangle$ ($i=1,\ldots,\Gamma$) at total momentum K and energy ω with asymptotic coefficients $\boldsymbol{q}_i(K,\omega)$. The asymptotic form of these eigenstates is thus a linear combination of all possible non-decaying solutions of the asymptotic problem:

$$
|\Upsilon_i(K,\omega)\rangle = \sum_{\gamma=1}^{2\Gamma} q_i^\gamma(K,\omega)\times\sum_{n>N}\sum_j \mathrm{e}^{i\kappa_\gamma n} v_\gamma^j(\kappa_\gamma)|\chi_{j,K}(n)\rangle \tag{3.25}
$$

where the coefficients are obtained from solving the local problem. The number of eigenstates equals half the number of oscillating modes that appear in the linear combination. With every oscillating mode γ we can associate a function $\omega_\gamma(\kappa)$ giving the energy of this mode as a function of the momentum κ_γ of the second particle at a fixed total momentum K. If γ corresponds to the two-particle mode with particles α_γ and β_γ, this function is given by $\omega_\gamma(\kappa)=\Delta_{\alpha_\gamma}(K-\kappa)+\Delta_{\beta_\gamma}(\kappa)$. The derivative of this function, which will prove of crucial importance, is $\omega'_\gamma(\kappa)=\Delta'_{\beta_\gamma}(\kappa)-\Delta'_{\alpha_\gamma}(K-\kappa)$. It can be interpreted as the difference in group velocity between the two particles, i.e. the relative group velocity in the center of mass frame.

Much like the proof of conservation of particle current in one-particle quantum mechanics, it can be shown that (see Appendix A.4), if (3.25) is to be the asymptotic form of an eigenstate, the coefficients $q_i^\gamma(K,\omega)$ should obey

$$\sum_{\gamma} \left|q_i^{\gamma}(K,\omega)\right|^2 \left(\frac{\mathrm{d}\omega_\gamma}{\mathrm{d}\kappa}(\kappa_\gamma)\right) = 0. \tag{3.26}$$

This equation can indeed be read as a form of conservation of particle current, with $\omega'_\gamma(\kappa_\gamma)$ playing the role of the (relative) group velocity of the asymptotic mode γ. As any linear combination of eigenstates with the same energy ω is again an eigenstate, this relation can be extended to

$$\sum_{\gamma} \overline{q_j^{\gamma}(K,\omega)} q_i^{\gamma}(K,\omega) \left(\frac{\mathrm{d}\omega_\gamma}{\mathrm{d}\kappa}(\kappa_\gamma)\right) = 0.$$

With this equation satisfied, we can define the two-particle S matrix $S(K,\omega)$. Firstly, the different modes are classified according to the sign of the derivative: the incoming modes have $\frac{\mathrm{d}\omega}{\mathrm{d}\kappa} > 0$ (two particles moving towards each other), the outgoing modes have $\frac{\mathrm{d}\omega}{\mathrm{d}\kappa} < 0$ (two particles moving away from each other), so that we have

$$\sum_{\gamma\in\Gamma_{\text{in}}} \overline{q_j^{\gamma}(K,\omega)} q_i^{\gamma}(K,\omega) \left|\frac{\mathrm{d}\omega_\gamma}{\mathrm{d}\kappa}(\kappa_\gamma)\right| = \sum_{\gamma\in\Gamma_{\text{out}}} \overline{q_j^{\gamma}(K,\omega)} q_i^{\gamma}(K,\omega) \left|\frac{\mathrm{d}\omega_\gamma}{\mathrm{d}\kappa}(\kappa_\gamma)\right|.$$

If we group the coefficients of all solutions in (square) matrices $Q_{\text{in}}(K,\omega)$ and $Q_{\text{out}}(K,\omega)$, so that the ith column is a vector with the coefficients q_i^{γ} for the in- and outgoing modes of the ith solution, we can rewrite this equation as

$$Q_{\text{in}}(K,\omega)^{\dagger} V_{\text{in}}^2(K,\omega) Q_{\text{in}}(K,\omega) = Q_{\text{out}}(K,\omega)^{\dagger} V_{\text{out}}^2(K,\omega) Q_{\text{out}}(K,\omega),$$

with $V_{\text{in,out}}(K,\omega)_{ij} = \delta_{ij} \left|\frac{\mathrm{d}\omega_\gamma}{\mathrm{d}\kappa}(\kappa_\gamma)\right|^{1/2}$ a diagonal matrix. As $Q_{\text{in}}(K,\omega)$ and $Q_{\text{out}}(K,\omega)$ should be connected linearly, we can define a unitary matrix $S(K,\omega)$ as

$$V_{\text{out}}(K,\omega) Q_{\text{out}}(K,\omega) = S(K,\omega) V_{\text{in}}(K,\omega) Q_{\text{in}}(K,\omega).$$

In Sect. 3.6 we will show that this definition corresponds to the S matrix that is known in standard scattering theory. Note, however, that $S(K,\omega)$ is only defined up to a set of phases. Indeed, since the vectors $\boldsymbol{v}_\gamma$ can only be determined up to a phase, the coefficient matrices C_{in} and C_{out} are only defined up to a diagonal matrix of phase factors. These arbitrary phase factors show up in the S matrix as well. We will show how to fix them in the case of the elastic scattering of two identical particles (Sect. 3.6); in the case where we have different outgoing channels only the square of the magnitude of the S matrix elements is physically well-defined.

This formalism allows to calculate the norm of the scattering states in an easy way. Indeed, the general overlap between two scattering states is given by

$$
\begin{aligned}
&\langle \Upsilon_{i'}(K',\omega')|\Upsilon_i(K,\omega)\rangle \\
&\quad = 2\pi\delta(K-K')\left(\sum_{\gamma,\gamma'} \overline{q_{i'}^{\gamma'}(K',\omega')} q_i^{\gamma}(K,\omega)\boldsymbol{v}_{\gamma'}^{\dagger}\boldsymbol{v}_{\gamma} \sum_{n,n'>N} \mathrm{e}^{i(\kappa_\gamma-\kappa'_{\gamma'})n} + \text{finite}\right) \\
&\quad = 2\pi\delta(K-K')\left(\sum_{\gamma,\gamma'} \overline{q_{i'}^{\gamma'}(K',\omega')} q_i^{\gamma}(K,\omega)\boldsymbol{v}_{\gamma'}^{\dagger}\boldsymbol{v}_{\gamma}\pi\delta(\kappa_\gamma(\omega)-\kappa'_{\gamma'}(\omega')) + \text{finite}\right).
\end{aligned}
$$

The δ factor for the momenta κ_γ is obviously only satisfied if $\omega = \omega'$, so we can transform this to a $\delta(\omega-\omega')$. Moreover, if $\kappa_\gamma(\omega) = \kappa'_{\gamma'}(\omega')$ for $\gamma \neq \gamma'$, then necessarily $\boldsymbol{v}_{\gamma'}^{\dagger}\boldsymbol{v}_{\gamma} = 0$, so we can reduce the double sum in γ, γ' to a single one. If we omit all finite parts, we have

$$
\langle \Upsilon_{i'}(K',\omega')|\Upsilon_i(K,\omega)\rangle = 2\pi\delta(K-K')\pi\delta(\omega-\omega')\sum_{\gamma} \overline{q_{i'}^{\gamma}(K',\omega')} q_i^{\gamma}(K,\omega)\left|\frac{\mathrm{d}\omega_\gamma}{\mathrm{d}\kappa}(\kappa_\gamma)\right|.
$$

With the $Q_{\text{in/out}}$ as defined above the overlap reduces to

$$
\begin{aligned}
\langle \Upsilon_{i'}(K',\omega')|\Upsilon_i(K,\omega)\rangle &= 2\pi\delta(K-K')2\pi\delta(\omega-\omega')\,[Q_{\text{in}}(K,\omega)]_{i'}^{\dagger}\,V_{\text{in}}^{2}(K,\omega)\,[Q_{\text{in}}(K,\omega)]_i \\
&= 2\pi\delta(K-K')2\pi\delta(\omega-\omega')\,[Q_{\text{out}}(K,\omega)]_{i'}^{\dagger}\,V_{\text{out}}^{2}(K,\omega)\,[Q_{\text{out}}(K,\omega)]_i\,.
\end{aligned}
$$

3.4.5 Bound States

In Sect. 3.3.6 we have seen how the one-particle ansatz can be extended to larger blocks in order to describe very broad excitations, a situation that arises when a bound state forms out of a two-particle continuum. We could, however, study these bound states with the two-particle ansatz as well. Specifically, the formation of a bound state out of a two-particle continuum should correspond to a continuous deformation of a two-particle wavefunction into a very broad, yet localized one-particle wavefunction.

In contrast to a scattering state the energy of a bound state is not known from the one-particle dispersions, so that we will have to scan a certain energy range in search of bound state solutions – of course, with the one-particle ansatz we can get a pretty good idea where to look. A bound state corresponds to solutions for the eigenvalue equation with only decaying modes in the asymptotic regime. In principle we should even be able to find bound state solutions within a continuum of scattering states (i.e. a stationary bound-state, not a resonance within the continuum) by the presence of additional localized solutions for the scattering problem.

3.5 *Intermezzo*: Excitations and the MPS Tangent Space

Before going on to explain how the two-particle wavefunction contains the information to characterize the particle interactions, this short section describes how our work falls under the concept of "post-MPS methods". This set of methods can be best introduced by again going back to Anderson's description of the concept of an elementary excitation:

> Another way of saying it is that the concept of elementary excitations is a way of *linearizing* the equations of the system about the true ground state rather than about some independent particle approximation [1, p. 102].

Indeed, the idea of linearizing the equations of motion is exactly contained in the concept of a *tangent space* around a variational ground state. Let us explain why.

Consider the set of all matrix product states in the thermodynamic limit at a fixed bond dimension $|\Psi(A)\rangle$. This set of states constitutes a smooth (complex) manifold $\mathcal{M}$ embedded in the full Hilbert space.[19] This manifold has a tangent space $\mathbb{T}$ in every point, where a tangent vector is given by

$$\begin{aligned} |\Phi^{(A)}(B)\rangle &= B^i \partial_i |\Psi(A)\rangle \\ &= \sum_{n=-\infty}^{\infty} \sum_{\{s\}=1}^{d} \boldsymbol{v}_L^\dagger \left[\prod_{m<n} A^{s_m} \right] B^{s_n} \left[\prod_{m>n} A^{s_m} \right] \boldsymbol{v}_{\boldsymbol{R}} |\{s\}\rangle . \end{aligned}$$

The tangent space is a linear vector space and can be visualized as a plane in Fig. 3.2. Crucially, the manifold $\mathcal{M}$ and the tangent space $\mathbb{T}$ contain completely different physics: the MPS describes the translation invariant bulk physics of ground states with intensive (scaling with system size) expectation values, while the tangent vectors parametrize extensive (finite quantities) properties on top of this ground state. Therefore, the tangent space contains the *low-energy dynamics* that is going on against the MPS vacuum state.

A perfect example of this idea is the time-dependent variational principle (TDVP) [56, 57], which dictates how to best approximate time evolution within a variational manifold. Take the time evolution of a state $|\Psi(A)\rangle$ within the manifold of matrix product states according to the time-dependent Schrödinger equation

$$i \frac{\partial}{\partial t} |\Psi(A)\rangle = H |\Psi(A)\rangle .$$

The right hand side is a vector that points out of the manifold; the best approximation within the manifold is obtained by projecting the full evolution onto the tangent space and walking in that direction. The TDVP equation for the time evolution within the MPS manifold then reads [36]

[19] In Ref. [37] this was shown with full mathematical rigour, with a number of caveats that do not need to worry us here.

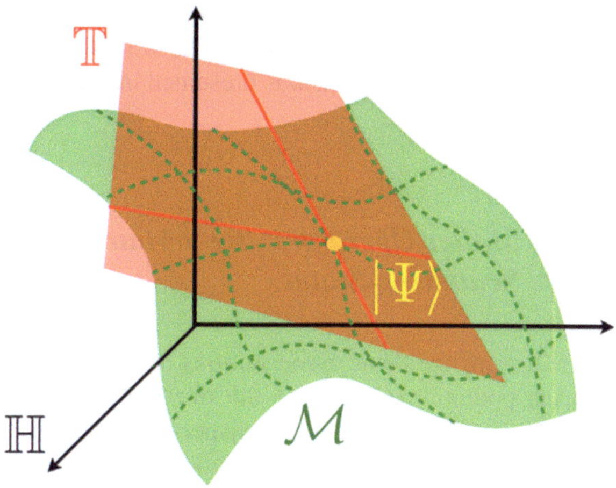

Fig. 3.2 The tangent space (*red*) around a point in the MPS manifold (*green*), both embedded in the physical Hilbert space. Whereas the MPS manifold is a highly non-linear subspace of Hilbert space that describes the bulk properties of ground states, the linear tangent space describes the low-energy dynamics on top of the ground state

$$i\frac{\partial}{\partial t}|\Psi(A(t))\rangle = \mathsf{P}_{\mathbb{T}} H |\Psi(A(t))\rangle.$$

Another example is the one-particle ansatz for elementary excitations that was introduced previously; the ansatz wavefunction has the form of a tangent vector, with a momentum factor added to capture the momentum dependence of the excitation. From this perspective, ground states and excitations belong to different regions of Hilbert space and ask for a different parametrization. This is the reason why studying low-energy excitations or dynamical correlations with direct matrix product techniques – i.e. without leaving the MPS manifold – is essentially misguided and can never work in the thermodynamic limit.

The set of MPS methods that are based on the tangent space has been coined "post-MPS methods" [40] or "post-DMRG methods" [59]. These methods stand in perfect analogy to post-Hartree–Fock methods [60, 61]; the Hartree–Fock manifold consists of Slater determinants [58, 62, 63], whereas the tangent vectors correspond to particle-hole excitations. Post-Hartree–Fock methods such as the random-phase approximation [64] and the Tamm–Dancoff approximation [65, 66] have been translated to the MPS/DMRG setting [40, 59] and stand in close connection with the techniques of this dissertation.[20]

3.6 Scattering Theory in One Dimension

We now discuss how the variational formulation of scattering theory using matrix product states, as developed in the previous section, relates to standard scattering theory. There the S matrix is typically defined from a dynamical point of view: Its

[20]These connections with other computational methods is worked out in more detail in Ref. [40].

elements are the overlaps of asymptotically free in and out states with respect to the full time-evolution operator. Because it is a priori not clear that this definition corresponds to the one that was presented in the previous sections, we show that this is indeed the case.

3.6.1 Møller Operators, the S Matrix and Stationary Scattering States

We will translate some basic notions of single particle scattering theory from an external potential [67, 68] to the one-dimensional case where we have different types of particles with general dispersion relations. The two-particle scattering in the many body Hilbert space considered in this manuscript can be mapped to this setting by taking out the conservation of total momentum and only looking at the relative wave function, which is encoded in the coefficients $c^j(n)$. For the remainder of this section, we assume to have a Hilbert space spanned by states $\{|x, j\rangle\}$ where x is a spatial coordinate that can be discrete ($x \in \mathbb{Z}$) or continuous ($x \in \mathbb{R}$) and $j = 1, \ldots, L$ labels different internal levels at every position (corresponding to different particle types). We assume we have some Hamiltonian H, which can be written as the sum of a free part H_0 and a potential V. The free Hamiltonian is translation invariant ($\langle x', j'|H_0|x, j\rangle = (\mathsf{A}_{x-x'})_{j',j}$ with $\mathsf{A}_x = (\mathsf{A}_{-x})^\dagger$) and also assumed to be short-ranged ($\mathsf{A}_{x-x'} = 0$ for $|x - x'| > M$). The potential is centered around $x = 0$ and goes to zero quickly, e.g. $\langle x', j'|V|x, j\rangle = 0$ for $|x| > M + N$ or $|x'| > M + N$. The free Hamiltonian is diagonalized in momentum space and describes the free propagation of a number of types of particles $\alpha = 1, \ldots, N$ with eigenvalues (dispersion relations) $E_\alpha(p)$. Indeed, by using the momentum states $|p, j\rangle = \int \mathrm{d}x e^{ipx}|x, j\rangle$ (an integral over x should be read as a sum for the discrete case), the free Hamiltonian H_0 is brought into block-diagonal form:

$$\langle p, j|H_0|p', j'\rangle = 2\pi\delta(p - p')(\mathsf{A}(p))_{j,j'} \tag{3.27}$$

where the $L \times L$ hermitian matrix $\mathsf{A}(p) = \int \mathrm{d}x e^{ipx}\mathsf{A}_x$ is an analytic function of p (since A_x vanishes for $|x| > M + N$). Its eigenvalues $E_\alpha(p)$ and corresponding eigenvectors $\boldsymbol{v}_\alpha(p)$ define the spectrum of H_0. Also note for further reference the relation

$$\boldsymbol{v}_\beta(p)^\dagger \frac{\mathrm{d}\mathsf{A}}{\mathrm{d}p}(p)\boldsymbol{v}_\alpha(p) = \frac{\mathrm{d}E_\alpha}{\mathrm{d}p}(p)\delta_{\alpha,\beta}. \tag{3.28}$$

We will henceforth denote the eigenvalues of the free Hamiltonian H_0 as $E(p_\alpha)$ and the corresponding eigenstates as $|p_\alpha\rangle$ with coordinate representation $\langle x, j|p_\alpha\rangle = v^j_\alpha(p)e^{ipx}$. By choosing $\boldsymbol{v}_\alpha(p)^\dagger\boldsymbol{v}_\beta(p) = \delta_{\alpha,\beta}$, the eigenstates $|p_\alpha\rangle$ of H_0 are normalized as $\langle p'_\beta|p_\alpha\rangle = 2\pi\delta(p' - p)\delta_{\alpha,\beta}$ and span the whole Hilbert space

$$1 = \sum_{\alpha} \int \frac{\mathrm{d}p}{2\pi} |p_\alpha\rangle\langle p_\alpha|.$$

The range of p determines whether we are dealing with a discrete or continuous system, and will not be specified. In order to describe scattering experiments, one should build wave packets from these momentum eigenstates

$$|\phi_\alpha\rangle = \int \frac{\mathrm{d}p}{2\pi}\, \phi(p)|p_\alpha\rangle.$$

Typically, we will be interested in wave packets $\phi(p)$ that are strongly centred around some momentum p_0, so that it makes sense to express scattering amplitudes (S matrix elements) in the basis of momentum eigenstates.

Let $U(t)$ and $U_0(t)$ denote the unitary evolution associated to respectively H and H_0. We now want to describe some orbit $U(t)|\psi\rangle$, which has an in-asymptote and an out-asymptote in the following sense

$$U(t)|\psi\rangle \to U_0(t)|\psi_{\mathrm{in}}\rangle \qquad \text{as} \quad t \to -\infty$$
$$U(t)|\psi\rangle \to U_0(t)|\psi_{\mathrm{out}}\rangle \qquad \text{as} \quad t \to +\infty.$$

For given $|\psi_{\mathrm{in}}\rangle$ or $|\psi_{\mathrm{out}}\rangle$, one can try to define

$$|\psi\rangle = \lim_{t\to-\infty} U(t)^\dagger U_0(t)|\psi_{\mathrm{in}}\rangle = \Omega_+|\psi_{\mathrm{in}}\rangle$$
$$|\psi\rangle = \lim_{t\to+\infty} U(t)^\dagger U_0(t)|\psi_{\mathrm{out}}\rangle = \Omega_-|\psi_{\mathrm{out}}\rangle$$

with $\Omega_\pm$ the Møller operators. The existence of these limits, and thus of the Møller operators, can be proven by studying wave packets and linear combinations thereof. For a quadratic dispersion relation, the dispersive behaviour of the wave packet is often sufficient to guarantee convergence. Since we are studying general dispersion relations $E_\alpha(p)$, a sufficient condition can be obtained by restricting to wave packets centred around momenta p_0 with non-zero group velocity $dE_\alpha/dp \neq 0$. Since they are the limit of unitary operators, the Møller operators $\Omega_\pm$ are isometries. Finally, we need the condition of asymptotic completeness (which is often harder to prove) to ensure that the range of Ω_+ and Ω_- is the same: they map every state to the space of scattering states and satisfy the intertwining relations

$$H\Omega_\pm = \Omega_\pm H_0.$$

The scattering operator or S matrix can then be defined as the operator mapping the in-asymptote to the out-asymptote

$$|\psi_{\mathrm{out}}\rangle = \Omega_-^\dagger \Omega_+|\psi_{\mathrm{in}}\rangle = S|\psi_{\mathrm{in}}\rangle \to S = \Omega_-^\dagger \Omega_+.$$

One can easily show that the free Hamiltonian commutes with S so it makes sense to represent the S matrix in the basis of free momentum states

$$\langle q_\beta | S | p_\alpha \rangle = 2\pi\delta(E(q_\beta) - E(p_\alpha)) \times S_{q_\beta, p_\alpha}.$$

If asymptotic completeness is obeyed the S matrix is unitary, which can be expressed in the momentum basis as

$$\langle q_\beta | S^\dagger S | p_\alpha \rangle = 2\pi\delta(p_\alpha - q_\beta)\delta_{\alpha\beta}. \tag{3.29}$$

We can translate this condition to the matrix elements S_{q_β, p_α} as

$$\begin{aligned}
&\langle q_\beta | S^\dagger S | p_\alpha \rangle \\
&= \sum_\gamma \int \frac{\mathrm{d}r}{2\pi} \langle q_\beta | S^\dagger | r_\gamma \rangle \langle r_\gamma | S | p_\alpha \rangle \\
&= \sum_\gamma \int \frac{\mathrm{d}r}{2\pi} 4\pi^2 \delta(E(p_\alpha) - E(r_\gamma))\delta(E(q_\beta) - E(r_\gamma)) \overline{S}_{r_\gamma, q_\beta} S_{r_\gamma, p_\alpha} \\
&= \sum_\gamma \int \frac{\mathrm{d}r}{2\pi} \overline{S}_{r_\gamma, q_\beta} S_{r_\gamma, p_\alpha} 2\pi\delta(E(q_\beta) - E(r_\gamma)) \left(\sum_{p_{\alpha'} \in A(p_\alpha)} \left| \frac{\mathrm{d}E}{\mathrm{d}p}(p_{\alpha'}) \right|^{-1} 2\pi\delta(p_{\alpha'} - r_\gamma) \right) \\
&= \left(\sum_{r_\gamma \in A(p_\alpha)} \overline{S}_{r_\gamma, q_\beta} S_{r_\gamma, p_\alpha} \left| \frac{\mathrm{d}E}{\mathrm{d}p}(r_\gamma) \right|^{-1} \right) \times 2\pi\delta(E(q_\beta) - E(p_\alpha)) \\
&= (\tilde{S}^\dagger \tilde{S})_{q_\beta p_\alpha} \times \left| \frac{\mathrm{d}E}{\mathrm{d}p}(q_\beta) \right|^{1/2} 2\pi\delta(E(q_\beta) - E(p_\alpha)) \left| \frac{\mathrm{d}E}{\mathrm{d}p}(p_\alpha) \right|^{1/2}
\end{aligned}$$

where $A(p_\alpha)$ is the set of momenta $\{q_\beta\}$ such that $E(q_\beta) = E(p_\alpha)$, and we have defined the matrix elements of $\tilde{S}$ as

$$\tilde{S}_{q_\beta, p_\alpha} = \left| \frac{\mathrm{d}E}{\mathrm{d}p}(q_\beta) \right|^{-1/2} S_{q_\beta, p_\alpha} \left| \frac{\mathrm{d}E}{\mathrm{d}p}(p_\alpha) \right|^{-1/2}. \tag{3.30}$$

Unitariness of the S matrix, Eq. (3.29), implies that $\tilde{S}_{q_\beta, p_\alpha}$ should be a unitary matrix.

There are different ways to calculate these S matrix elements; one way is to construct the stationary scattering states, i.e. the eigenstates of the full Hamiltonian $H = H_0 + V$. One first introduces the Green's operators as

$$\begin{aligned}
G_0(z) &= (z - H_0)^{-1} \\
G(z) &= (z - H)^{-1},
\end{aligned}$$

which are related through the relation

$$G(z) = G_0(z) + G_0(z)VG(z)$$
$$= G_0(z) + G(z)VG_0(z).$$

The T operator is defined as

$$T(z) = V + VG(z)V$$

for which we can easily derive the Lippman–Schwinger equation [69]

$$T(z) = V + VG_0(z)T(z),$$

and thc equations

$$G_0(z)T(z) = (G_0(z) + G_0(z)VG(z))V = G(z)V$$
$$T(z)G_0(z) = V(G_0(z) + G(z)VG_0(z)) = VG(z).$$

The Lippman–Schwinger equation can be rewritten as an integral equation for the matrix elements of $T(z)$

$$\langle q_\beta|T(z)|p_\alpha\rangle = \langle q_\beta|V|p_\alpha\rangle + \sum_\gamma \int \frac{\mathrm{d}r_\gamma}{2\pi} \frac{\langle q_\beta|V|r_\gamma\rangle}{z - E(r_\gamma)} \langle r_\gamma|T(z)|p_\alpha\rangle.$$

One can derive a related equation for the Møller operators

$$\begin{aligned}
\Omega_+|\phi\rangle &= \lim_{t\to-\infty} U(t)^\dagger U_0(t)|\phi\rangle \\
&= |\phi\rangle - i\int_{-\infty}^{0} \mathrm{d}\tau U(\tau)^\dagger V U_0(\tau)|\phi\rangle \\
&= |\phi\rangle - i\int_{-\infty}^{0} \mathrm{d}\tau \mathrm{e}^{\epsilon\tau} U(\tau)^\dagger V U_0(\tau)|\phi\rangle \\
&= |\phi\rangle - i\sum_\alpha \int \frac{\mathrm{d}p}{2\pi} \int_{-\infty}^{0} \mathrm{d}\tau \mathrm{e}^{\epsilon\tau} U(\tau)^\dagger V U_0(\tau)|p_\alpha\rangle\langle p_\alpha|\phi\rangle \\
&= |\phi\rangle + \sum_\alpha \int \frac{\mathrm{d}p}{2\pi} G(E(p_\alpha) + i0)\, V|p_\alpha\rangle\langle p_\alpha|\phi\rangle,
\end{aligned}$$

where we have introduced the time-dependent damping factor to the potential $V \to V\mathrm{e}^{-\epsilon t}$, which is allowed for $\epsilon \to 0$ according to the adiabatic theorem. The S matrix

$$\langle q_\beta|S|p_\alpha\rangle = \langle q_\beta|\Omega_-^\dagger \Omega_+|p_\alpha\rangle = \lim_{t\to\infty} \langle q_\beta|\mathrm{e}^{iH_0t}\mathrm{e}^{-2iHt}\mathrm{e}^{iH_0t}|p_\alpha\rangle$$

can be worked out by writing it as the integral of its derivative

$$
\begin{aligned}
\langle q_\beta|S|p_\alpha\rangle &= \langle q_\beta|p_\alpha\rangle - i\int_0^\infty \mathrm{d}t \langle q_\beta|\left(\mathrm{e}^{iH_0t}V\mathrm{e}^{-2iHt}\mathrm{e}^{iH_0t} + \mathrm{e}^{iH_0t}\mathrm{e}^{-2iHt}V\mathrm{e}^{iH_0t}\right)|p_\alpha\rangle \\
&= \langle q_\beta|p_\alpha\rangle - i\lim_{\epsilon\to 0}\int_0^\infty \mathrm{d}t \\
&\quad \langle q_\beta|\left(V\mathrm{e}^{i(E(q_\beta)+E(p_\alpha)+i\epsilon-2H)t} + \mathrm{e}^{i(E(q_\beta)+E(p_\alpha)+i\epsilon-2H)t}V\right)|p_\alpha\rangle \\
&= \langle q_\beta|p_\alpha\rangle + \frac{1}{2}\lim_{\epsilon\to 0}\langle q_\beta|\left(VG\left(\frac{1}{2}(E(p_\alpha)+E(q_\beta))+i\epsilon\right)\right. \\
&\quad \left. + G\left(\frac{1}{2}(E(p_\alpha)+E(q_\beta))+i\epsilon\right)V\right)|p_\alpha\rangle \\
&= \langle q_\beta|p_\alpha\rangle + \lim_{\epsilon\to 0}\left(\frac{1}{E(q_\beta)-E(p_\alpha)+i\epsilon} + \frac{1}{E(p_\alpha)-E(q_\beta)+i\epsilon}\right) \\
&\quad \times \langle q_\beta|T\left(\frac{1}{2}(E(p_\alpha)+E(q_\beta))+i\epsilon\right)|p_\alpha\rangle \\
&= 2\pi\delta(q_\beta - p_\alpha)\delta_{\beta\alpha} - 2\pi\delta(E(q_\beta)-E(p_\alpha))\, i\langle q_\beta|T(E(p_\alpha)+i0)|p_\alpha\rangle.
\end{aligned}
$$

The off-diagonal elements of the S matrix are given by the on-shell T-matrix elements. We define the amplitudes f

$$
f(q_\beta \leftarrow p_\alpha) = -i\left|\frac{\mathrm{d}E}{\mathrm{d}p}(p_\alpha)\right|^{-1/2}\langle q_\beta|T(E(p_\alpha)+i0)|p_\alpha\rangle\left|\frac{\mathrm{d}E}{\mathrm{d}p}(q_\beta)\right|^{-1/2},
$$

which are the off-diagonal elements of $\tilde{S}$ as defined in the unitary matrix (3.30).

We can now define the scattering states

$$
|p_\alpha\pm\rangle = \Omega_\pm|p_\alpha\rangle, \qquad H|p_\alpha\pm\rangle = E(p_\alpha)|p_\alpha\pm\rangle
$$

which, through the Lippmann–Schwinger equation for the Møller operators, obey the relation

$$
|p_\alpha\pm\rangle = |p_\alpha\rangle + G(E(p_\alpha)\pm i0)V|p_\alpha\rangle = |p_\alpha\rangle + G_0(E(p_\alpha)\pm i0)V|p_\alpha\pm\rangle.
$$

Another important relation is

$$
\begin{aligned}
\langle q_\beta|T(E(p_\alpha)\pm i0)|p_\alpha\rangle &= \langle q_\beta|(V + VG(E(p_\alpha)\pm i0)V)|p_\alpha\rangle \\
&= \langle q_\beta|V|p_\alpha\pm\rangle. \qquad (3.31)
\end{aligned}
$$

An explicit expression for the asymptotic wave functions of the scattering states can thus be obtained:

$$
\begin{aligned}
&\langle x, j|p_\alpha+\rangle \\
&= \langle x, j|p_\alpha\rangle + \sum_{j'} \int \mathrm{d}x' \langle x, j|G_0(E(p_\alpha)+i0)|x', j'\rangle\langle x', j'|V|p_\alpha+\rangle \\
&= \mathrm{e}^{ip_\alpha x} v_\alpha^j(p) + \int \mathrm{d}x' \sum_{j'} \langle x, j|\frac{1}{E(p_\alpha)-H_0+i0}|x', j'\rangle\langle x', j'|V|p_\alpha+\rangle.
\end{aligned}
\tag{3.32}
$$

Since we know the exact eigenvalues and eigenvectors of H_0, we will now first introduce a resolution of the identity

$$
\sum_j \int \frac{\mathrm{d}q}{2\pi} |q, j\rangle\langle q, j|
$$

which brings the Green's function in block diagonal form

$$
\begin{aligned}
\langle x, j|p_\alpha+\rangle &= \mathrm{e}^{ipx} v_\alpha^j(p) \\
&+ \int \mathrm{d}x' \sum_{j'} \int \frac{\mathrm{d}q}{2\pi} \left(\frac{1}{E(p_\alpha)-\mathsf{A}(q)+i0}\right)_{j,j'} e^{iq(x-x')}\langle x', j'|V|p_\alpha+\rangle
\end{aligned}
$$

with the matrix $\mathsf{A}(q)$ an analytic function of q, as defined at the beginning of this section. The integral over q can be calculated with the residue theorem. For continuous systems, where q ranges over the real axis, we will have to close the contour in the upper or lower half plane depending on the whether $x - x' > 0$ or $x - x' < 0$. A first set of poles will be close to the real axis and can be obtained from the eigenvalue decomposition of $\mathsf{A}(q)$. Together with the analytic dependence on q and Eq. (3.28), we obtain

$$
\mathsf{A}(q \pm i0) = \sum_\beta \left(E(q_\beta) \pm i0\frac{\mathrm{d}E}{\mathrm{d}p}(q_\beta)\right) \boldsymbol{v}_\beta(q)\boldsymbol{v}_\beta(q)^\dagger. \tag{3.33}
$$

We should therefore separate the set $A(p_\alpha)$ of all solutions q_β for which $E(q_\beta) = E(p_\alpha)$ into two parts $A^\pm(p_\alpha)$ corresponding to solutions for which the energy derivative $\frac{\mathrm{d}E}{\mathrm{d}p}(q_\beta)$ is positive (+) or negative (−). We then find a first set of poles of $\left(\frac{1}{E(p_\alpha)-\mathsf{A}(q)+i0}\right)_{j,j'}$ which are of the form $q_\beta + i0$ for $q_\beta \in A^+(p_\alpha)$ and of the form $q_\beta - i0$ for $q_\beta \in A^-(p_\alpha)$. The corresponding residues are given by

$$
\begin{aligned}
&\lim_{\substack{q\to q_\beta \pm i0 \\ q_\beta \in A^\pm(p_\alpha)}} (q-(q_\beta \pm i0))\left(\frac{1}{E(p_\alpha)-\mathsf{A}(q)+i0}\right)_{j,j'} e^{iq(x-x')} \\
&\qquad = -\left(\frac{\mathrm{d}E}{\mathrm{d}p}(q_\beta)\right)^{-1} v_\beta^j(p)\overline{v}_\beta^{j'}(p)e^{iq_\beta(x-x')}.
\end{aligned}
$$

Aside from those solutions, there could be other solutions $q = i\lambda_\gamma$ further away from the real axis (Re$\lambda \neq 0$). These correspond to values of λ where the analytically continued (but non-hermitian) matrix $\mathsf{A}(i\lambda)$ has a real eigenvalue $E_\gamma(i\lambda) = E(i\lambda_\gamma)$ that equals $E(p_\alpha)$; we denote the corresponding left and right eigenvectors as $\tilde{\boldsymbol{w}}_\gamma(\lambda)^\dagger$ and $\boldsymbol{w}_\gamma(\lambda)$ (which will in general not be related by hermitian conjugation). The corresponding residue is then given by

$$-\frac{\mathrm{d}E}{\mathrm{d}p}(\lambda_\gamma) w_\gamma^j \overline{\tilde{w}}_\gamma^{j'} e^{-\lambda_\gamma(x-x')},$$

or, more generally,

$$-\frac{\mathrm{d}E}{\mathrm{d}p}(\lambda_\gamma) \mathsf{P}_{j,j'}(i\lambda_\gamma) e^{-\lambda_\gamma(x-x')}$$

with $\mathsf{P}(i\lambda_\gamma)$ the corresponding eigenspace projector.

Let us now return to the evaluation of the integral over q. Depending on the sign of $x - x'$, we will close the contour in the upper or lower half plane and pick up the contributions of the poles in those respective domains. Since we also have an integral over x', it seems we will need to split this into the two regions $x < x'$ and $x > x'$. However, we can make use of the locality of the potential to conclude that $\langle x', j'|V|p_\alpha+\rangle$ is only nonzero for $|x'| \leq M + N$. Thus, if $|x| > M + N$, then $x - x'$ will have a fixed sign throughout the integral over x'. For e.g. $x - x' > 0$, we will need to sum up the contributions of all the poles in the upper half plane, corresponding to $q_\beta + i0$ for $p_\beta \in A^+(p_\alpha)$ and all $i\lambda_\gamma$ with Re$\lambda_\gamma > 0$. The latter contributions will actually vanish if we now take the limit $x \to \infty$. We can then write the asymptotic wave function as

$$\langle x, j|p_\alpha+\rangle \approx v_\alpha^j(p)\mathrm{e}^{ip_\alpha x} - i \begin{cases} \sum_{q_\beta \in A^-(p_\alpha)} v_\beta^j(q) \mathrm{e}^{iq_\beta x} \left|\frac{\mathrm{d}E}{\mathrm{d}p}(q_\beta)\right|^{-1} \langle q_\beta|V|p\pm\rangle & x \to -\infty \\ \sum_{q_\beta \in A^+(p_\alpha)} v_\beta^j(q) \mathrm{e}^{iq_\beta x} \left|\frac{\mathrm{d}E}{\mathrm{d}p}(q_\beta)\right|^{-1} \langle q_\beta|V|p\pm\rangle & x \to +\infty \end{cases}$$

and with Eq. (3.31)

$$\langle x, j|p_\alpha+\rangle \approx v_\alpha^j(p)\mathrm{e}^{ip_\alpha x} - i \sum_{q_\beta \in A^\pm(p_\alpha)} v_\beta^j(q) \mathrm{e}^{iq_\beta x} \left|\frac{\mathrm{d}E}{\mathrm{d}p}(q_\beta)\right|^{-1} \times \langle q_\beta|T(E(p_\alpha) + i0)|p_\alpha\rangle \qquad x \to \pm\infty.$$

The coefficients that appear are the amplitudes that were defined earlier, so we have the nice final result

$$\langle x, j|p_\alpha+\rangle = v_\alpha^j(p)\mathrm{e}^{ip_\alpha x} + \sum_{q_\beta \in A^\pm(p_\alpha)} \left|\frac{\mathrm{d}E}{\mathrm{d}p}(p_\alpha)\right|^{1/2} \times f(q_\beta \leftarrow p_\alpha) \left|\frac{\mathrm{d}E}{\mathrm{d}p}(q_\beta)\right|^{-1/2} \mathrm{e}^{iq_\beta x} v_\beta^j(q), \qquad x \to \pm\infty. \tag{3.34}$$

For discrete systems, we can proceed in a similar way. Momentum integrals now range from 0 to 2π and we automatically obtain a contour around the unit circle in the complex plane by going to the complex variable $\mu = e^{iq}$ (for $x - x' > 0$) or $\mu = e^{-iq}$ (for $x - x' < 0$). The derivation then follows analogously.

3.6.2 The Connection with Variational Scattering States

In order to connect the variational scattering states of Sect. 3.4 to the above form for the stationary scattering states, we have to make a few modifications. First of all, we can reformulate the two-particle scattering problem as a one-particle problem by factoring out the conservation of total momentum and only focusing on the matrix elements between different relative momenta. At every value of the total momentum, we can define relative momentum states $|p_\gamma\rangle$ with dispersions $\omega(p_\gamma)$, which are solutions of the free Hamiltonian H_0. This free Hamiltonian corresponds to the effective two-particle Hamiltonian matrix in the asymptotic regime (3.18) and the states $|p_\gamma\rangle$ are the asymptotic modes (3.21).

Secondly, our "one-particle" Hilbert space is only defined on a half-infinite line, because the particles are essentially bosonic. The way around this consists of artificially assigning particle labels and distinguishing the situation where particle 1 (2) is on the left (right), and the opposite situation; the relative coordinate $n = n_2 - n_1$ now ranges over the positive and negative integers. Alternatively, one could add to the free Hamiltonian H_0 a potential V which is infinite everywhere on the negative real line, making this a forbidden region. Scattering theory would still work (existence of the Møller operators etc.), provided that we restrict the "in" states to momenta for which $\frac{\mathrm{d}\omega}{\mathrm{d}p}(p) < 0$ and the out states to momenta for which $\frac{\mathrm{d}\omega}{\mathrm{d}p}(p) > 0$. This corresponds exactly to how we defined the incoming and outgoing modes in Sect. 3.4.4.

Translating the expression for the asymptotic wave function of the scattering states $|p_\alpha+\rangle$ to the framework of Sect. 3.4 amounts to the following form for the wave function $\boldsymbol{c}_\alpha(n)$

$$\boldsymbol{c}_\alpha(n) \to \left|\frac{\mathrm{d}\omega}{\mathrm{d}\kappa}(\kappa_\alpha)\right|^{-1/2} \boldsymbol{v}_\alpha e^{ip_\alpha n} + \sum_{\gamma\in A^+(\kappa_\alpha)} f(\kappa_\gamma \leftarrow \kappa_\alpha) \left|\frac{\mathrm{d}\omega}{\mathrm{d}\kappa}(\kappa_\gamma)\right|^{-1/2} \boldsymbol{v}_\gamma e^{i\kappa_\gamma n} \tag{3.35}$$

for every incoming mode $\alpha = 1, \ldots, \Gamma$. In this representation, we choose one incoming mode α that couples only to all outgoing modes $\{\gamma \in A^+(\kappa_\alpha)\}$. The coefficient matrix for the incoming modes that was defined earlier takes on the form

$$(Q_{\text{in}})_{\gamma,\alpha} = \delta_{\gamma\alpha} \left|\frac{\mathrm{d}\omega}{\mathrm{d}\kappa}(\kappa_\alpha)\right|^{-1/2}$$

while the coefficients for the outgoing modes are given by

$$(Q_{\text{out}})_{\gamma,\alpha} = \left|\frac{\mathrm{d}\omega}{\mathrm{d}\kappa}(\kappa_\gamma)\right|^{-1/2} f(\kappa_\gamma \leftarrow \kappa_\alpha).$$

The S matrix $S(K,\omega)$ that was defined takes on the simple form (as $V_{\text{in}}Q_{\text{in}} = 1$ in this representation)

$$\begin{aligned} S(K,\omega) &= V_{\text{out}}Q_{\text{out}} \\ &= f(\kappa_\gamma \leftarrow \kappa_\alpha). \end{aligned}$$

Through this identification the unitariness of the S matrix $S(K,\omega)$ that was proven in the previous section is indeed equivalent to the unitary S matrix defined through the Møller operators as $S = \Omega_-^\dagger \Omega_+$.

3.6.3 Scattering Phase, Scattering Length and Bound States

Let us make things more concrete by working out the case where the one-particle spectrum consists of just one type of particle with dispersion relation $\Delta(\kappa)$. Suppose we have only one combination of momenta κ_1 and κ_2 such that they add up to total momentum $K = \kappa_1 + \kappa_2$ and total energy $\omega = \Delta(\kappa_1) + \Delta(\kappa_2)$. There are two asymptotic modes – one mode with κ_1 on the left and κ_2 on the right, and one mode with the momenta interchanged – that combine into one scattering state with the asymptotic form

$$\boldsymbol{c}(n) \to q^1 \mathrm{e}^{i\kappa_2 n}\boldsymbol{v_1} + q^2 \mathrm{e}^{i\kappa_1 n}\boldsymbol{v_2}.$$

The conservation equation takes on the simple form

$$|q^1|^2 = |q^2|^2$$

because $\omega'(\kappa_1) = -\omega'(\kappa_2)$. As we mentioned above in the general case, the relative phase of the two vectors $\boldsymbol{v_1}$ and $\boldsymbol{v_2}$ can be chosen arbitrarily. However, since the two modes correspond to the interchanging of two identical particles, it makes sense to fix the phase such that $\boldsymbol{v_2}^\dagger \boldsymbol{v_1} > 0$. Due to the momentum dependence of the one-particle solutions, this overlap will be slightly smaller than one. In this setting, the S matrix reduces to a phase factor and can be labelled by the two individual momenta κ_1 and κ_2

$$S(\kappa_1,\kappa_2) = S(K,\omega) = \frac{q^2}{q^1}.$$

and the asymptotic wave function takes the form

$$|\Upsilon(\kappa_1,\kappa_2)\rangle \to \sum_{n_1<2} e^{i(\kappa_1 n_1+\kappa_2 n_2)}\left[B_{\kappa_1}\text{at } n_1, B_{\kappa_2} \text{ at } n_2\right]$$
$$+ S(\kappa_1,\kappa_2)e^{i(\kappa_2 n_1+\kappa_1 n_2)}\left[B_{\kappa_2} \text{ at } n_1, B_{\kappa_1} \text{ at } n_2\right]. \qquad (3.36)$$

From simple arguments [70] one can argue that in one dimension the S matrix should have the universal limiting value for low-energy scattering [71, 72]

$$S(\kappa_1,\kappa_2) \to -1 \quad \text{as} \quad |\kappa_1-\kappa_2| \to 0.$$

We define the scattering phase θ as the phase shift of the S matrix relative to its universal low-energy value $S(\kappa_1,\kappa_2) = -e^{i\theta(\kappa_1,\kappa_2)}$.

In general, the scattering phase depends on the two individual momenta κ_1 and κ_2 in a very non-trivial way. When one is only interested in low-energy scattering, however, the dispersion of the scattering phase $\theta(\kappa_1,\kappa_2)$ can typically be captured in one characteristic quantity called the scattering length a. It is defined as the coefficient of the first-order correction of the scattering phase as a function of the relative momentum $\kappa_1-\kappa_2$,

$$\theta(\kappa_1,\kappa_2) \approx -a(\kappa_1-\kappa_2), \qquad \kappa_1,\kappa_2 \to 0. \qquad (3.37)$$

The physical meaning of this scattering length becomes clear when looking at the wave function in this limit of vanishing relative momentum[21]

$$\begin{aligned} c(n) &\to e^{i\kappa_2 n}\left(1-e^{i\theta(\kappa_1,\kappa_2)}e^{i(\kappa_1-\kappa_2)n}\right) \\ &\propto 1-e^{i(\kappa_1-\kappa_2)(n-a)} \\ &\approx (\kappa_1-\kappa_2)(n-a), \qquad \kappa_1-\kappa_2 \to 0 \end{aligned}$$

up to some unimportant prefactors. The scattering length thus appears as the size of an effective infinitely hard wall, where the relative wavefunction vanishes.

In principle, the scattering length is defined in the low-energy limit where both individual momenta are close to the minimum of their dispersion relation. One can, however, define a generalized scattering length $a(K)$, depending on the total momentum $K=\kappa_1+\kappa_2$, which gives the slope of the scattering length at a fixed total momentum (Fig. 3.3)

$$a(K) = -\frac{\theta(\kappa_1,\kappa_2)}{\kappa_1-\kappa_2}, \qquad \begin{cases} \kappa_1-\kappa_2 \to 0 \\ K=\kappa_1+\kappa_2 \end{cases}.$$

This momentum resolved scattering length will prove crucial for detecting the formation of bound states. Think of a bound state splitting off from the two-particle scattering continuum at a certain value of the total momentum K. In the regime

[21] We have omitted the vectors $\boldsymbol{v}_1$ and $\boldsymbol{v}_2$, because they should become equal when the two momenta approach.

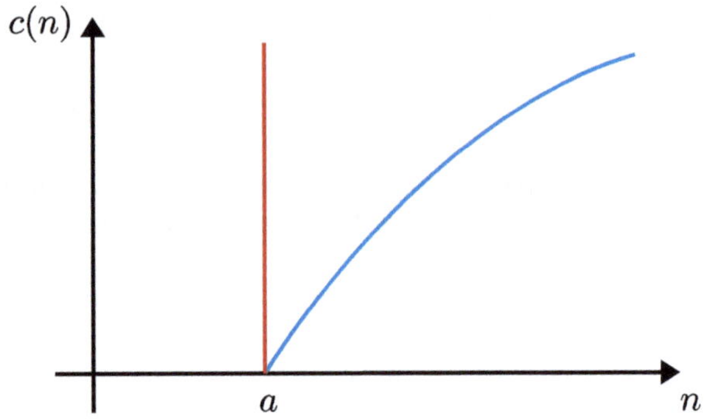

Fig. 3.3 The scattering length shows up as the node of the relative wavefunction $c(n)$ in the limit for $\kappa_1, \kappa_2 \to 0$

where the bound state is just below the continuum, the asymptotic form of the bound state wave function should look like

$$c(n) \propto \mathrm{e}^{-\lambda n} \approx 1 - \lambda n$$

with $\lambda \to 0$ as the bound state comes closer to the continuum and gets wider. We expect that, as the bound state eventually gets absorbed into the continuum and transforms into a scattering state, that this wavefunction smoothly goes over into the above form for the scattering wavefunction

$$(\kappa_1 - \kappa_2)(n - a) \rightsquigarrow 1 - \lambda n$$

which can only happen if the scattering length diverges. This means that the formation of a bound state out of a scattering continuum at a certain momentum should be accompanied by a diverging scattering length, a feature which presents us with a perfect tool to detect the formation of a bound state in the two-particle S matrix.

3.6.4 Dynamical Scattering Theory

In real-life scattering experiments we need localized particles that collide and, after the scattering process, are again recorded as localized particles. This can be modeled by building wave packets as

$$|\psi\rangle = \int \frac{\mathrm{d}p}{2\pi} \phi(p) |p\rangle.$$

In this expression, we assume that $\phi(p)$ is a real-valued function, but it is clear that $|p\rangle$ contains an arbitrary momentum-dependent phase factor. For simplicity, however, we will assume that $\phi(p)$ is strongly peaked around a certain momentum $\bar{p}$, so that the momentum eigenstates for $p \approx \bar{p}$ can be written as

$$|p\rangle = \sum_n e^{ipn} |\chi(n)\rangle,$$

where $|\chi(n)\rangle$ represents a momentum-independent perturbation of the ground-state wave function centred around site n. The best choice for $\phi(p)$ is of course a Gaussian distribution centred around $\bar{p}$ with a width $1/\sigma$:

$$\phi(p) = \frac{\sqrt{2\pi}}{\sigma} e^{-ip\bar{n}} \exp\left(\frac{-(p-\bar{p})^2}{2\sigma^2}\right).$$

The time evolution of a wave packet is determined by the momentum dispersion of the particle, which we expand to second order as

$$E(p) = \omega + v(p-\bar{p}) + \frac{c}{2}(p-\bar{p})^2,$$

such that the time evolution of the wave packet is given by

$$\begin{aligned} |\psi_1(t)\rangle &= \int \frac{dp}{\sqrt{2\pi}\sigma} \exp\left(-ip\bar{n} - iE(p)t - \frac{(p-\bar{p})^2}{2\sigma^2}\right) |p\rangle \\ &= \frac{\sigma(t)}{\sigma(0)} \sum_n e^{i(\bar{p}(n-\bar{n})-\omega t)} \exp\left(-\frac{\sigma(t)^2}{2}(n-\bar{n}-vt)^2\right) |\chi(n)\rangle. \end{aligned}$$

This represents a one-particle wave packet centred around $\bar{n}+vt$ and spreading out with a time-dependent width

$$\frac{1}{\text{Re}\,\sigma(t)^2} = \frac{1}{\sigma(0)^2} + c^2t^2\sigma(0)^2$$

As compared to the linear movement with velocity v, the spreading of the wave packet is a second-order effect, so we can certainly find a time scale where, during the scattering process, the wave packets remain well-localized. Let us therefore take the following two-particle initial state and neglect all second-order effects ($c \approx 0$):

$$|\psi_2(0)\rangle = \int \frac{dp_1}{2\pi} \int \frac{dp_2}{2\pi} \phi(p_1,p_2) |p_1,p_2\rangle$$

with

$$\begin{aligned} \phi(p_1,p_2) &= \phi_1(p_1) \times \phi_2(p_2) \\ &= \frac{\sqrt{2\pi}}{\sigma_1} e^{ip_1\bar{n}_1} \exp\left(\frac{-(p_1-\bar{p}_1)^2}{2\sigma_1^2}\right) \frac{\sqrt{2\pi}}{\sigma_2} e^{ip_2\bar{n}_2} \exp\left(\frac{-(p_2-\bar{p}_2)^2}{2\sigma_2^2}\right) \end{aligned}$$

and the asymptotic form of the two-particle eigenstate

$$|p_1, p_2\rangle \to \sum_{n_1<n_2} \left(e^{ip_1n_1}e^{ip_2n_2} + S(p_1, p_2)e^{ip_2n_1}e^{ip_1n_2}\right)|\chi(n_1, n_2)\rangle.$$

We write the S matrix in first order in $p_i - \bar{p}_i$ as

$$S(p_1, p_2) = S(\bar{p}_1, \bar{p}_2) \times e^{-ia_1(p_1-\bar{p}_1)}e^{-ia_2(p_2-\bar{p}_2)}.$$

The time evolution of this state can be computed similarly as before,

$$\begin{aligned}|\psi_2(t)\rangle = \sum_{n_1<n_2} e^{-i(\omega_1+\omega_2)t}\Big(&e^{i\bar{p}_1(n_1-\bar{n}_1)}e^{i\bar{p}_2(n_2-\bar{n}_2)}\\ &\quad\times e^{-\frac{\sigma_1^2}{2}(n_1-\bar{n}_1-v_1t)^2}e^{-\frac{\sigma_2^2}{2}(n_2-\bar{n}_2-v_2t)^2}\\ &+S(\bar{p}_1, \bar{p}_2)e^{i\bar{p}_2(n_1-\bar{n}_2-a_2)}e^{i\bar{p}_1(n_2-\bar{n}_1-a_2)}\\ &\quad\times e^{-\frac{\sigma_2^2}{2}(n_1-\bar{n}_2-a_2-v_2t)^2}e^{-\frac{\sigma_1^2}{2}(n_2-\bar{n}_1-a_1-v_1t)^2}\Big)|\chi(n_1, n_2)\rangle.\end{aligned}$$

The physical set-up is designed as $\bar{n}_1 \ll \bar{n}_2$, such that the first term dominates for short times. In the long-time limit only the second part survives under the condition that $v_1 > v_2$; otherwise, the two wave packets never see each other and the first term will dominate for all times. In between these two time limits, the scattering process occurs and we need the local part of the stationary wave functions. We can summarize the asymptotic situations by tracking the movement of the center of the left and right wave packets as

$$\text{before collision:}\qquad \begin{cases} n_L = \bar{n}_1 + v_1t \\ n_R = \bar{n}_2 + v_2t \end{cases}$$

$$\text{after collision:}\qquad \begin{cases} n_L = \bar{n}_2 - a_2 + v_2t \\ n_R = \bar{n}_1 - a_1 + v_1t \end{cases},$$

which shows that (i) the velocities are interchanged, and (ii) the scattering process induces a displacement given by the derivatives of the scattering phases. The derivative is of course the scattering length that we defined earlier, showing again that it is this quantity which provides the non-trivial physical properties at lowest order.

In the case where there is more than one type of particle and different scattering channels are opened, we can play the same game, using generalized wave packets

$$|\psi_{1,\alpha}(t=0)\rangle = \sum_{\alpha'} \int \frac{dp_{\alpha'}}{2\pi}\phi_{\alpha'}(p)|p_{\alpha'}\rangle.$$

From the asymptotic form of the stationary wave function [Eq. (3.35)] we can construct a state with two well-defined wave packets of types α_1 and α_2, which will time-evolve towards a state that is the superposition of different states with wave packets of types α_1' and α_2', where the relative magnitudes are determined by the scattering amplitudes $f(\cdots)$.

3.7 Finite Density of Excitations: The Approximate Bethe Ansatz

In this section we will develop a method to describe a finite density of excitations based on the coordinate Bethe ansatz. Obviously the vacuum state without any particles is the ground state of the original Hamiltonian. Therefore, in order to generate a finite density of excitations in equilibrium,[22] the Hamiltonian needs to be perturbed or the temperature turned on. In the former scenario, the easiest case is just the application of a chemical potential for the particles. In the original spin Hamiltonian this typically corresponds to an external magnetic field that is applied to the spin chain. In this chapter we will assume that the simplest case of a chemical potential h or a temperature T can indeed be realized; the exact realization for spin chains will be the subject of the next chapter. We will show that the density of particles can be treated as a bose gas for which, based on the information on the one- and two-particle excitations, the thermodynamic properties can be computed with the Bethe ansatz. Moreover, based on the mapping to a Luttinger liquid, the power-law behaviour of the correlation functions is inferred.

For simplicity, we will restrict for the remainder of this section to the case of one type of particle – making the consistency conditions for factorized scattering (Yang–Baxter equation) trivial – but the framework can be extended to multicomponent situations [73]. We will interpret the strongly correlated MPS ground state as a vacuum state on which we can build N-particle states, described by a N-particle wave function $\Psi(x_1, \ldots, x_N)$. Although in general we have no particle conservation in the system, we will argue that the first-quantized approach gives a good approximation at low densities. Indeed, particle-number violating processes involve three or more particles and can be neglected at low densities. In Sect. 3.9 we will discuss how to develop a second-quantization approach.

3.7.1 *The Bethe Ansatz Wavefunction*

We start with one particle. We can link the one-particle excitation $|\Phi_\kappa[B]\rangle$ with dispersion $\Delta(\kappa)$ in an obvious way with a one-particle wave function $\Psi_1(x)$ in first quantization as

$$\Psi_1(x) = \mathrm{e}^{i\kappa x}.$$

[22] In the outlook we will discuss how our particle picture can be applied to non-equilibrium physics such as quantum quenches and transport processes.

The energy of this particle under a chemical potential equals $\epsilon = \Delta(\kappa) - h$. Adding a second particle can be done by only taking account of the asymptotic part of the two-particle wave function ($x_1 < x_2$)

$$\Psi_2(x_1, x_2) = \mathrm{e}^{i(\kappa_1 x_1 + \kappa_2 x_2)} + S(\kappa_1, \kappa_2)\mathrm{e}^{i(\kappa_2 x_1 + \kappa_1 x_2)}.$$

The two-particle S matrix $S(\kappa_1, \kappa_2)$ is computed with the variational methods of the previous sections. As we are working with identical particles, the wave function in the other sector ($x_1 > x_2$) has to be determined from the statistics of the particles. On the level of the spin system, the addition of a particle is a local operation, so we will work with bosonic many-particle wave functions.

The addition of a third particle can only be done approximately. Indeed, a three-particle wave function has the general form [73]

$$\begin{aligned}\Psi_3(x_1, x_2, x_3) = \mathrm{e}^{i(\kappa_1 x_1 + \kappa_2 x_2 + \kappa_3 x_3)} &+ S(\kappa_1, \kappa_2)\mathrm{e}^{i(\kappa_2 x_1 + \kappa_1 x_2 + \kappa_3 x_3)} + \cdots \\ &+ \iiint \mathrm{d}\kappa_1' \mathrm{d}\kappa_2' \mathrm{d}\kappa_3' \; S(\kappa_1' \kappa_2' \kappa_3') \; \mathrm{e}^{i(\kappa_1' x_1 + \kappa_2' x_2 + \kappa_3' x_3)} \\ &+ \text{other particle numbers.}\end{aligned} \tag{3.38}$$

The first terms represent a sum over all six permutations of the three momenta, with the S matrices for all possible two-particle scattering processes as prefactors. The next term is the diffractive part, which accounts for the three-particle scattering. For these scattering processes, the two conservation laws are not enough to preserve individual momenta and a whole continuum of other momenta is generated. The last term accounts for the non-particle preserving scattering processes, which can generate two- or four-particle states as well. As a result, it is no longer possible to assign a set of individual momenta $\{\kappa_1, \kappa_2, \kappa_3\}$ (or even a particle number) to this wave function, because they are completely mixed up with all other possible sets of momenta that are compatible with conservation of total energy and momentum.

The crucial approximation of our approach is that we neglect the two last terms in Eq. (3.38): every many-particle scattering event can be decomposed into two-particle scatterings that preserve particle number and individual momenta. This implies that three-particle eigenstates can be labeled by three individual momenta and that the three-particle wave function is given by the permutation terms only. The absence of diffractive scattering is the hallmark of integrability [73], so we are essentially assuming that our many-particle system is integrable [75–77].[23]

[23] Our approach is assumed to be valid at low densities, and should be connected to the low-density approximation in terms of virial coefficients. In particular, the fact that the second virial coefficient can be written as a function of the two-particle phase shift only [86, 87] and the idea of computing the thermodynamics of a gas directly with the S matrix, seem both to be closely related [88]. On the other hand, as the solution of the Yang–Yang equation is supposed to be the analytic continuation of the full virial expansion [73], our method seems to correspond to an approximate continuation, that is valid beyond first order.

If this approximation proves to be valid, we can apply the Bethe ansatz machinery [48, 73, 74] – see also Sect. 2.5. The first-quantized wave function of an integrable N-particle system, unambiguously defined by a set of momenta $\{\lambda_1, \ldots, \lambda_M\}$, is a sum of plane waves with all possible permutations of the momenta

$$\Psi(x_1, \ldots, x_N) = \sum_{\mathcal{P}} A(\mathcal{P}) e^{i(\lambda_{\mathcal{P}1} x_1 + \cdots + \lambda_{\mathcal{P}N} x_N)} \tag{3.39}$$

where $A(\mathcal{P})/A(\mathcal{P}') = S(\lambda_i, \lambda_j)$ if the permutations $\mathcal{P}$ and $\mathcal{P}'$ differ by the interchange of the momenta λ_i and λ_j.

By imposing periodic boundary conditions on the Bethe wave function in the thermodynamic limit, we arrive at a description of the ground state as a Fermi sea of "pseudo-momenta" filled up to a certain Fermi level q. In contrast to the free-fermion case, the density of occupied modes is not constant but given by the function $\rho(\lambda)$ such that $\rho(\lambda) = 0$ for $|\lambda| > q$. The energy of the modes $\epsilon(\lambda)$ can be determined from the integral equation

$$\epsilon(\lambda) - \frac{1}{2\pi} \int_{-q}^{q} K(\lambda, \mu) \epsilon(\mu) \mathrm{d}\mu = \epsilon_0(\lambda) \tag{3.40}$$

where $\epsilon_0(\lambda)$ is the "bare energy" of the particle, i.e. the energy an isolated particle with momentum λ would have in an infinite system. The kernel of the integral equation is given by the derivative of the scattering phase $K(\lambda, \mu) = \partial_\lambda \theta(\lambda, \mu)$. The value of the Fermi level is computed self-consistently from this equation and the requirement that $\epsilon(\pm q) = 0$. Once q has been determined, the density $\rho(\lambda)$ is the solution of a similar integral equation [78, 79]

$$\rho(\lambda) - \frac{1}{2\pi} \int_{-q}^{q} K(\lambda, \mu) \rho(\mu) \mathrm{d}\mu = \frac{1}{2\pi}. \tag{3.41}$$

The total density and energy density are given by

$$D = \int_{-q}^{q} \rho(\lambda) \mathrm{d}\lambda \quad \text{and} \quad E = \frac{1}{2\pi} \int_{-q}^{q} \epsilon(\lambda) \mathrm{d}\lambda. \tag{3.42}$$

The excitation spectrum is easily characterized in terms of the pseudo-particles of the Bethe ansatz. We can construct two types of elementary excitations: either we take one particle with momentum $|\lambda| < q$ out of the Fermi sea (hole excitation) or we add one particle with momentum $|\lambda| > q$ (particle excitation). These elementary particle and hole excitations have a topological nature [48], so that the physical excitations – the ones having a finite overlap with a local operator – consist of an even number of particles and holes [79]. This gives rise to the physical excitation spectrum as shown in Fig. 3.4.

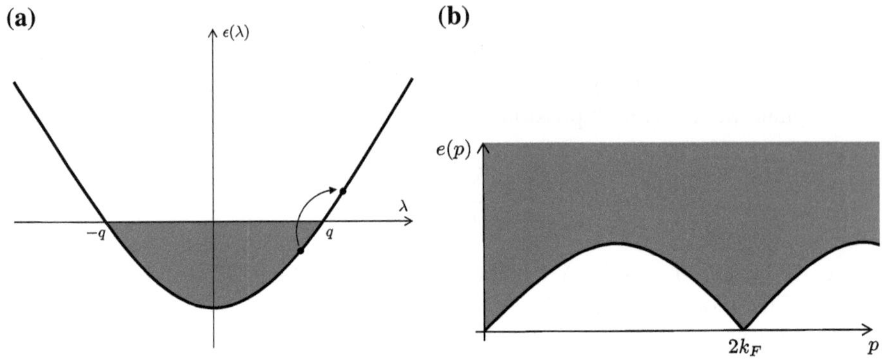

Fig. 3.4 **a** The Fermi sea of pseudo momenta, filled up to the Fermi level q. Physical excitations can be pictured as particle-hole excitations close to the Fermi-level. **b** The physical excitation spectrum, the *grey area* represents a continuum of states. Because of the fact that physical excitations always come in pairs, the spectrum has its minima at momentum 0 and $2k_F$. The slope of the dispersion relation at these momenta is the Fermi velocity u

3.7.2 Correlation Functions: The Luttinger Liquid

The excitation spectrum in Fig. 3.4 shows that we are dealing with a critical one-dimensional bose gas, which can be described as a Luttinger liquid (LL) [81, 82, 84]. A first important quantity is the Fermi momentum k_F, the physical momentum of the gapless particle and hole excitations. It is given by the dressed momentum of the Fermi level,

$$k_F = p(q) = \pm q \pm \int_{-q}^{q} \theta(q, \mu)\rho(\mu)\mathrm{d}\mu,$$

where we have used Eq. (2.4). The integral can be evaluated by integrating Eq. (2.3) with respect to λ:

$$\begin{aligned} D &= \frac{q}{\pi} + \frac{1}{2\pi}\int_{-q}^{q} \mathrm{d}\lambda \int_{-q}^{q} \mathrm{d}\mu \; K(\lambda, \mu)\rho(\mu) \\ &= \frac{q}{\pi} + \frac{1}{\pi}\int_{q}^{q} \mathrm{d}\mu \; \theta(q, \mu)\rho(\mu) \end{aligned}$$

so that we obtain the simple result

$$k_F = \pi D.$$

Since we have gapless excitations at 0 and $\pm 2k_F$, correlation functions will have their oscillation periods at these values. The slope of the dispersion relation is the Fermi velocity u and can be calculated from the Bethe ansatz. The third important characteristic quantity is the LL parameter K which determines the power-law decay of correlation functions. In order to calculate it, we define the function $S_R(\lambda)$ as (h is a chemical potential for the particles)

$$S_R(\lambda) = -\frac{\partial \epsilon(\lambda)}{\partial h}$$

which (from Eq. (3.40)) follows the integral equation

$$S_R(\lambda) - \frac{1}{2\pi} \int_{-q}^{q} K(\lambda, \mu) S_R(\mu) \mathrm{d}\mu = 1. \tag{3.43}$$

In the context of a dilute gas of magnons $S_R(q)$ can be interpreted as the renormalized spin of the magnon close to the Fermi surface. With the low-energy excitations just above the Fermi sea behaving as free fermions [89] (i.e. their S matrix is -1), one can show that the LL parameter K is related to $S_R(q)$ as [90]

$$K = S_R(q)^2. \tag{3.44}$$

By thus making the connection between the approximate Bethe ansatz and the LL description, we can infer information on the critical correlations in a system where a finite density of excitations forms on top of a strongly-correlated vacuum state. More specifically, we can infer the long-range behaviour of one-particle and pair correlation functions as [83, 84]

$$\begin{aligned} g_1(x) &= A_0 \frac{1}{x^{1/2K}} - A_1 \frac{\cos(2\pi D x)}{x^{2K+1/2K}} + \cdots \\ D_2(x) &= D^2 - \frac{K}{2\pi^2 x^2} + B_1 \frac{\cos(2\pi D x)}{x^{2K}} + \cdots \end{aligned} \tag{3.45}$$

where D is the density, A_0, A_1, and B_1 are non-universal constants and the dots denote higher order terms. These correlation functions can be used to infer the behaviour of spin correlations of the original spin chain: depending on whether the spin operator targets a particle excitation or a pair of particles, the corresponding correlation functions will decay according to one of these two forms.

3.7.3 Limiting Cases

The Bethe ansatz equations of the previous section can be greatly simplified if we assume that we work at very low densities. Indeed, assuming that only the lowest pseudo-momentum states are occupied, we can approximate the full dispersion relation by its quadratic form $\epsilon_0(\lambda) \approx c\lambda^2 - h$, and the full two-particle S matrix by its limiting value of $S(\theta, \mu) \approx -1$. With the kernel of the integral equation zero, we easily find the density and the (physical) Fermi momentum

$$D = \frac{1}{\pi}\sqrt{\frac{h}{c}}, \qquad k_F = \sqrt{\frac{h}{c}}$$

and the LL parameters

$$u = 2\pi c D, \qquad K = 1.$$

Upon increasing the density, the limiting value of the S matrix will no longer apply. From Sect. 3.6.3 we know that the first order correction to the scattering phase is given by the scattering length, so we can insert the form (3.37) into the Bethe equations, while still assuming a quadratic dispersion relation. Since the kernel of the linear equation is constant, i.e. $K(\lambda, \nu) = \partial_\lambda \theta(\lambda, \mu) = -a$ – another example of the physical significance of the scattering length – the integral equation for the dressed energy reduces to

$$\epsilon(\lambda) + \frac{a}{2\pi}\int_{-q}^{q} \epsilon(\mu)\mathrm{d}\mu = c\lambda^2 - h. \tag{3.46}$$

One can show that the the first order correction to q with respect to the free-fermion case is linear in the scattering length

$$q = q_{\mathrm{FF}} + \delta q = \sqrt{\frac{h}{c}} - \frac{ah}{3\pi c} + \mathcal{O}(a^2) \tag{3.47}$$

so that the correction to the density is given by

$$D = \frac{1}{\pi}\sqrt{\frac{h}{c}} - \frac{4ha}{3\pi^2 c} + \mathcal{O}(a^2). \tag{3.48}$$

This result coincides with the one in Ref. [91]. The LL parameters in first order in a are given by [92, 93]

$$u = 2c\sqrt{\frac{h}{c}} + \frac{4ah}{3\pi} + \mathcal{O}(a^2)$$

and

$$K = 1 - 2aD + \mathcal{O}(a^2).$$

3.7.4 *Thermodynamic Bethe Ansatz*

As we have said, we can also treat a thermally excited density of particles with the approximate Bethe ansatz through the Yang–Yang equation [94, 95], a non-linear integral equation for the dressed energy function $\epsilon(\lambda)$

$$\epsilon(\lambda) = \epsilon_0(\lambda) - \frac{T}{2\pi}\int_{-\infty}^{+\infty} K(\lambda, \mu) \ln\left(1 + \mathrm{e}^{-\epsilon(\mu)/T}\right) \mathrm{d}\mu,$$

After solving this equation numerically, a generalized distribution $\theta(\lambda))$ for the quasi-momentum states is obtained,

$$\theta(\lambda) = \frac{1}{1 + \mathrm{e}^{\epsilon(\lambda)/T}},$$

which determines the integral equation for the density $\rho(\lambda)$

$$\rho(\lambda) = \frac{\theta(\lambda)}{2\pi}\left(1 + \int_{-\infty}^{+\infty} K(\lambda, \mu)\rho(\mu)\mathrm{d}\mu\right),$$

such that the total density can be calculated as

$$D = \int_{-\infty}^{+\infty} \rho(\lambda)\mathrm{d}\lambda.$$

Other thermodynamic quantities can be computed similarly.

3.7.5 *Effective Integrable Field Theories*

Another way of dealing with a finite density of excitations, based on information on the one-particle dispersion and the two-particle S matrix, consists of mapping the system to an effective integrable field theory. The parameters in this effective theory should be tuned to fit the variational information as good as possible. This approach has the advantage that integrability is exact for the effective model, but the mapping is typically only valid in some small region (e.g. low density and/or low temperature).

One possible field theory is obtained by making the approximation that the particles interact through a contact potential [77, 96], so that we end up with a Lieb–Liniger model [78, 79]. The first-quantized Hamiltonian for a collection of N bosons is given by

$$H = -\frac{1}{2m}\sum_{j=1}^{N}\frac{\partial^2}{\partial x_j^2} + 2c\sum_{j<k=1}^{N}\delta(x_j - x_k) \tag{3.49}$$

with the mass m of the bosons and the interaction strength c as the two tunable parameters. The two-boson S matrix is given by $S(\lambda_1, \lambda_2) = -\mathrm{e}^{i\theta(\lambda_1 - \lambda_2)}$ with

$$\theta(\lambda) = 2\arctan\left(\frac{\lambda}{c}\right), \tag{3.50}$$

so that the scattering length for a δ potential is $a_\delta = -2/c$. The boson dispersion relation is just quadratic, i.e. $\Delta(\lambda) = \lambda^2/(2m)$. By variationally calculating the dispersion relation and the scattering length of the relevant excitations, we can fix the two parameters and map the density of excitations to a Lieb–Liniger model. At low densities, we expect that this mapping is a good approximation.

Another possibility is the non-linear sigma model, which has proven to capture the qualitative behaviour of Haldane-gapped spin chains such as the spin-1 Heisenberg model [80] or two-leg spin-1/2 ladders. In contrast to the Lieb–Liniger model, however, we can not tune any parameters to fit the exactly known [97] two-particle S matrix. The universal behaviour of e.g. the magnon condensation of a gapped spin chain in a magnetic field [90], can nonetheless be captured with this mapping.

3.8 Dynamical Correlations

The third way to characterize elementary particle excitations and their interactions is by the contribution of one- and two-particle excitations to dynamical correlation functions or spectral functions. The central quantity of this section will be the momentum-frequency resolved dynamical correlation function

$$S(q,\omega) = \int \mathrm{d}t\, \mathrm{e}^{i\omega t} \sum_n \mathrm{e}^{-iqn} \langle O_n^\dagger(t) O_0(0)\rangle,$$

with $O_n(t)$ an operator at site n in the Heisenberg picture. The expectation value $\langle \ldots \rangle$ is with respect to the ground state (zero temperature) or a Gibbs distribution (finite temperature).

3.8.1 *Spectral Functions at Zero Temperature*

At zero temperature, the spectral function $S(q,\omega)$ is straightforwardly evaluated by projecting the time evolution on the exact low-energy excitations. Let us therefore define a projector on the one- and two-particle subspaces

$$P_{1\text{p},2\text{p}} = \int \frac{\mathrm{d}\kappa}{2\pi} \sum_{\alpha\in\Gamma_1(\kappa)} |\Phi_\alpha(\kappa)\rangle\langle\Phi_\alpha(\kappa)| + \int \frac{\mathrm{d}K}{2\pi} \int \frac{\mathrm{d}\omega}{2\pi} \sum_{\gamma\in\Gamma_2(K,\omega)} |\Upsilon_\gamma(K,\omega)\rangle\langle\Upsilon_\gamma(K,\omega)|$$

where Γ_1 (Γ_2) is the set of all types of one-particle (two-particle) states at that momentum (momentum-energy). The states are orthonormalized as

$$\langle\Phi_{\gamma'}(\kappa')|\Phi_\gamma(\kappa)\rangle = 2\pi\delta(\kappa'-\kappa)\delta_{\gamma\gamma'}$$
$$\langle\Upsilon_{\gamma'}(K',\omega')|\Upsilon_\gamma(K,\omega)\rangle = 4\pi^2\delta(K'-K)\delta(\omega'-\omega)\delta_{\gamma\gamma'}$$

so that we obtain the Lehmann representation [98] for the spectral function up to two-particle contributions

$$\begin{aligned} S(q,\omega) = &\sum_{\alpha\in\Gamma_1(q)} 2\pi\delta(\Delta_\alpha(q)-\omega)\,|\langle\Phi_\alpha(q)|O_0|\Psi_0\rangle|^2 \\ &+ \sum_{\gamma\in\Gamma_2(q,\omega)} \left|\langle\Upsilon_\gamma(q,\omega)|O_0|\Psi_0\rangle\right|^2 \\ &+\cdots \end{aligned}$$

In gapped systems, the one- and two-particle contributions saturate the full spectral function below the three-particle threshold, while contributions from higher-particle excitations might become important for higher energies. Yet, it appears that typically the one- and two-particle sectors already contain the largest portion of the spectral function, see e.g. Ref. [99]. The one- and two-particle form factors appearing in the above expression are calculated explicitly in Sect. 3.3.4 and Appendix A.3.4.

To get a quantitative estimate of how well the spectral function is approximated, we look at the zeroth and first frequency moment at a certain momentum, which are defined as

$$s_0(\kappa) = \int \frac{\mathrm{d}\omega}{2\pi} S(\kappa,\omega) \quad \text{and} \quad s_1(\kappa) = \int \frac{\mathrm{d}\omega}{2\pi} \omega S(\kappa,\omega).$$

These quantities follow the sum rules [100]

$$\begin{aligned} s_0(\kappa) &= \int \frac{\mathrm{d}\omega}{2\pi} \langle\Psi_0|O_{-\kappa}^\dagger 2\pi\delta(\omega-H)O_0(0)|\Psi_0\rangle \\ &= \langle\Psi_0|O_{-\kappa}^\dagger O_0(0)|\Psi_0\rangle \end{aligned}$$

and

$$\begin{aligned} s_1(\kappa) &= \int \frac{\mathrm{d}\omega}{2\pi} \omega\langle\Psi_0|O_{-\kappa}^\dagger 2\pi\delta(\omega-H)O_0(0)|\Psi_0\rangle \\ &= \langle\Psi_0|O_{-\kappa}^\dagger H O_0(0)|\Psi_0\rangle. \end{aligned}$$

If the ground state is taken to be an MPS, these quantities can be calculated exactly. Note that s_0 is just the static correlation function and that the ratio of the two is equal to the single mode approximation for the dispersion relation [101]

$$\Delta_{\mathrm{SMA}}(\kappa) = \frac{s_1(\kappa)}{s_0(\kappa)} = \frac{\langle\Psi_0|O_{-\kappa}^\dagger H O_0(0)|\Psi_0\rangle}{\langle\Psi_0|O_{-\kappa}^\dagger O_0(0)|\Psi_0\rangle}.$$

By comparing the one- and two-particle contributions for s_0 and s_1 to the exact values, we can get an idea of how well these eigenstates capture the effect of the operators working on the ground state and, consequently, how well the spectral function is approximated by only looking at these contributions.

3.8.2 Divergences in the Density of States

The spectral function in a two-particle continuum can contain divergences of a purely kinematic origin: due to a non-trivial dispersion relation of the particles, there are lines in the two-particle continuum (q, ω) that have a diverging number of states. This effect is captured in the density-of-states function $\rho(q, \omega)$.

The origin of divergences in this density of states can be explained by considering the case of just one type of particle with dispersion relation $\Delta(\kappa)$ over the full Brillouin zone. We can construct a continuum of two-particle states by combining one-particle states with individual momenta κ_1 and κ_2; the total momentum and energy of this two-particle state is $q = \kappa_1 + \kappa_2$ and $\omega = \Delta(\kappa_1) + \Delta(\kappa_2)$. The density of states within this two-particle continuum is not uniform but given by [102]

$$\rho(q, \omega) = \frac{1}{|\Delta'(\kappa_1) - \Delta'(\kappa_2)|},$$

where $\Delta'(\kappa)$ denotes the derivative of the dispersion relation, corresponding to the group velocity of the particle. This formula arises from transforming the uniform density of states as parametrized by the individual momenta $\rho(\kappa_1, \kappa_2)$ to the non-uniform $\rho(q, \omega)$. The derivative of the dispersion relation is the particle's group velocity, so the density of states diverges whenever the group velocity of the two particles is equal – the difference between individual group velocities can be interpreted as the relative velocity of the two-particle state in the center-of-mass frame.

The form of this divergence can be derived as follows. Assume we have a combination of individual momenta κ_1^* and κ_2^* that gives rise to a diverging density of states (i.e. we have $\Delta'(\kappa_1^*) = \Delta'(\kappa_2^*)$) in the point $q = \kappa_1^* + \kappa_2^*$ and $\omega^* = \Delta(\kappa_1^*) + \Delta(\kappa_2^*)$. For fixed q, we look at small deviations of the individual momenta, i.e.

$$\begin{array}{l} \kappa_1 = \kappa_1^* + \epsilon \\ \kappa_2 = \kappa_2^* - \epsilon \end{array} \rightarrow \begin{array}{l} \Delta'(\kappa_1) = \Delta'(\kappa_1^*) + \epsilon\Delta''(\kappa_1^*) \\ \Delta'(\kappa_2) = \Delta'(\kappa_2^*) - \epsilon\Delta''(\kappa_2^*) \end{array}$$

The shift in total energy ω as a function of ϵ is computed as

$$\begin{aligned} \omega &= \Delta(\kappa_1^*) + \epsilon\Delta'(\kappa_1^*) + \frac{\epsilon^2}{2}\Delta''(\kappa_1^*) + \Delta(\kappa_2^*) - \epsilon\Delta'(\kappa_2^*) + \frac{\epsilon^2}{2}\Delta''(\kappa_2^*) \\ &= \omega^* + \frac{\epsilon^2}{2}(\Delta''(\kappa_1^*) + \Delta''(\kappa_2^*)) \end{aligned}$$

so that we obtain an inverse square-root dependence close to the divergence

$$\rho(q, \omega_\epsilon) = \left|\Delta''(\kappa_1^*) + \Delta''(\kappa_2^*)\right|^{-1/2} \times \frac{1}{\sqrt{|\omega - \omega^*|}}$$

Note that these square-root divergences always occur at the edges of the two-particle continuum. If there is an inflection point in the one-particle dispersion relation, there are regions in the two-particle continuum (so-called folding regions) where different combinations of individual momenta lead to the same total momentum and energy. At the boundaries of these folding regions, one again expects a diverging density of states.

The spectral function $S(q, \omega)$ is obtained by multiplying the density of states with the spectral weight of the corresponding two-particle eigenstate. Consequently, the spectral function does not have to diverge on the lines of diverging $\rho(q, \omega)$, because the spectral weight typically vanishes on these edges.

3.8.3 Adding an Imaginary Frequency

In experiments the resolution on the spectral function is never perfect. This is often modelled by introducing a small imaginary frequency. To introduce it consistently, the spectral function is written in terms of the dynamical susceptibility

$$S(q, \omega) = -2\mathrm{Im}\chi(q, \omega)$$

with

$$\chi(q, \omega) = i \int_0^\infty \mathrm{d}t \sum_n \mathrm{e}^{i(\omega + i\eta)t} \mathrm{e}^{-iqn} \langle O_n(t)^\dagger O_0(0) \rangle .$$

By again inserting a projector on the one- and two-particle subspace, this is given by

$$\begin{aligned} \chi(q, \omega) = & \sum_{\alpha \in \Gamma_1(q)} \frac{1}{\omega + i\eta - \Delta_\alpha(q)} \left| \langle \Phi_\alpha(q) | O_0 | \Psi_0 \rangle \right|^2 \\ & + \sum_{\gamma \in \Gamma_2(q,\omega)} \int \frac{\mathrm{d}\omega'}{2\pi} \frac{1}{\omega + i\eta - \omega'} \left| \langle \Upsilon_\gamma(q, \omega) | O_0 | \Psi_0 \rangle \right|^2 \end{aligned}$$

and, taking the imaginary part,

$$\begin{aligned} S(q, \omega) = & \sum_{\alpha \in \Gamma_1(q)} \frac{2\eta}{(\omega - \Delta_\alpha(q))^2 + \eta^2} \left| \langle \Phi_\alpha(q) | O_0 | \Psi_0 \rangle \right|^2 \\ & + \sum_{\gamma \in \Gamma_2(q,\omega)} \int \frac{\mathrm{d}\omega'}{2\pi} \frac{2\eta}{(\omega - \omega')^2 + \eta^2} \left| \langle \Upsilon_\gamma(q, \omega) | O_0 | \Psi_0 \rangle \right|^2 \end{aligned}$$

In the limit $\eta \to 0$ the original result is retrieved because

$$\lim_{\eta \to 0} \frac{2\eta}{\omega^2 + \eta^2} = 2\pi\delta(\omega).$$

3.8.4 Spectral Functions at Finite Temperature

At finite temperatures, the thermally excited density of excitations already present in the thermal state destroys the perfect coherence of one-particle contributions to spectral functions: the δ peaks at zero temperature will get smeared out in finite temperature spectral functions. This thermal broadening of the one-particle response can be treated in a semi-classical approximation [103], but it appears that the broadening depends heavily on the interactions between the particles requiring a full quantum-mechanical treatment [105, 106].

At zero temperature the spectral function $S(q,\omega)$ can be expressed in terms of the spectral weights of the low-energy excitations of the system. At finite temperatures, this is no longer true as we generally need form factors corresponding to states with arbitrarily high energies. In gapped integrable systems – where the higher energy states can be labeled with a particle number n and have an energy of the order $n\Delta$ – the multi-particle form factors are suppressed with a Boltzmann factor $\mathcal{O}(\mathrm{e}^{-n\Delta/T})$, so one can restrict to low-particle form factors at low enough temperatures (compared to the gap) [104–106].

In the above we have shown that, even in non-integrable systems, the particle picture remains valid at low densities (low temperatures); this implies that we can also apply the the low-temperature expansion in $\mathcal{O}(\mathrm{e}^{-\Delta/T})$ to the case of non-integrable models.[24] So we can associate a particle number to higher excitations and we can write down the finite temperature expression for the spectral function in the Lehmann representation. Let us express the spectral function in terms of the dynamical susceptibility

$$S(q,\omega) = \frac{2}{1-\mathrm{e}^{-\beta\omega}} \mathrm{Im}\chi(q,\omega)$$

with

$$\chi(q,\omega) = i\int_0^\infty \mathrm{d}t \mathrm{e}^{i(\omega+i\eta)t} \sum_n \mathrm{e}^{-iqn} \langle O_n(t) O_0(0)\rangle,$$

where we have introduced a small imaginary frequency η, which will be put to zero at the end. By inserting a double resolution of the identity, we obtain the spectral representation

$$\chi(q,\omega) = \frac{1}{Z}\sum_{r,s} G_{r,s}(q,\omega)$$

with

$$G_{r,s}(q,\omega) = \sum_{\{\alpha_r\}\{\beta_s\}} \int \frac{\mathrm{d}\kappa_1 \ldots \mathrm{d}\kappa_r}{2\pi} \int \frac{\mathrm{d}\kappa_1' \ldots \mathrm{d}\kappa_s'}{2\pi} 2\pi\delta(q-K_r+K_s)$$
$$\frac{\mathrm{e}^{-\beta E_r}}{\omega+i\eta-E_r+E_s} \left|\langle\Phi_{\alpha_1,\ldots,\alpha_r}(\kappa_1,\ldots,\kappa_r)|O|\Phi_{\beta_1,\ldots,\beta_s}(\kappa_1',\ldots,\kappa_s')\rangle\right|^2.$$

[24] See also Ref. [107] for a similar expansion for non-integrable systems.

Computing this form-factor expansion is not straightforward, because the form factors contain a number of divergences which only cancel after a careful analysis [39]. It starts with the partition function

$$\begin{aligned}
Z &= \sum_{m=0}^{\infty} Z_m \\
&= \langle \Psi_0 | \Psi_0 \rangle + \sum_{\alpha \in \Gamma} \int \frac{\mathrm{d}\kappa}{2\pi} \mathrm{e}^{-\beta \Delta_\alpha(\kappa)} \langle \Phi_\alpha(\kappa) | \Phi_\alpha(\kappa) \rangle \\
&\quad + \sum_{\alpha\beta} \int \frac{\mathrm{d}\kappa_1}{2\pi} \frac{\mathrm{d}\kappa_2}{2\pi} \mathrm{e}^{-\beta(\Delta_\alpha(\kappa_1)+\Delta_\beta(\kappa_2))} \langle \Upsilon_{\alpha\beta}(\kappa_1, \kappa_2) | \Upsilon_{\alpha\beta}(\kappa_1, \kappa_2) \rangle \\
&\quad + \cdots ,
\end{aligned}$$

which shows that Z_m formally carries a $[2\pi\delta(0)]^m$ divergence. At this point, we make use of the fact that the temperature is smaller than the gap in the system, such that the Boltzmann factor $\mathrm{e}^{-\beta\Delta}$ is a small parameter. Expanding the $1/Z$ factor in this small parameter and grouping all terms in its powers, we have

$$\chi(q, \omega) = \sum_{s=0}^{\infty} C_s(q, \omega)$$

with the first terms given by

$$\begin{aligned}
C_0(q, \omega) &= \sum_{m=0}^{\infty} G_{0,m}(q, \omega) \\
C_1(q, \omega) &= G_{1,0} + \sum_{m=1}^{\infty} \left(G_{1,m} - Z_1 G_{0,m-1} \right) .
\end{aligned}$$

The first term is the zero-temperature result. The diverging (disconnected) contributions in $E_{1,m}$ are exactly cancelled by the ones in the partition function Z_1, such that these terms are all finite quantities in the thermodynamic limit. Still, all terms in this expression contains a divergence when the zero-temperature one-particle energy is approached (e.g. the zero-temperature result is still present). On physical grounds, we expect the zero-temperature δ peak to broaden into a strongly peaked, yet non-singular curve around the one-particle energy, which can be modelled by giving the one-particle mode a self-energy $\Sigma(q, \omega)$ as a result of the interactions with the thermally excited particle density. Accordingly, the susceptibility $\chi(q, \omega)$ can be rewritten according to a Dyson equation

$$\chi(q, \omega) = \frac{D(q, \omega)}{1 - \Sigma(q, \omega) D(q, \omega)}$$

where $D(q, \omega)$ is the "free propagator" or the dynamical correlation function at zero temperature. This Dyson equation can again be expanded as

$$\chi(q, \omega) = D(q, \omega) + D^2(q, \omega)\Sigma(q, \omega) + \cdots$$

and matched to the above expansion.

Let us focus on the thermal broadening of the δ peak corresponding to an isolated particle α with dispersion $\omega = \Delta_\alpha(q)$. As explained, we can associate a self-energy $\Sigma_\alpha(q, \omega)$ with this mode. By only taking into account the terms that have a divergence close to the dispersion $\omega = \Delta_\alpha(q)$, one has in lowest order in $e^{-\beta\Delta}$

$$\Sigma_\alpha(q, \omega) = \frac{1}{G_{0,1}(q, \omega)^2}\left(G_{1,2}(q, \omega) - Z_1 G_{0,1}(q, \omega)\right). \tag{3.51}$$

The $G_{1,2}(q, \omega)$ contains form factors linking the one- and two-particle sector, and represent the physical process of creating a particle α by the operator O on a state where another particle is already present. The thermal lineshape corresponding to this particle mode α becomes

$$S_\alpha(q, \omega) = \frac{-2}{1 - e^{-\beta\omega}} \mathrm{Im}\left[\frac{G_{0,1}(q, \omega)}{1 - G_{0,1}(q, \omega)\Sigma_\alpha(q, \omega)}\right].$$

These expressions still contain a finite η, which is now extrapolated to zero. Because of the finite self-energy $\Sigma_\alpha(q, \omega)$ this will not give rise to a δ function, but will result in a thermal line (not necessarily symmetric).

The computation of the self-energy $\Sigma_\alpha(q, \omega)$ from Eq. (3.51) requires a careful isolation of the divergent (disconnected) contributions in the partition function and form factors, but should in principle be feasible. The explicit expressions, however, are left for future work.

3.9 Effective Field Theory

In the previous sections, the interactions between effective particles in spin chains has been approached from different sides: scattering processes, applying a chemical potential and temperature and dynamical correlation functions. These characterizations should in principle be derivable from one effective many-particle theory that describes both the dynamics and the interactions of the particles in second quantization [40, 108, 109]. Therefore, we introduce momentum space creation and annihilation operators that act on the ground state as

$$c_\alpha^\dagger(\kappa)|\Psi[A]\rangle = |\Phi_\alpha(\kappa)\rangle$$
$$c_\alpha(\kappa)|\Psi[A]\rangle = 0$$

and write down an effective interacting theory

$$\begin{aligned} H = &\sum_{\alpha} \int \frac{d\kappa}{2\pi} \Delta_{\alpha}(\kappa) c_{\alpha}^{\dagger}(\kappa) c_{\alpha}(\kappa) \\ &+ \sum_{\alpha'\beta'\alpha\beta} \int \frac{d\kappa}{2\pi} \frac{d\kappa_1}{2\pi} \frac{d\kappa_2}{2\pi} V_{\alpha'\beta',\alpha\beta}(\kappa, \kappa_1, \kappa_2) \\ &\qquad \times c_{\alpha'}^{\dagger}(\kappa_1 + \kappa_2 - \kappa) c_{\beta'}^{\dagger}(\kappa) c_{\beta}(\kappa_2) c_{\alpha}(\kappa_1). \end{aligned} \tag{3.52}$$

Since we only have explicit access to the operator acting on the ground state and not the operator itself, it is a priori not clear how to determine the $c_{\alpha}^{\dagger}(\kappa)$ and $c_{\alpha}(\kappa)$ in a unique way. Moreover, there seems to be no trivial way for imposing the correct commutation relations. Thirdly, because these operators will be momentum-dependent, the transition to a local, real-space representation of the Fock operators might not be well-defined. The construction of Wannier states out of the momentum eigenstates might provide a good starting point [109], although it is still not clear how to find the unique real-space operators that are essential for computing the interaction term in Eq. (3.52). These obstacles have made the systematic construction of an effective field theory non-feasible.

A different approach can be taken by starting from a free theory of particles with generalized statistics that match the two-particle S matrix. Within the approximate Bethe ansatz, we assume that the full spectrum of a Hamiltonian consists of scattering states of particle excitations. We have seen [Sect. 2.5.4] that the excitations are formally created by the Faddeev–Zamolodchikov (FZ) operators, so that the low-energy part of a Hamiltonian can be expressed as

$$H_0 = \sum_{\alpha} \int \frac{d\kappa}{2\pi} \Delta_{\alpha}(\kappa) Z_{\alpha}^{\dagger}(\kappa) Z_{\alpha}(\kappa)$$

This naive way of writing this down enables to consider a perturbation to the Hamiltonian and express it in terms of the FZ operators. Indeed, when applying a non-commuting perturbation, we could have a new Hamiltonian of the form
$H = H_0 +$

$$\sum_{\alpha\beta} \int \frac{d\kappa}{2\pi} \left(M_p^{\alpha\beta} Z_{\alpha}^{\dagger}(\kappa) Z_{\beta}(\kappa) + M_n^{\alpha\beta} Z_{\alpha}(-\kappa) Z_{\beta}(\kappa) + h.c. \right) \tag{3.53}$$

where

$$\begin{aligned} M_p^{\alpha\beta}(\kappa) &\sim \langle \Phi_{\alpha}(\kappa) | M | \Phi_{\beta}(\kappa) \rangle \\ M_n^{\alpha\beta}(\kappa) &\sim \langle \Psi(A) | M | \Upsilon_{\beta\alpha}(\kappa, -\kappa) \rangle \end{aligned}$$

are the particle preserving, resp. particle non-preserving parts of the perturbation up to quadratic order in the FZ operators. Because there seems to be no straightforward way of diagonalizing this Hamiltonian – a canonical Bogoliubov for FZ operators is not possible in general [110] – the expression only has a formal meaning. For small perturbations, however, we can assume that only small momentum states will be occupied and that the S matrix is approximately -1. In that case, the FZ operators reduce to fermion creation and annihilation operators and we can diagonalize the Hamiltonian [Eq. (3.53)] with a Bogoliubov rotation (see Sect. 4.5). This will be carried out in Sect. 4.5 in more detail.

3.10 Non-equilibrium Processes

In this last section we discuss some of the extensions of our framework to out-of-equilibrium processes.

Quantum Quenches and Thermalization

When studying the time evolution of integrable systems, the occupation numbers $n_\alpha(\kappa) = Z_\alpha^\dagger(\kappa)Z_\alpha(\kappa)$ corresponding to the FZ operators are integrals of motion [111]. For non-integrable systems this is no longer the case, although the observation of so-called prethermalization plateaus might point to the fact that they are almost preserved. Indeed, the mode occupation numbers $n_\alpha(\kappa)$ provide a way to distinguish a thermal Gibbs ensemble from a generalized Gibbs ensemble [112, 113]. Consequently, by finding an explicit (real-space) representation of the FZ operators we could follow the occupation numbers $n_\alpha(\kappa)$ through time, also when starting from an interacting theory.

Transport Properties

One particular context where the effects of integrability-breaking perturbations are still under debate is transport properties at finite temperature [114]. The question is whether in the frequency-dependent conductivity

$$\sigma(\omega) = 2\pi D\delta(\omega) + \sigma_{\text{reg}}(\omega),$$

the divergence at $\omega \to 0$ or "Drude weight" D survives as the temperature is turned on, corresponding to dissipationless transport. Indeed, a Drude weight can be expressed in terms of an asymptotic, homogeneous current as

$$D = \lim_{t\to\infty} \lim_{N\to\infty} \frac{\text{Re}\langle J(t)J(0)\rangle}{2NT},$$

where J is some current operator $J = \sum_n j_n$. Due to the local conservation laws D remains non-zero for integrable systems, but the conductivity is expected to be regularized ($D = 0$) by integrability-breaking perturbations [115]. On the other hand, some perturbations seem to maintain a non-zero Drude weight [116].

In the case of a simple setup where two thermalized leads at temperatures T_L and T_R are coupled at time $t = 0$, a non-zero Drude weight can be understood from the simple form the homogeneous asymptotic current takes on [116]

$$\lim_{t\to\infty} \langle j_n(t)\rangle = f(T_L) - f(T_R), \tag{3.54}$$

where $f(T)$ is a distribution for a thermal excitation of particles, to be computed with the thermodynamic Bethe ansatz.

Regarding the regular part of the of the conductivity, gapped spin ladder systems have exhibited very large thermal conductivities due to spin excitations (magnons), but it remains a matter for debate whether the magnons exhibit standard diffusive dynamics. In particular, it seems that the connection between the magnons and an analysis in terms of mean-free paths asks for better studies [117].

References

1. P.W. Anderson, *Concepts in Solids: Lectures on the Theory of Solids* (World Scientific, Singapore, 1963)
2. R.P. Feynman, Atomic theory of the λ transition in helium. Phys. Rev. **91**, 1291 (1953). doi:10.1103/PhysRev.91.1291
3. R.P. Feynman, Atomic theory of liquid helium near absolute zero. Phys. Rev. **91**, 1301 (1953). doi:10.1103/PhysRev.91.1301
4. R.P. Feynman, Atomic theory of the two-fluid model of liquid helium. Phys. Rev. **94**, 262 (1954). doi:10.1103/PhysRev.94.262
5. R.P. Feynman, M. Cohen, Energy spectrum of the excitations in liquid helium. Phys. Rev. **102**, 1189 (1956). doi:10.1103/PhysRev.102.1189
6. M. Cohen, R.P. Feynman, Theory of inelastic scattering of cold neutrons from liquid helium. Phys. Rev. **107**, 13 (1957). doi:10.1103/PhysRev.107.13
7. D. Pines, P. Nozières, *Theory of Quantum Liquids* (Perseus Books Publishing, Cambridge, 1966)
8. L.D. Landau, Theory of the superfluidity of helium II, J. Phys. U.S.S.R **5**, 356 (1941). doi:10.1103/PhysRev.60.356
9. A. Bijl, J. de Boer, A. Michels, Properties of liquid helium II. Physica **8**, 655 (1941). doi:10.1016/S0031-8914(41)90422-6
10. B.I. Lundqvist, Single-particle spectrum of the degenerate electron gas. Physik der Kondensierten Materie **6**, 193 (1967). doi:10.1007/BF02422716
11. A.W. Overhauser, Simplified theory of electron correlations in metals. Phys. Rev. B **3**, 1888 (1971). doi:10.1103/PhysRevB.3.1888
12. S.M. Girvin, A. MacDonald, P. Platzman, Collective-excitation gap in the fractional quantum Hall effect. Phys. Rev. Lett. **54**, 581 (1985). doi:10.1103/PhysRevLett.54.581
13. S.M. Girvin, A. MacDonald, P. Platzman, Magneto-roton theory of collective excitations in the fractional quantum Hall effect. Phys. Rev. B **33**, 2481 (1986). doi:10.1103/PhysRevB.33.2481
14. D.P. Arovas, A. Auerbach, F.D.M. Haldane, Extended Heisenberg models of antiferromagnetism: analogies to the fractional quantum Hall effect. Phys. Rev. Lett. **60**, 531 (1988). doi:10.1103/PhysRevLett.60.531
15. I. Affleck, T. Kennedy, E.H. Lieb, H. Tasaki, Rigorous results on valence-bond ground states in antiferromagnets. Phys. Rev. Lett. **59**, 799 (1987). doi:10.1103/PhysRevLett.59.799

16. I. Affleck, T. Kennedy, E.H. Lieb, H. Tasaki, Valence bond ground states in isotropic quantum antiferromagnets. Commun. Math. Phys. **115**, 477 (1988). doi:10.1007/BF01218021
17. M. Takahashi, Excitation spectra of S=1 antiferromagnetic chains. Phys. Rev. B **50**, 3045 (1994). doi:10.1103/PhysRevB.50.3045
18. E. Sorensen, I. Affleck, S(k) for Haldane-gap antiferromagnets: large-scale numerical results versus field theory and experiment. Phys. Rev. B **49**, 13235 (1994). doi:10.1103/PhysRevB.49.13235
19. E. Sorensen, I. Affleck, Equal-time correlations in Haldane-gap antiferromagnets. Phys. Rev. B **49**, 15771 (1994). doi:10.1103/PhysRevB.49.15771
20. S. Östlund, S. Rommer, Thermodynamic limit of the density matrix renormalization for the spin-1 Heisenberg chain. Phys. Rev. Lett. **75**, 13 (1995). doi:10.1103/PhysRevLett.75.3537
21. S. Rommer, S. Östlund, Class of ansatz wave functions for one-dimensional spin systems and their relation to the density matrix renormalization group. Phys. Rev. B **55**, 2164 (1997). doi:10.1103/PhysRevB.55.2164
22. B. Pirvu, J. Haegeman, F. Verstraete, Matrix product state based algorithm for determining dispersion relations of quantum spin chains with periodic boundary conditions. Phys. Rev. B **85**, 13 (2012). doi:10.1103/PhysRevB.85.035130
23. J. Haegeman, B. Pirvu, D.J. Weir, J.I. Cirac, T.J. Osborne, H. Verschelde, F. Verstraete, Variational matrix product ansatz for dispersion relations. Phys. Rev. B **85**, 100408 (2012). doi:10.1103/PhysRevB.85.100408
24. E. Bartel, A. Schadschneider, J. Zittartz, Excitations of anisotropic spin-1 chains with matrix product ground state. Eur. Phys. J. B - Condens. Matter **31**, 209 (2003). doi:10.1140/epjb/e2003-00025-7
25. S.G. Chung, L. Wang, Entanglement perturbation theory for the elementary excitation in one dimension. Phys. Lett. A **373**, 2277 (2010), http://www.sciencedirect.com/science/article/pii/S0375960109005106
26. M.E. Peskin, D.V. Schroeder, *An Introduction to Quantum Field Theory* (Westview Press, Boulder, 1995)
27. R.F. Streater, A.S. Wightman, *PCT, Spin and Statistics, and All That* (W.A. Benjamin Inc, New York, 1964)
28. R. Haag, *Local Quantum Physics: Fields, Particles, Algebras* (Springer, Berlin, 1996)
29. R. Haag, Quantum field theories with composite particles and asymptotic conditions. Phys. Rev. **112**, 669 (1958). doi:10.1103/PhysRev.112.669
30. D. Ruelle, On asymptotic condition in quantum field theory. Helv. Phys. Acta **35**, 147 (1962)
31. E.H. Lieb, D.W. Robinson, The finite group velocity of quantum spin systems. Commun. Math. Phys. **28**, 251 (1972). doi:10.1007/BF01645779
32. V. Zauner, D. Draxler, L. Vanderstraeten, M. Degroote, J. Haegeman, M.M. Rams, V. Stojevic, N. Schuch, F. Verstraete, Transfer matrices and excitations with matrix product states. New J. Phys. **17**, 053002 (2015). doi:10.1088/1367-2630/17/5/053002
33. I.P. McCulloch, Infinite size density matrix renormalization group, revisited (2008), arXiv:0804.2509
34. G. Vidal, Efficient simulation of one-dimensional quantum many-body systems. Phys. Rev. Lett. **93**, 4 (2004). doi:10.1103/PhysRevLett.93.040502
35. G. Vidal, Classical simulation of infinite-size quantum lattice systems in one spatial dimension. Phys. Rev. Lett. **98**, 5 (2007). doi:10.1103/PhysRevLett.98.070201
36. J. Haegeman, J.I. Cirac, T.J. Osborne, I. Pizorn, H. Verschelde, F. Verstraete, Time-dependent variational principle for quantum lattices. Phys. Rev. Lett. **107**, 070601 (2011). doi:10.1103/PhysRevLett.107.070601
37. J. Haegeman, M. Mariën, T.J. Osborne, F. Verstraete, Geometry of matrix product states: metric, parallel transport, and curvature. J. Math. Phys. **55**, 021902 (2014). doi:10.1063/1.4862851
38. J. Haegeman, C. Lubich, I. Oseledets, B. Vandereycken, F. Verstraete, Unifying time evolution and optimization with matrix product states (2014), arXiv:1408.5056
39. F.H.L. Essler, R.M. Konik, Finite-temperature dynamical correlations in massive integrable quantum field theories. J. Stat. Mech.: Theory Exp. **2009**, P09018 (2009). doi:10.1088/1742-5468/2009/09/P09018

40. J. Haegeman, T.J. Osborne, F. Verstraete, Post-matrix product state methods: to tangent space and beyond. Phys. Rev. B **88**, 075133 (2013). doi:10.1103/PhysRevB.88.075133
41. J. Haegeman, S. Michalakis, B. Nachtergaele, T.J. Osborne, N. Schuch, F. Verstraete, Elementary excitations in gapped quantum spin systems. Phys. Rev. Lett. **111**, 080401 (2013). doi:10.1103/PhysRevLett.111.080401
42. S. Bachmann, W. Dybalski, P. Naaijkens, Lieb-Robinson bounds, Arveson spectrum and Haag-Ruelle scattering theory for gapped quantum spin systems (2014), arXiv:1412.2970
43. W. Zimmermann, On the bound state problem in quantum field theory. Il Nuovo Cimento **10**, 597 (1958). doi:10.1007/BF02859796
44. E.H. Lieb, The Bose fluid, *Lecture Notes in Theoretical Physics*, vol. VIIC (University of Colorado Press, Boulder, 1965), pp. 175–224
45. T. Davison, E. Feenberg, Variance of H in the Bijl-Feynman description of an elementary excitation. Phys. Rev. **171**, 221 (1968). doi:10.1103/PhysRev.171.221
46. R. Rajaraman, *Solitons and Instantons: An Introduction to Solitons and Instantons in Quantum Field Theory* (North-Holland, Amsterdam, 1996)
47. L.D. Faddeev, V. Korepin, Quantum theory of solitons. Phys. Rep. **42**, 1 (1978). doi:10.1016/0370-1573(78)90058-3
48. V.E. Korepin, N.M. Bogoliubov, A.G. Izergin, *Quantum Inverse Scattering Method and Correlation Functions* (Cambridge University Press, Cambridge, 1997)
49. L.D. Faddeev, L.A. Takhtajan, What is the spin of a spin wave? Phys. Lett. A **85**, 375 (1981). doi:10.1016/0375-9601(81)90335-2
50. Y. Tang, A.W. Sandvik, Method to characterize spinons as emergent elementary particles. Phys. Rev. Lett. **107**, 157201 (2011). doi:10.1103/PhysRevLett.107.157201
51. Y. Tang, A.W. Sandvik, Quantum Monte Carlo studies of spinons in one-dimensional spin systems. Phys. Rev. B **92**, 184425 (2015). doi:10.1103/PhysRevB.92.184425
52. K. Okunishi, N. Maeshima, Spinon excitation and Möbius boundary condition in S=1/2 antiferromagnetic Heisenberg spin ladder with zigzag structure. Phys. Rev. B **64**, 212406 (2001). doi:10.1103/PhysRevB.64.212406
53. S. Mandelstam, Soliton operators for the quantized sine-Gordon equation. Phys. Rev. D **11**, 3026 (1975). doi:10.1103/PhysRevD.11.3026
54. R.P. Feynman, *Statistical Mechanics: A Set of Lectures* (W.A. Benjamin, Inc., Reading, 1972)
55. W. Kohn, Variational scattering theory in momentum space I. Central field problems. Phys. Rev. **84**, 495 (1951). doi:10.1103/PhysRev.84.495
56. J. Frenkel, *Wave Mechanics, Advanced General Theory* (Oxford Clarendon Press, Oxford, 1934)
57. P.A.M. Dirac, Note on exchange phenomena in the Thomas atom. Math. Proc. Camb. Philos. Soc. **26**, 376 (1930). doi:10.1017/S0305004100016108
58. J.C. Slater, The theory of complex spectra. Phys. Rev. **34**, 1293 (1929). doi:10.1103/PhysRev.34.1293
59. S. Wouters, N. Nakatani, D. Van Neck, G.K.-L. Chan, Thouless theorem for matrix product states and subsequent post density matrix renormalization group methods. Phys. Rev. B **88**, 075122 (2013). doi:10.1103/PhysRevB.88.075122
60. P. Ring, P. Schuck, *The Nuclear Many-Body Problem* (Springer, Berlin, 1980)
61. T. Helgaker, P. Jorgensen, J. Olsen, *Molecular Electronic-Structure Theory* (Wiley, New York, 2014)
62. W. Heisenberg, Mehrkörperproblem und Resonanz in der Quantenmechanik. Zeitschrift für Physik **38**, 411 (1926). doi:10.1007/BF01397160
63. P.A.M. Dirac, On the theory of quantum mechanics. Proc. R. Soc. Lond. Ser. A **112**, 661 (1926). doi:10.1098/rspa.1926.0133
64. D. Bohm, D. Pines, A collective description of electron interactions: III. Coulomb interactions in a degenerate electron gas. Phys. Rev. **92**, 609 (1953). doi:10.1103/PhysRev.92.609
65. I.E. Tamm, Relativistic interaction of elementary particles. J. Phys. USSR **9**, 449 (1945)
66. S.M. Dancoff, Non-adiabatic meson theory of nuclear forces. Phys. Rev. **78**, 382 (1950). doi:10.1103/PhysRev.78.382

67. J.R. Taylor, *Scattering Theory: The Quantum Theory on Nonrelativistic Collisions* (Wiley, New York, 1972)
68. M. Reed, B. Simon, *Scattering Theory* (Academic Press, New York, 1972)
69. B. Lippmann, J. Schwinger, Variational principles for scattering processes. I. Phys. Rev. **79**, 469 (1950). doi:10.1103/PhysRev.79.469
70. S. Sachdev, *Quantum Phase Transitions* (Cambridge University Press, Cambridge, 2011)
71. S. Sachdev, K. Damle, Low temperature spin diffusion in the one-dimensional quantum O(3) nonlinear σ Model. Phys. Rev. Lett. **78**, 943 (1997). doi:10.1103/PhysRevLett.78.943
72. K. Damle, S. Sachdev, Nonzero-temperature transport near quantum critical points. Phys. Rev. B **56**, 8714 (1997). doi:10.1103/PhysRevB.56.8714
73. B. Sutherland, *Beautiful Models: 70 Years of Exactly Solved Quantum Many-Body Problems* (World Scientific, Singapore, 2004)
74. H. Bethe, Zur Theorie der Metalle. I. Eigenwerte und Eigenfunktionen der linearen Atomkette. Zeitschrift für Physik **71**, 205 (1931). doi:10.1007/BF01341708
75. W. Krauth, Bethe ansatz for the one-dimensional boson Hubbard model. Phys. Rev. B **44**, 9772 (1991). doi:10.1103/PhysRevB.44.9772
76. H. Kiwata, Y. Akutsu, Bethe-Ansatz approximation for the S=1 antiferromagnetic spin chain. J. Phys. Soc. Jpn. **63**, 3598 (1994). doi:10.1143/JPSJ.63.3598
77. K. Okunishi, Magnetization process of bilinear-biquadratic spin chains at finite temperature. Phys. Rev. B **60**, 4043 (1999). doi:10.1103/PhysRevB.60.4043
78. E.H. Lieb, W. Liniger, Exact analysis of an interacting Bose gas. I. The general solution and the ground state. Phys. Rev. **130**, 1605 (1963). doi:10.1103/PhysRev.130.1605
79. E.H. Lieb, Exact analysis of an interacting Bose gas. II. The excitation spectrum. Phys. Rev. **130**, 1616 (1963). doi:10.1103/PhysRev.130.1616
80. F.D.M. Haldane, Continuum dynamics of the 1-D Heisenberg antiferromagnet: identification with the O(3) nonlinear sigma model. Phys. Lett. A **93**, 464 (1983). doi:10.1016/0375-9601(83)90631-X
81. F.D.M. Haldane, Demonstration of the Luttinger liquid character of Bethe-ansatz-soluble models of 1-D quantum fluids. Phys. Lett. A **81**, 153 (1981), http://www.sciencedirect.com/science/article/pii/0375960181900499
82. T. Giamarchi, *Quantum Physics in One Dimension* (Oxford University Press, Oxford, 2004)
83. F.D.M. Haldane, Effective harmonic-fluid approach to low-energy properties of one-dimensional quantum fluids. Phys. Rev. Lett. **47**, 1840 (1981). doi:10.1103/PhysRevLett.47.1840
84. M.A. Cazalilla, R. Citro, T. Giamarchi, E. Orignac, M. Rigol, One dimensional bosons: from condensed matter systems to ultracold gases. Rev. Mod. Phys. **83**, 1405 (2011). doi:10.1103/RevModPhys.83.1405
85. D. Pérez-García, F. Verstraete, M.M. Wolf, J.I. Cirac, Matrix product state representations. Quantum Inf. Comput. **7**, 401 (2007), arXiv:quant-ph/0608197
86. E. Beth, G.E. Uhlenbeck, The quantum theory of the non-ideal gas. II. Behaviour at low temperatures. Physica **4**, 915 (1937). doi:10.1016/S0031-8914(37)80189-5
87. K. Huang, *Statistical Mechanics* (Wiley, New York, 1963)
88. R. Dashen, S.-K. Ma, H. Bernstein, S-matrix formulation of statistical mechanics. Phys. Rev. **187**, 345 (1969). doi:10.1103/PhysRev.187.345
89. F. Lesage, H. Saleur, Correlations in one-dimensional quantum impurity problems with an external field or a temperature. Nucl. Phys. B **490**, 543 (1997). doi:10.1016/S0550-3213(97)00024-2
90. R.M. Konik, P. Fendley, Haldane-gapped spin chains as Luttinger liquids: correlation functions at finite field. Phys. Rev. B **66**, 144416 (2002). doi:10.1103/PhysRevB.66.144416
91. J. Lou, S. Qin, T.-K. Ng, Z. Su, I. Affleck, Finite-size spectrum, magnon interactions, and magnetization of S=1 Heisenberg spin chains. Phys. Rev. B **62**, 3786 (2000). doi:10.1103/PhysRevB.62.3786
92. I. Affleck, Luttinger liquid parameter for the spin-1 Heisenberg chain in a magnetic field. Phys. Rev. B **72**, 132414 (2005). doi:10.1103/PhysRevB.72.132414

93. I. Affleck, W. Hofstetter, D.R. Nelson, U. Schollwöck, Non-Hermitian Luttinger liquids and flux line pinning in planar superconductors. J. Stat. Mech.: Theory Exp. **2004**, P10003 (2004). doi:10.1088/1742-5468/2004/10/P10003
94. C.N. Yang, C.P. Yang, Thermodynamics of a one-dimensional system of bosons with repulsive delta-function interaction. J. Math. Phys. **10**, 1115 (1969). doi:10.1063/1.1664947
95. M. Takahashi, *Thermodynamics of One-Dimensional Solvable Models* (Cambridge University Press, Cambridge, 2005)
96. K. Okunishi, Y. Hieida, Y. Akutsu, δ-function Bose-gas picture of S=1 antiferromagnetic quantum spin chains near critical fields. Phys. Rev. B **59**, 6806 (1999). doi:10.1103/PhysRevB.59.6806
97. A.B. Zamolodchikov, A.B. Zamolodchikov, Factorized S-matrices in two dimensions as the exact solutions of certain relativistic quantum field theory models. Ann. Phys. **120**, 253 (1979). doi:10.1016/0003-4916(79)90391-9
98. H. Lehmann, Über Eigenschaften von Ausbreitungsfunktionen und Renormierungskonstanten quantisierter Felder. Il Nuovo Cimento **11**, 342 (1954). doi:10.1007/BF02783624
99. J.-S. Caux, J. Mossel, I.P. Castillo, The two-spinon transverse structure factor of the gapped Heisenberg antiferromagnetic chain. J. Stat. Mech.: Theory Exp. **2008**, P08006 (2008). doi:10.1088/1742-5468/2008/08/P08006
100. P. Hohenberg, W. Brinkman, Sum rules for the frequency spectrum of linear magnetic chains. Phys. Rev. B **10**, 128 (1974). doi:10.1103/PhysRevB.10.128
101. D.P. Arovas, S.M. Girvin, Exact questions to some interesting answers in many body physics, in *Recent Progress in Many-Body Theories*, ed. by T.L. Ainsworth, C.E. Campbell, B.E. Clements, E. Krotscheck (Springer, Boston, 1992). doi:10.1007/978-1-4615-3466-2
102. T. Barnes, Boundaries, cusps, and discontinuities in the multimagnon continua of one-dimensional quantum spin systems. Phys. Rev. B **67**, 024412 (2003). doi:10.1103/PhysRevB.67.024412
103. K. Damle, S. Sachdev, Spin dynamics and transport in gapped one-dimensional Heisenberg antiferromagnets at nonzero temperatures. Phys. Rev. B **57**, 8307 (1998). doi:10.1103/PhysRevB.57.8307
104. R.M. Konik, Haldane-gapped spin chains: exact low-temperature expansions of correlation functions. Phys. Rev. B **68**, 104435 (2003). doi:10.1103/PhysRevB.68.104435
105. F.H.L. Essler, R.M. Konik, Finite-temperature lineshapes in gapped quantum spin chains. Phys. Rev. B **78**, 100403 (2008). doi:10.1103/PhysRevB.78.100403
106. D.A. Tennant, B. Lake, A.J.A. James, F.H.L. Essler, S. Notbohm, H.-J. Mikeska, J. Fielden, P. Kögerler, P.C. Canfield, M.T.F. Telling, Anomalous dynamical line shapes in a quantum magnet at finite temperature. Phys. Rev. B **85**, 014402 (2012). doi:10.1103/PhysRevB.85.014402
107. B. Fauseweh, J. Stolze, G.S. Uhrig, Finite-temperature line shapes of hard-core bosons in quantum magnets: a diagrammatic approach tested in one dimension. Phys. Rev. B **90**, 024428 (2014). doi:10.1103/PhysRevB.90.024428
108. L. Vanderstraeten, J. Haegeman, T.J. Osborne, F. Verstraete, S matrix from matrix product states. Phys. Rev. Lett. **112**, 257202 (2014). doi:10.1103/PhysRevLett.112.257202
109. F. Keim, G.S. Uhrig, Effective one-dimensional models from matrix product states. Eur. Phys. J. B **88**, 154 (2015). doi:10.1140/epjb/e2015-60188-0
110. S. Sotiriadis, Zamolodchikov-Faddeev algebra and quantum quenches in integrable field theories. J. Stat. Mech.: Theory Exp. **2012**, P02017 (2012), http://iopscience.iop.org/1742-5468/2012/02/P02017
111. F.H.L. Essler, G. Mussardo, M. Panfil, Generalized Gibbs ensembles for quantum field theories. Phys. Rev. A **91**, 051602 (2015). doi:10.1103/PhysRevA.91.051602
112. F.H.L. Essler, S. Kehrein, S.R. Manmana, N.J. Robinson, Quench dynamics in a model with tuneable integrability breaking. Phys. Rev. B **89**, 165104 (2014). doi:10.1103/PhysRevB.89.165104
113. B. Bertini, F.H.L. Essler, S. Groha, N.J. Robinson, Prethermalization and thermalization in models with weak integrability breaking. Phys. Rev. Lett. **115**, 180601 (2015). doi:10.1103/PhysRevLett.115.180601

114. R. Vasseur, J.E. Moore, Nonequilibrium quantum dynamics and transport: from integrability to many-body localization. J. Stat. Mech.: Theory Exp. **2016**, 064010 (2016). doi:10.1088/1742-5468/2016/06/064010
115. Y. Huang, C. Karrasch, J.E. Moore, Scaling of electrical and thermal conductivities in an almost integrable chain. Phys. Rev. B **88**, 115126 (2013). doi:10.1103/PhysRevB.88.115126
116. C. Karrasch, R. Ilan, J.E. Moore, Nonequilibrium thermal transport and its relation to linear response. Phys. Rev. B **88**, 195129 (2013). doi:10.1103/PhysRevB.88.195129
117. C. Karrasch, D.M. Kennes, F. Heidrich-Meisner, Spin and thermal conductivity of quantum spin chains and ladders. Phys. Rev. B **91**, 115130 (2015). doi:10.1103/PhysRevB.91.115130

Chapter 4
Effective Particles in Quantum Spin Chains: Applications

Throughout this dissertation, we have made the claim that the low-energy degrees of freedom in quantum spin chains are effective particles that move against a strongly-correlated background. These quasiparticles cannot be understood as dressed versions of the microscopic degrees of freedom – as is the case in e.g. Fermi-liquid theory – but they are collective modes that often have exotic emergent properties. The framework of the previous chapter allows to target these effective degrees of freedom: starting from the microscopic Hamiltonian, we arrive at an effective particle description of the low-energy physics.

In Sect. 2.3 we have already explained that the qualitative features of elementary excitations in one-dimensional spin chains is more or less understood. Yet, despite this general understanding a lot of questions still remain; in fact, previous breakthroughs have shown that one-dimensional spin systems show an extremely rich variety of physical phenomena in the low-energy sector, a variety that can be traced back to the non-trivial properties of the elementary excitation spectra. Precisely because the elementary excitations have a strong collective nature, their emergent properties are no longer connected to the microscopic degrees of freedom in the system. Instead, they should be pictured as effective particles living on a strongly-correlated ground state, where the correlations in this background states gives rise to their exotic properties. Examples of this rich variety include:

- spectral functions that are dominated by either (i) a strong peak at finite frequencies, corresponding to a gapped one-particle excitation, or (ii) a broad continuum, corresponding to multi-particle excitations; the latter can be explained by excitations that have fractional quantum numbers and can only be created in pairs, hence the absence of a strong one-particle signal in the spectral function
- the linear growth of entanglement entropy of a system after a quench in the Hamiltonian, caused by strongly-entangled quasiparticle excitations propagating through the system [1, 2]
- the absence of thermalization in integrable systems after a quantum quench, a consequence of the factorized scattering of the particle excitations [3, 4]

L. Vanderstraeten, *Tensor Network States and Effective Particles for Low-Dimensional Quantum Spin Systems*, Springer Theses,
DOI 10.1007/978-3-319-64191-1_4

- bose condensation of magnons in gapped spin-rotation invariant spin chains under a magnetic field [5–7]
- the emergence of signal wave fronts after local and global quenches, where the speed by which a wave front travels through the system is determined by the quasiparticle velocity [8, 9].

What makes these topics especially fascinating is that their physical realizations have come within reach of experiment, and the point has been reached where quantitative comparisons between theory and experiment are possible. More specifically, non-trivial properties of the excitations can be probed with a high resolution in a lot of experiments, and an accurate understanding asks for a detailed account of the quasiparticles and their interactions. The most direct probe for low-energy spectra is inelastic neutron scattering, where multi-particle effects are becoming important [10, 11]. The thermal broadening of one-particle signals can only be accounted for by many-particle processes [12]. Also, the critical properties of magnetized spin chains and ladders can only be understood by identifying it as a condensed gas of interacting magnons on a strongly-correlated background state [5–7, 13]. Lastly, the properties of quasiparticles can be probed in cold-atom experiments [8, 14, 15] and their interactions are important [16, 17].

On the theoretical side, a quantitative numerical simulation of the low-energy dynamics remains, despite a profound theoretical understanding, a huge challenge. Exact diagonalization is the most straightforward and unbiased approach [18–22], but, due to its exponential scaling, limited to small system sizes. When frustration is present in the system, quantum Monte Carlo methods are plagued by the sign problem. In recent years, the scope of DMRG has been extended to study time evolution and, through a Fourier transform, spectral functions [23]. This method has the advantage of being completely generic and has an efficient scaling in system size, but is limited in its resolution; it is the growth in entanglement during real-time evolution that limits the resolution of the spectral functions. Other DMRG approaches, based on Lanczos-methods [24] or Chebyshev expansions [25], have similar defects. Yet, if one is only interested in low-energy dynamics, this entanglement growth seems counter-intuitive as all states in this low-energy sector are characterized by a small entanglement entropy [26–28].

A different strategy consists in targeting the low-lying particle excitations explicitly. For integrable systems, all excited states can be constructed exactly and it has been shown how to compute their spectral weights in an efficient way [29–31]. Alternatively, if the system can be connected perturbatively to a trivial non-interacting limit, perturbative continuous unitary transformations provide access to the non-trivial properties of low-lying excitation spectra [32–35]. Both approaches have the advantage that they have a built-in particle picture, but, for these particles to be well-defined, they need an extensive number of conserved quantities or a well-defined non-interacting limit to perturb from.

In the previous chapter, however, we have argued extensively that a particle picture should hold for any gapped quantum lattice system and that the particles can be simulated with a set of efficient variational methods. In this chapter, we will apply this framework to a number of spin chain models in order to show its versatility. This will involve the computation of the elementary excitation spectrum, the two-particle S matrix, the formation of bound states, the low-energy part of the spectral function, and the response to an external magnetic field or temperature. In these simulations, it will become manifest that the interactions between quasiparticles can exhibit very non-trivial behaviour, which the aforementioned alternative methods are not able to resolve.

Note that our formalism builds on previous work. First of all, in the paper where it was first introduced, the one-particle excitation ansatz was used to compute the spectrum of the spin-1 Heisenberg chain and the spin-1/2 XXZ chain [36]. These results were extended to compute the spectrum of the bilinear-biquadratic spin-1 chain [37] and a number of other spin chain models [38] and quantum field theories [39]. The ansatz has a natural translation to the framework of continuous matrix product states and has led to the computation of the spectrum of one-dimensional quantum field theories in Ref. [40]. Lastly, the ansatz was used to compute the mass spectrum for one-dimensional gauge theories in Refs. [41–43]. Also note that all the calculations in this chapter require a good MPS approximation for the ground state, for which we have always applied tangent-space methods [Sect. 2.6.4].

4.1 The Spin-1 Heisenberg Chain

Let us first look at the spin-1 Heisenberg chain, defined by the Hamiltonian

$$H = \sum_i \vec{S}_i \cdot \vec{S}_{i+1}$$

with $\vec{S}_i = (S_i^x, S_i^y, S_i^z)$ the three spin-1 operators at site i. The model has been of great theoretical and numerical interest because of the radically different behaviour between integer and half-odd-integer spin chains, which was first conjectured by Haldane in 1983 [44, 45]. More specifically, Haldane suggested that integer spin chains have a finite gap, in sharp contrast to half-odd-integer spin chains that are rigorously known to be gapless and long-range ordered by the Lieb–Schultz–Mattis theorem [46] and its extension [47].

The argument is based on a mapping of the spin chain onto the O(3) non-linear sigma model,[1] which is supposed to capture the low-energy physics of the system [44, 45, 48, 49]. This mapping becomes exact for large spin, but it is expected to hold for lower spins as well. The O(3) non-linear sigma model is believed to have a triplet of massive bosons with an exact S matrix [50]. In this effective description, the boson triplet has the relativistic dispersion relation

---

[1] An alternative field theory treatment of the Heisenberg spin-1 chain was proposed using Majorana fermions, giving similar qualitative predictions.

$$\omega(k) = \sqrt{\Delta^2 + c^2 k^2}$$

and the exact S matrix is given by

$$S_{\alpha\beta}^{\alpha'\beta'}(k_1, k_2) = \sigma_1(\theta)\,\delta_{\alpha\beta}\delta_{\alpha'\beta'} + \sigma_2(\theta)\delta_{\alpha\alpha'}\,\delta_{\beta\beta'} + \sigma_3(\theta)\,\delta_{\alpha\beta'}\delta_{\beta\alpha'}$$

with

$$\sigma_1(\theta) = \frac{2\pi i\theta}{(\theta + \pi i)(\theta - 2\pi i)}$$
$$\sigma_2(\theta) = \frac{\theta(\theta - \pi i)}{(\theta + \pi i)(\theta - 2\pi i)}$$
$$\sigma_3(\theta) = \frac{-2\pi i(\theta - \pi i)}{(\theta + \pi i)(\theta - 2\pi i)}$$

and $\theta = \theta_1 - \theta_2$ with the rapidities defined as $k_{1,2} = (\Delta/c)\sinh(\theta_{1,2})$.

When applied to the Heisenberg chain, this mapping predicts how the spectrum looks like around momenta 0 and π. The massive boson triplet appears as a threefold degenerate magnon branch in the spectrum around momentum π, while the lowest excitations around zero momentum are two-magnon scattering states. No bound states should exist in the system and the magnon-magnon S matrix should behave more or less according to the above expressions. Interpolating between the isolated magnon branch and the two-magnon continuum, one expects that the magnon emerges out of the continuum at some intermediate momentum. Although these predictions are only strictly valid for large spin and around the minima of the spectrum, it remains an open question to what extent they are correct for the full spectrum.

For a part, these predictions have been tested numerically. For example, in a number of numerical studies [51–55] it has been established that the elementary excitation is indeed a triplet (magnon) with a gap $\Delta = 0.4107$ [55–58] at momentum π. In this section, we will quantitatively study the other properties of the low-energy spectrum.

Elementary Magnon Excitation

With our methods, we can get an estimate for the gap $\Delta = 0.41047924871(1)$. Around the minimum we expect that the magnon dispersion follows a Lorentz-invariant dispersion relation with a characteristic velocity v. In Fig. 4.1 the magnon dispersion relation is plotted, showing that it follows the Lorentz-invariant form in quite a large region. Our estimate for the characteristic velocity is $v = 2.4668(1)$, in correspondence with Ref. [59]. Interestingly, this form of the dispersion relation implies that the equal-time correlation functions in momentum space or structure factor follows a Lorentzian shape [60, 61],[2]

[2]This relation between the dispersion relation and the structure factor follows from the Källén–Lehmann representation [62, 63] of two-point correlation functions in relativistic quantum field theory [64], see also [38].

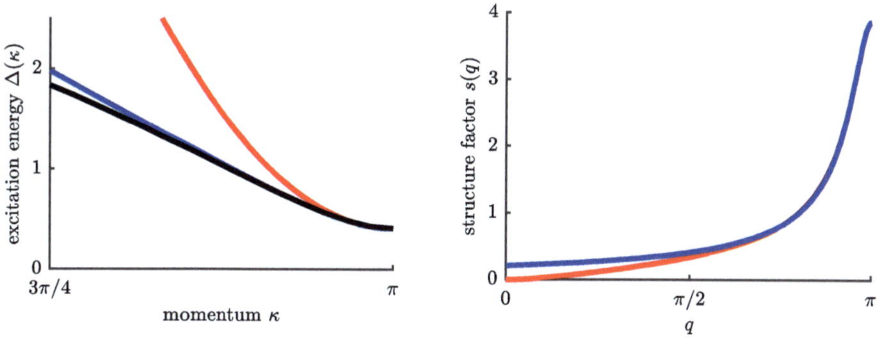

Fig. 4.1 Lorentz invariance of the magnon dispersion. *Left* variational calculation of the magnon dispersion relation (*black*), compared with Lorentz-invariant form (*blue*) and a quadratic form (*red*). *Right* the structure factor (*red*) compared to the Lorentzian (*blue*)

$$s(q) \propto \frac{1}{\sqrt{\Delta^2 + v^2 q^2}}$$

around momentum π. In Fig. 4.1 the exact structure factor was compared to this form, showing a good correspondence over a large region in momentum space. This form directly implies the Ornstein–Zernike form for the equal-time correlations in real space, i.e.

$$C(x) \propto \frac{\exp - |x|/\xi}{\sqrt{|x|/\xi}}$$

with $\xi = v/\Delta$ the correlation length of the system completely determined by the magnon gap and velocity. From our previous estimates, we arrive at $\xi = v/\Delta = 6.0097$ in very good agreement with other numerical results. Surprisingly, these estimates can be done accurately at low bond dimensions, for which a direct calculation of the correlation length – by means of the second largest eigenvalue of the transfer matrix – is a lot less accurate.

The accuracy of the excitation ansatz can be tested by computing the variance of the excited state with respect to the Hamiltonian. Figure 4.2 shows that the ansatz is already quite accurate at bond dimensions of the order $D = 30$, and that the accuracy increases exponentially. In fact, as the variance is negative for low bond dimension, the variational error is due to the ground-state approximation; a negative variance shows that in the region where the magnon is located, the excited state is more accurate than the ground state. Moreover, as the ratio r between the ground state variance and the one of the magnon is approximately constant around the value of $r = 20$ for different bond dimensions, it shows that the magnon size – the number of sites over which it is a "better eigenstate" – is more or less of that magnitude. For very high bond dimensions the ground state error, which should go to zero for gapped systems, drops below the error made by the one-particle ansatz which saturates at a finite, yet very small, value.

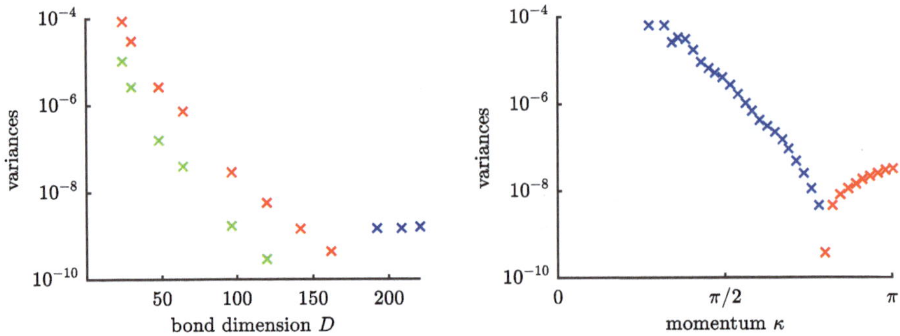

Fig. 4.2 Variance of the one-particle ansatz. For the log-plot absolute values have been taken; *red*, resp. *blue* crosses are negative, resp. positive values; *green* crosses correspond to the ground state variance per site (which is positive). *Left* variance of the magnon at momentum π as a function of bond dimension D. *Right* variance of the magnon as a function of momentum at a bond dimension $D = 100$. ed, resp. crosses are negative

Two-Magnon S Matrix

Just as in the non-linear sigma model, the magnons are now interpreted as interacting bosons with a non-trivial S matrix. Since the magnon is threefold degenerate, 9 incoming and outgoing two-magnon states can be composed and the S matrix is 9×9 dimensional unitary matrix. Because of $SU(2)$ invariance of the magnon interactions – both the ground state and the Hamiltonian are spin-rotation invariant – all scattering processes preserve the total spin and projection of the two-magnon state, such that the S matrix is diagonalized in this coupled basis and constant within every sector of total spin[3]:

$$S(\kappa_1, \kappa_2) = \begin{pmatrix} -e^{i\theta_0(\kappa_1,\kappa_2)}\mathbb{1}_{1\times 1} & & \\ & -e^{i\theta_1(\kappa_1,\kappa_2)}\mathbb{1}_{3\times 3} & \\ & & -e^{i\theta_2(\kappa_1,\kappa_2)}\mathbb{1}_{5\times 5} \end{pmatrix}.$$

The minus signs are chosen such that the scattering phases $\theta_{1,2,3}$ approach zero as $\kappa_{1,2} \to 0$.

In Fig. 4.3 the three scattering phases are plotted for total momentum 0 and as a function of relative momentum $\kappa_1 - \kappa_2$ (the momenta are relative to the magnon

[3]The S matrix was defined in Sects. 3.4.4 and 3.6.2; in our setting we have three types of particles (the three components of the magnon triplet) and they all have the same dispersion relation. This implies that, for every combination of total momentum K and total energy ω within the two-magnon continuum, we can build 9 scattering states. The relative coefficients of the asymptotic modes in these scattering states give rise to a (9×9) unitary S matrix (the group velocities will factor out, as all particles have the same dispersion). Furthermore, instead of labeling these scattering states with momentum and energy (K, ω), we can equally well label them with total and relative momentum $(K, \kappa_1 - \kappa_2)$ where κ_1 and κ_2 are the two momenta that show up in the asymptotic modes (there is still an ambiguity in the ordering of the momenta, we will always take the convention that $\kappa_1 > \kappa_2$, i.e. positive relative momentum).

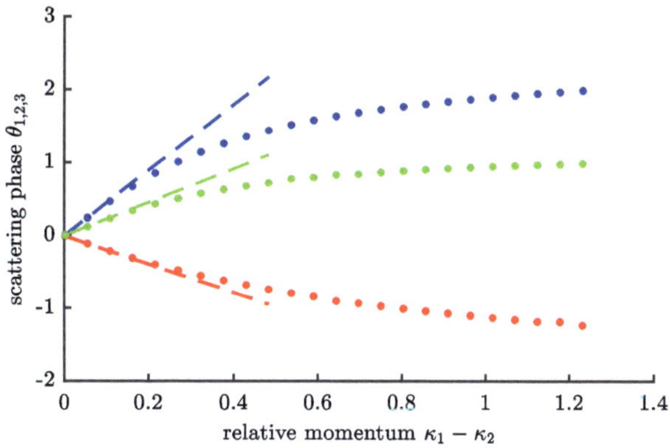

Fig. 4.3 The angle of the S-matrix elements for the sectors of total spin $S = 2$ (*green*), $S = 1$ (*blue*) and $S = 0$ (*red*) at total momentum $\kappa = 0$ and different relative momenta $\kappa_{\text{rel}} = \kappa_1 - \kappa_2$. The linear regime for small relative momenta is clearly visible, as well as the deviations from that regime at higher relative momenta. Calculations were done with bond dimension $D = 64$. The striped lines indicate the linear region and only serve as a guide to the eye

minimum at π). We can clearly observe a linear regime for small relative momenta, with the slope giving us a direct measure of the scattering length in the different sectors. We find the following values for the scattering lengths

$$a_0 = 1.945 \qquad a_1 = -4.515 \qquad a_2 = -2.306.$$

The signs of these scattering lengths are in agreement with the predictions of the non-linear sigma model. In the $S = 2$ sector we have excellent agreement with Ref. [59], while for the other sectors we have found no previous quantitative estimates. When we go to larger relative momenta, the curve loses its linearity.

For a Lorentz or Galilean invariant system the S matrix should depend only on the relative rapidity resp. momentum. In a lattice setting, however, both of these symmetries are broken so that the scattering phases depend on the total momentum of the two-particle state as well. Still, in Fig. 4.4 it is shown that, close to the minimum of the dispersion relation, the S matrix is nearly Galilean independent, and larger deviations are observed for larger total momenta.

The Full Spectrum

What about the rest of the spectrum? Far away from the momenta 0 and π, the mapping to the non-linear sigma model makes no predictions. A simple interpolation, however, implies that at some intermediate value of the momentum the one-magnon branch merges into the two-magnon continuum. From Fig. 4.5, where the full excitation spectrum is plotted, one can indeed see that this is the correct scenario. The exact point where the magnon dives into the continuum is estimated at $\kappa \approx 0.222\pi$, a slightly smaller value than the estimate 0.23–0.24π as reported in [23]. As the magnon comes closer to the continuum, the one-particle ansatz is expected to become less accurate

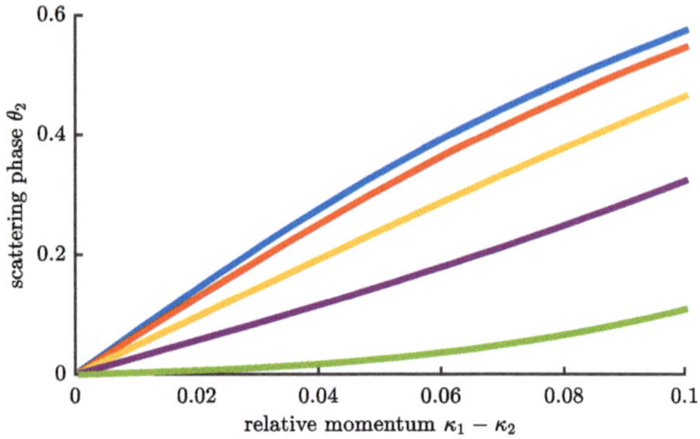

Fig. 4.4 The scattering phase in the $S = 2$ sector as a function of relative momentum for equally spaced values of the total momentum between 0 (*lower line*) and $\pi/5$ (*upper line*)

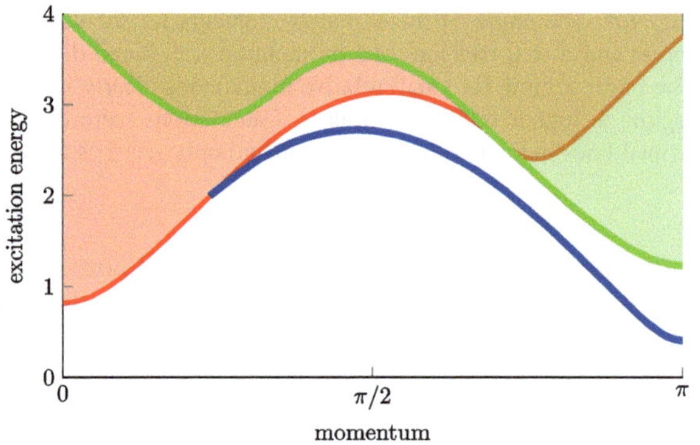

Fig. 4.5 The full excitation spectrum of the Heisenberg spin-1 chain. The elementary magnon branch (*blue line*) merges into the two-magnon continuum (*red*). At momentum π the three-magnon continuum (*green*) is the lowest multi-particle band

– remember that the localized nature of excitations was dependent on the gap below and above the isolated branch. This behaviour is confirmed by Fig. 4.2, where the variance is plotted as a function of the momentum. This plot clearly shows that the variance quickly goes up as the continuum comes closer.[4]

The formation of the isolated magnon branch is visible in the two-magnon S matrix. Indeed, very close to the continuum, the isolated magnon can be seen as a two-magnon bound state that gains stability as it leaves the two-particle band [23]. This scenario is confirmed in Fig. 4.6, where the scattering lengths (i.e. the slope of the scattering phase as a function of relative momentum for fixed total momentum) in the three sectors are plotted as a function of total momentum. The divergence of the scattering length in the $S = 1$ sector signals a formation of a bound state in that sector, corresponding to the formation of the isolated magnon branch. This two-magnon bound state continuously evolves into a well-defined one-magnon state closer to the dispersion's minimum.

[4] In order to attain a good accuracy on the momentum where the gap to the continuum vanishes, it was necessary to extend the one-particle ansatz to bigger blocks. Only by systematically growing the block size, it was possible to obtain the reported value of $\kappa \approx 0.222\pi$.

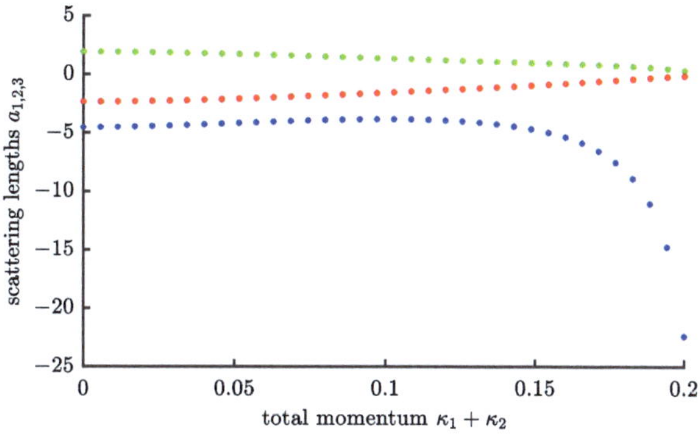

Fig. 4.6 The scattering length in the three sectors of total spin $S = 2$ (*green*), $S = 1$ (*blue*) and $S = 0$ (*red*) as a function of total spin. The divergence in the $S = 1$ sector is clearly visible

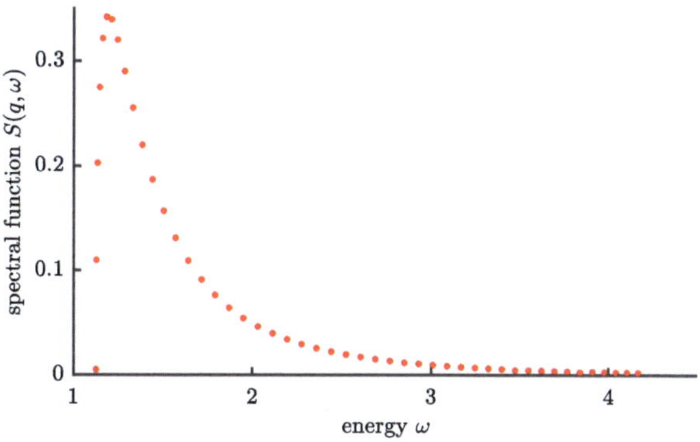

Fig. 4.7 The two-particle contribution for the spectral function $S(q, \omega)$ at momentum $\kappa = \frac{\pi}{10}$. Calculations were done with bond dimension $D = 48$

The full excitation spectrum can be probed in experiment through a direct measurement of the spectral function defined as[5]

$$S(q, \omega) = \sum_{n=-\infty}^{+\infty} \mathrm{e}^{-iqn} \int_{-\infty}^{+\infty} \mathrm{d}t \, \mathrm{e}^{i\omega t} \langle \Psi_0 | S_n^y(t) S_0^y(0) | \Psi_0 \rangle .$$

The action of the spin operators on the ground state can be projected onto the low-lying excited states; for low enough energies only the one- and two-particle excitations give contributions. Around momentum π the spectral function is dominated by a single-particle δ peak, whereas the spectral function has a continuous two-particle contribution around momentum 0. Note that $S(q, \omega) \to q^2$ as $q \to 0$, because the ground state is a spin singlet state. In Fig. 4.7 the spectral function was plotted as a function of energy ω at fixed momentum $q = \pi/10$, inside the region where it is expected that the two-particle contributions dominate. This is the shape one expects from the non-linear sigma model [65] and compares nicely with DMRG results [23].

We can get an idea of how well the full spectral function is reproduced by looking at its zeroth and first frequency moment, i.e. $s_0(\kappa) = \int \frac{\mathrm{d}\omega}{2\pi} S(\kappa, \omega)$ and

[5]Because of rotation invariance it doesn't matter in what spin direction the correlations are measured.

$s_1(\kappa) = \int \frac{d\omega}{2\pi}\omega S(\kappa,\omega)$. As the former is equal to the static structure factor and the latter can be written as the expectation value of a simple double commutator [66], both can be easily calculated with the MPS ground state. It appears that the two-particle contribution in Fig. 4.7 approaches the exact values up to 98.7 and 96.4%, showing indeed that the two-particle sector carries the dominant contribution of the spectral function at this momentum.

Magnon Condensation

The physical properties of the magnons (mass, velocity, S matrix, ...) can be brought to the fore by applying a magnetic field h to the spin chain. Indeed, as the ground state is a spin singlet, it does not feel the field. The magnons, however, have spin one and will shift in energy according to their spin component along the magnetic field. This Zeeman splitting will lower the energy of one of the triplet components linearly, until it reaches zero. At that point, the magnons condense in the ground state and a finite density of them is created. This magnetized spin chain is characterized as a Luttinger liquid with a Luttinger parameter K that varies continuously with the magnetization [67–69].

The physical picture of this condensation can be understood from the approximate Bethe ansatz that was developed in Sect. 3.7. Indeed, once it crosses the gap, the magnetic field serves as a chemical potential for the $+1$ component of the magnon triplet.[6] The above information on the magnon dispersion relation and the magnon-magnon S matrix will allow us to compute both thermodynamic properties and correlation functions for the magnetized chain.

We start very close to the phase transition, where the particles can be approximated as free fermions (only the momenta around the minimum will be occupied, for which the S matrix is approximately -1). The magnetization, i.e. the density of condensed magnons, will be given by [58, 70, 71]

$$m(h) = \frac{\sqrt{2\Delta}}{\pi v}\sqrt{(h - h_c)}. \tag{4.1}$$

When more pseudo-momentum levels are filled up, the two-particle S matrix will deviate from its limiting value of -1 and the free-fermion approximation will no longer hold. As a first order correction, we can assume a linear scattering phase with the scattering length a as the slope (and still a quadratic dispersion). From Eq. (3.48) it follows that the correction to the magnetization curve is given by

$$m(h) = \frac{\sqrt{2\Delta}}{\pi v}\sqrt{(h - h_c)} - \frac{8\Delta a}{3\pi^2 v^2}(h - h_c), \tag{4.2}$$

a result which was obtained in Ref. [72] by a similar reasoning.

When even higher momenta are occupied, these approximations – quadratic dispersion relation, linear scattering phase and Galilean invariance – will get worse and

[6] The other components remain gapped, so we will not consider them in our calculations.

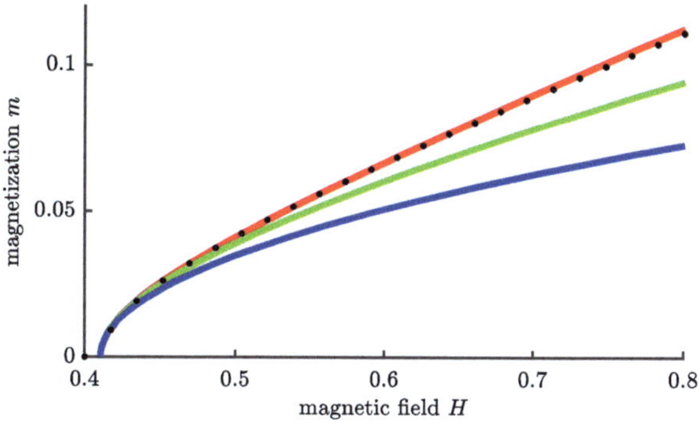

Fig. 4.8 The magnetization m versus applied magnetic field h for the spin-1 Heisenberg chain. Our results (*red*, bond dimension $D = 64$) are compared to the hard-core boson square-root dependence (*blue*) [71] and first order corrections by the scattering length a_2 (*green*) [72]. The direct MPS calculations (*black dots*) were done at the same bond dimension of $D = 64$. The *bottom-right* inset provides a close-up of the phase transition

only a full approximate Bethe ansatz calculation will give the correct magnetization curve; the result is plotted in Fig. 4.8.

Next we look at correlation functions of the magnetized ladder. With our methods, we have no direct access to these correlation functions, but we can infer their form by combining the Luttinger liquid formalism with the thermodynamic properties as computed from the approximate Bethe ansatz. Indeed, since the S^x and S^y operator essentially creates a magnon out of the vacuum at momentum π and the S^z operator creates a two-magnon state at momentum 0, we can translate the expressions for the bose gas correlators [Eq. (3.45)] to the magnetized ladder as [68]

$$\begin{aligned} \langle S_n^x S_{n'}^x \rangle \approx \langle S_j^y S_0^y \rangle &\propto \frac{(-1)^{n-n'}}{|n-n'|^{1/2K}} \\ \langle S_j^z S_0^z \rangle &\approx m^2 + \frac{K}{2\pi^2 |n-n'|^2} + A_z \frac{\cos(2\pi m(n-n'))}{|n-n'|^{2K}}. \end{aligned}$$

The power-law decay of these correlation functions is controlled by the Luttinger parameter K. In Fig. 4.22 we have plotted K as a function of the magnetization m. This behaviour can again be explained by starting with the free-fermion limit at very low densities. In Sect. 3.7.3 we have shown that $K = 1$ in this case. The first order correction on this value is determined by the magnon-magnon scattering length; in first order in m the LL parameter is given by [68]

$$K(m) = 1 - 2am. \tag{4.3}$$

For larger values of the magnetization, again only the full approximate Bethe ansatz gives the results in Fig. 4.9, in correspondence with the direct DMRG simulations from Ref. [69].

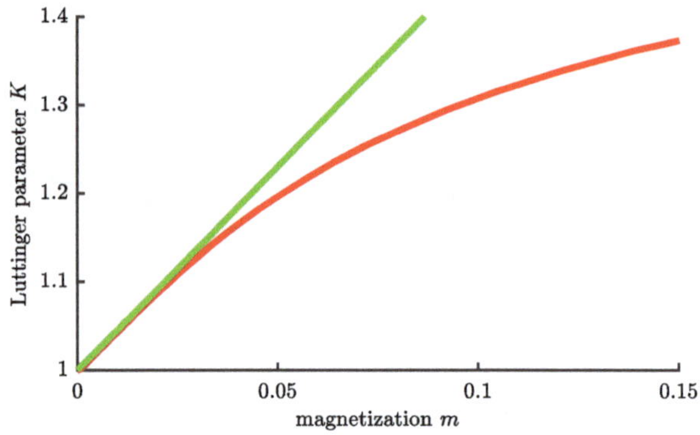

Fig. 4.9 The *top-left* inset provides our result for the LL parameter K as a function of the magnetization m (*red*), compared to the linear relation Eq. (4.3) based on the scattering length (*green*)

4.2 The Spin-1/2 Heisenberg Antiferromagnetic Ladder

As a second application, we will study the spin-1/2 Heisenberg antiferromagnetic (HAF) two-leg ladder in a magnetic field, defined by the Hamiltonian

$$H = \sum_{i,l} \vec{S}_{i,l} \cdot \vec{S}_{i+1,l} + \gamma \sum_{i} \vec{S}_{i,1} \cdot \vec{S}_{i,2} - h \sum_{i,l} S^z_{i,l}$$

where $l = 1, 2$ denote the two legs of the ladder and $\vec{S}_{i,l}$ denotes the spin operator at site i in the l'th leg (see Fig. 4.10).

The two-leg HAF ladder and its excitation spectrum have been studied intensively for many reasons. First of all, it is the first step to study the transition from one-dimensional systems to higher-dimensional versions. Secondly, the excitation spectrum has a lot of interesting features, such as the presence of a gap [44] and the existence of bound states, and can be studied with a variety of methods depending on the parameter regime. These features can also be observed experimentally [73–76], so that ladders provide an ideal test for these theoretical methods [77, 78]. Finally, the experimental realization of magnetized spin ladders provides an ideal quantitative test of the Luttinger liquid model [5–7, 79].

In this section we will test our variational method on the two-leg ladder. An MPS approximation for the ground state can be found by first blocking two spins on a rung into one four-level system and applying an MPS optimization algorithm. In this

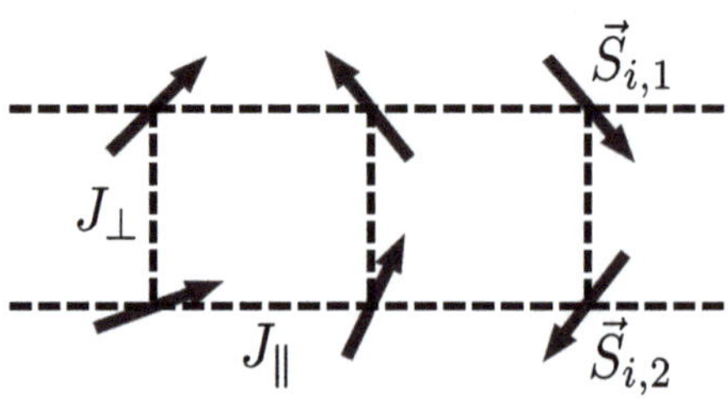

Fig. 4.10 The ladder geometry with $J_\parallel$ and $J_\perp$ the couplings along the leg, resp. rung. We will always put $J_\parallel = 1$ and define the coupling ratio $\gamma = J_\perp / J_\parallel$

representation (to every rung there corresponds one MPS tensor A) we find a ground state that is invariant under translations over one site in the leg direction; all momenta in the following subsections are defined with respect to this translation operator. The Hamiltonian and the ground state are invariant under the reflection operator $\mathcal{P}$ which flips the two legs of the ladder. We impose no additional symmetries (e.g. $SU(2)$ invariance) on the MPS, but our variational solution will of course have the right symmetries to high precision.

In the first three subsections we will investigate the low-lying spectrum of the ladder without magnetic field. In the following two subsections we will apply the approximate Bethe ansatz to the magnetization process, at zero and finite temperature.

One-Particle Excitations: Elementary Spectrum and Bound States

The nature of the elementary excitations in the ladder can be understood starting from the following limits.

At zero coupling ($\gamma \to 0$), we have two independent spin-1/2 Heisenberg chains where the elementary excitations are spinons (carrying spin 1/2). These spinons are topologically non-trivial excitations and can only be created in pairs by the action of a local operator. Upon coupling the chains, the spinons are confined into magnons carrying integer spin. This picture has been studied with bosonization techniques [80], showing that the interchain coupling opens up a gap to a triplet of massive magnons (triplons) and a higher-up singlet.

At infinite coupling ($\gamma \to \infty$) we have a collection of independent rungs with antiferromagnetic interaction. In the ground state all rungs are in a singlet state and an elementary excitation is constructed by promoting one rung to a triplet state. When the leg coupling $J_\perp$ is turned on, this triplet obtains a kinetic energy and we get a non-trivial dispersion. This qualitative picture survives for intermediate couplings: through perturbative continuous unitary transformations an effective particle picture can be established and very accurate results on e.g. the elementary dispersion relation and bound states can be obtained [32, 34, 81–84].

In Fig. 4.11 we have plotted the gap as a function of the interchain coupling. One can observe that the gap goes to zero in the weak-coupling limit, while it grows to the

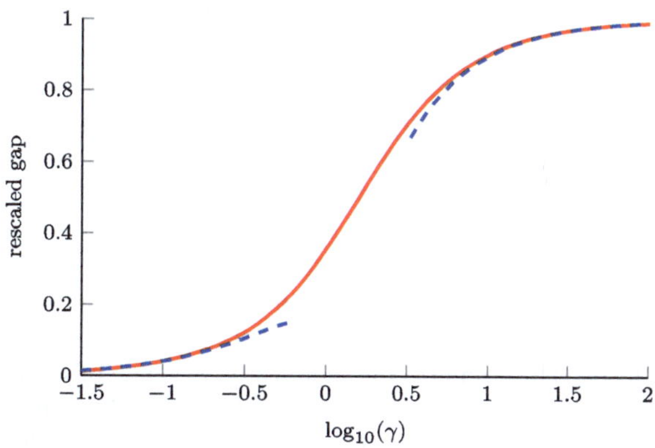

Fig. 4.11 The rescaled gap $\Delta/\sqrt{1+\gamma^2}$ as a function of the interchain coupling γ (*red*). The *blue dashed lines* are the first order correction from the strong-coupling limit ($\gamma \to \infty$) and results from bosonization in the weak-coupling limit ($\gamma \to 0$) [80, 85]

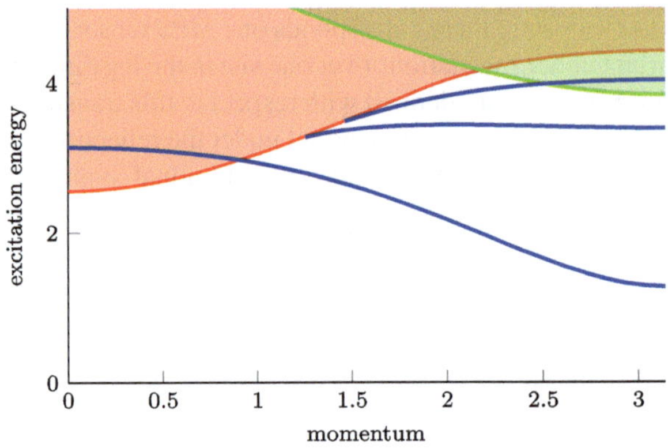

Fig. 4.12 The one-particle spectrum consists of a triplet (magnon) which is stable over the whole Brillouin zone (*lowest lying blue curve*), an singlet (bound state) which is stable for momenta between $\kappa_{\text{BS1}} \approx 0.39\pi$ and π (*second blue curve*), and a triplet (bound state) which is stable for momenta between $\kappa_{\text{BS2}} \approx 0.46\pi$ and π (*third blue curve*). Note that the determination of κ_{BS1} and κ_{BS2} is not very precise because the one-particle ansatz is not accurate near the transition. The *red* region is the two-magnon continuum, the *green* region is the three-magnon continuum; the other continua (e.g. triplet-singlet continuum) are not shown

constant value that one expects from a strong-coupling expansion. Our variational results smoothly interpolates between these two limits.

A typical excitation spectrum in the intermediate region ($\gamma = 2$) is shown in Fig. 4.12. The lowest-energy state is an elementary triplet excitation (magnon) with a minimum at momentum π. The magnon has odd parity under the reflection operator $\mathcal{P}$. The lowest-energy state around momentum zero is a two-magnon scattering state and has even parity. Because the one- and two-magnon state have different parity, the elementary magnon cannot decay and is stable in the whole Brillouin zone. From Fig. 4.13, where we have plotted the variance of the excitation ansatz, we can indeed see that the magnon is a bona fide particle excitation for all momenta. Note that under a parity-breaking interaction the stability of the magnon inside the continuum breaks down [86] and it might prove an interesting question whether we can capture its decay within our framework.

The elementary excitation spectrum at $\gamma = 2$ has two more elementary particle excitations, a singlet and a triplet, which are stable in a limited region around momentum π. Both are even under the parity operator $\mathcal{P}$. From the strong-coupling expansion, we can interpret them as two-magnon bound states [32], hence the even parity (without a well-defined particle number, we cannot make this interpretation, so we regard these branches as elementary particles). The variance of the bound states is sufficiently small in the stable region, but it grows larger as the momentum approaches the continuum. From the previous chapter we know that the localized nature of an elementary excitation is related to the gap below and above the excitation branch, so we expect the bound state to become wider as the gap to the continuum closes. This explains the increasing variance of the bound states in Fig. 4.13. Upon entering the continuum, the bound state has become completely delocalized and no longer exists as a stationary eigenstate of the Hamiltonian.

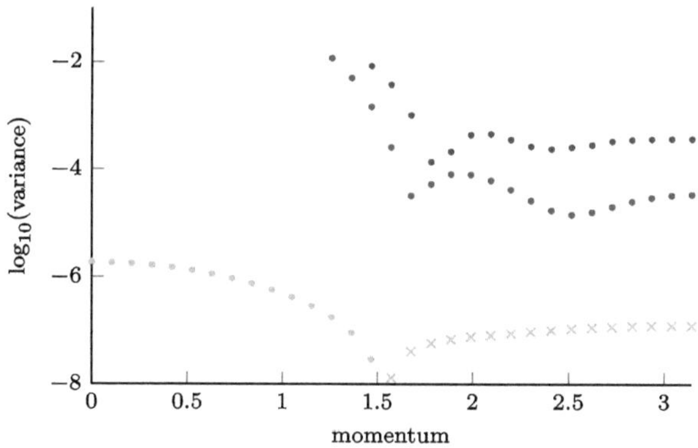

Fig. 4.13 The ($\log_{10}$ of the modulus of the) variance of the one-particle excitations; *dots*, resp. crosses are positive, resp. negative variances (see Appendix A.2 for the meaning of a negative variance). The magnon (*green*) is clearly a well-defined particle excitation in the whole Brillouin zone. The singlet (*red*) and triplet (*blue*) get larger variances as they come closer to the two-particle band, until they actually dive in and are no longer stable. Calculations were done at $\gamma = 2$ with bond dimension $D = 30$; the ground state variance density is 2.27×10^{-8} at that bond dimension

Table 4.1 Excitation energy and variance of the first 6 solutions of the one-particle problem for the HAF ($\gamma = 0.2$) at momentum π with bond dimension $D = 108$. The variance density of the ground state is $9.28.10^{-6}$. The first triplet has negative variance, which shows that this excitation is closer to an exact eigenstate locally than the ground state (see Appendix A.2). The fourth solution is also a true one-particle (singlet) excitation. All other solutions have a considerably larger variance and correspond to artificial two-particle states. Further up in the continuum, however, we have another triplet with quite small variance, although it is difficult to say whether this corresponds to a true bound state

Energy	Variance
0.081841224772803	−0.000178252361115
0.081841224779434	−0.000178252351941
0.081841224792513	−0.000178252347304
0.331378942771407	0.000337897356458
0.367322866763615	0.029803975299627
0.410460620351393	0.044970779553592
...	...
0.513408977989184	0.014052233372105
0.513408978649963	0.014052233100514
0.513408978939573	0.014052232922150
...	...

As a last illustration of the one-particle ansatz we have included Table 4.1, where we have listed the excitation energies and variances in the weak-coupling region, showing the elementary triplet and singlet excitations that we expect from a bosonization calculation. We observe that the variances are some orders of magnitude larger in this weak-coupling region. Since the gaps above and below these excitations are a lot smaller at small γ, this is not unexpected. Note that both the energies and the variances have the right degeneracies, even though we never imposed the corresponding symmetries explicitly.

Two-Particle S Matrix

In this section we will look at the two-magnon S matrix; the scattering of, e.g., an elementary magnon with a bound state will not be considered. Just as in the case of the spin-1 Heisenberg model [Sect. 4.1], the general expression for the magnon-magnon S matrix in this representation should reduce to a diagonal form where all information is in the three phases for every sector of total spin. In our simulations, we always found this reduced form to high precision, so in the following we can restrict to plotting these three phases.

In Fig. 4.14 we have plotted the S matrix as a function of the relative momentum $\kappa_1 - \kappa_2$ for total momentum $K = 0$. One can observe (i) the limit $S = -\mathbb{1}$ for the relative momentum going to zero, and (ii) the linear region around this limit (the slope is the scattering length). The sign of the phase is positive for all three sectors (although this does not have to be the case, see Figs. 4.15 and 4.16).

In Fig. 4.15 we have plotted the S matrix in the $S = 2$ sector for different values of the total momentum. We observe that the S matrix depends strongly on K in a non-trivial way, but there seems to be a small region around $K = 0$ where it is quasi-constant. This points to the presence of a region around the minimum of the dispersion relation where the interaction is Galilean invariant (note that the dispersion should be quadratic in this region). At larger momenta, this Galilean invariance is broken, as one expects in a lattice system.

Even more spectacular things can happen when we vary the total momentum, such as the formation of a bound state. In Fig. 4.16 we have plotted the scattering lengths in all three sectors as a function of the total momentum. We can see that the scattering lengths in the $S = 0$ and $S = 1$ sectors diverge, signalling the formation of the singlet and triplet bound states (in agreement with the discussion in Sect. 3.6.3).

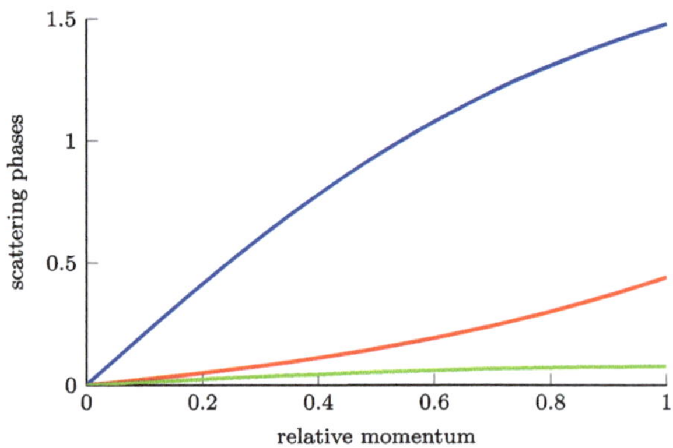

Fig. 4.14 The S matrix as a function of relative momentum $\kappa_1 - \kappa_2$ at total momentum $K = 0$. Plotted are the phases of the S matrix in the $S = 0$ (*red*), $S = 1$ (*blue*) and $S = 2$ (*green*) sector. Calculations were done at $\gamma = 2$ and with bond dimension $D = 32$

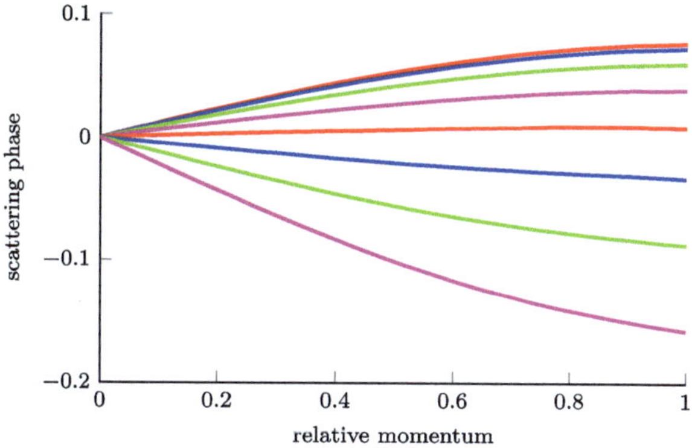

Fig. 4.15 The scattering phase in the $S = 2$ sector for 8 equally spaced values of the total momentum between $K = 0$ (*upper line*) and $K = \pi/3$ (*lower line*). Around $K = 0$ there is a region where the S matrix is independent of total momentum, which points to some Galilean invariance around the minimum of the dispersion relation. Calculations were done at $\gamma = 2$ and with bond dimension $D = 32$

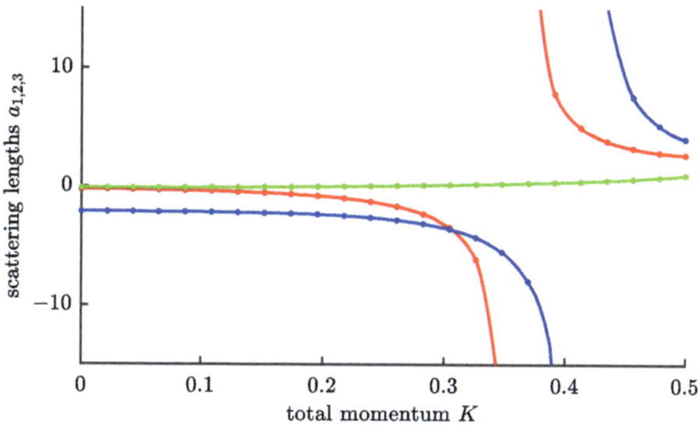

Fig. 4.16 The scattering lengths a_0 (*red*), a_1 (*blue*) and a_2 (*green*) as a function of the total momentum K. In the $S = 2$ sector nothing spectacular happens, although it does change sign. In the other sectors we see a divergence at the momentum where a bound state forms. The plotted range does not show all data points around the divergences, the *full lines* are a guide to the eye and give an indication on where the other points are situated. Calculations were done at $\gamma = 2$ and bond dimension $D = 32$

Spectral Function

Since we have a two-leg ladder system, we can look at spectral functions with transversal momentum q equal to 0 or π. We define the two rung operators (defined on rung i)

$$(S_0^z)_i = S_{i,1}^z + S_{i,2}^z \tag{4.4}$$

$$(S_\pi^z)_i = S_{i,1}^z - S_{i,2}^z. \tag{4.5}$$

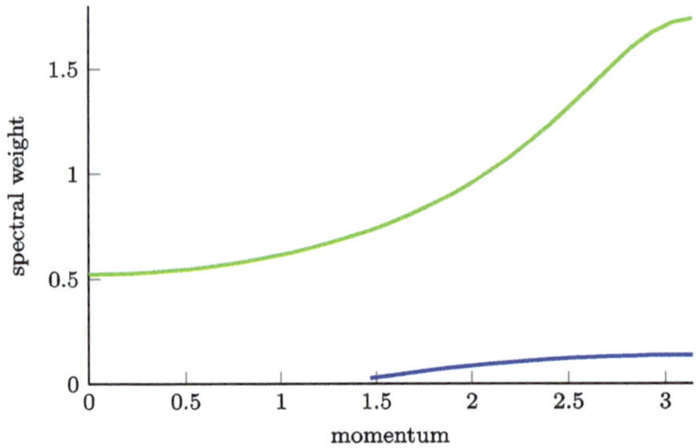

Fig. 4.17 The one-particle spectral weights; these appear in the spectral functions $S_{0/\pi}(\kappa, \omega)$ as the prefactor of the $2\pi\delta(\omega - \Delta(\kappa))$ function (where $\Delta(\kappa)$ is the dispersion relation of the particle). We have plotted the magnon weights w.r.t. to the odd operator (*green*) and the weight of the triplet bound state w.r.t. to the even operator (*blue*). All the other one-particle spectral weights are identically zero. These results are in accordance with Ref. [84]. Note that the one-particle description of the bound state gets worse when coming closer to the continuum, so that the calculation of its spectral weight loses accuracy in this region. It is nevertheless clear that the spectral weight goes to zero as the bound state loses stability

These operators have even, resp. odd parity under the action of the reflection operator $\mathcal{P}$. We will look at spectral functions $S_{0/\pi}(\kappa, \omega)$ with respect to these two operators,

$$S_{0/\pi}(\kappa, \omega) = \sum_n \int \mathrm{d}t \; \mathrm{e}^{i(\omega t - \kappa n)} \langle \Psi_0 | \mathrm{e}^{-iHt} (S^z_{0/\pi})^\dagger_n \mathrm{e}^{iHt} (S^z_{0/\pi})_0 | \Psi_0 \rangle,$$

where $\sum_n$ represents a sum over rungs.

Let us first look at the one-particle contributions. Since the elementary magnon is odd under $\mathcal{P}$, it can only carry spectral weight with respect to the odd operator. From $SU(2)$ symmetry we know that the singlet bound state does not carry any spectral weight with respect to both operators (they are both spin-1 operators). Lastly, the triplet bound state is even under $\mathcal{P}$, so it only contributes to the even operator spectral function $S_0(\kappa, \omega)$. These considerations lead to the picture in Fig. 4.17. One can see that the spectral weight of the bound state goes to zero as it approaches the continuum.

Next we look at the two-magnon contribution, which has only overlap with the even parity operator. In Fig. 4.18 we have plotted different momentum slices of the spectral function. At momentum zero, the spectral function is identically zero (the ground state is a singlet) and grows for small momenta as $\propto \kappa^2$ (cfr. Ref. [65]). For larger momenta, we see that the spectral function gets strongly peaked at some value for κ, after which the peak again disappears. The origin of this resonance is of course the formation of the bound state: before it is stable, the bound state is already visible in the spectral function as a resonance.

To further confirm this picture, we have plotted the maximum of the peak as a function of the momentum in Fig. 4.19. One can see the resonance clearly diverging

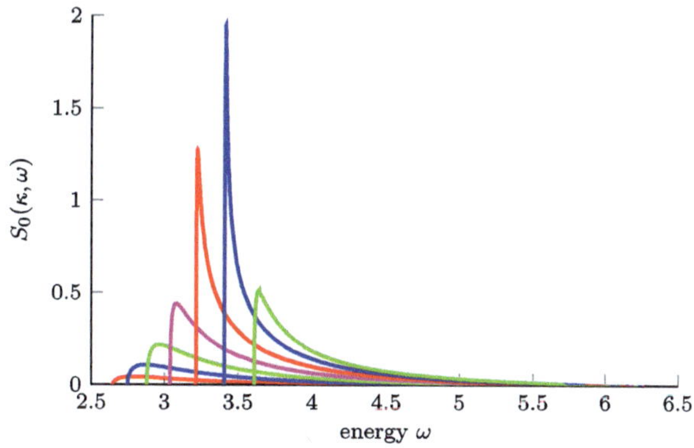

Fig. 4.18 The two-particle contribution to the spectral function $S_0(\kappa, \omega)$ for equally spaced values of the momentum between $\kappa = 0$ and $\kappa = \pi/2$. The $\kappa = 0$ curve is not shown as it is equal to zero everywhere. Calculations were done at $\gamma = 2$ with bond dimension $D = 32$

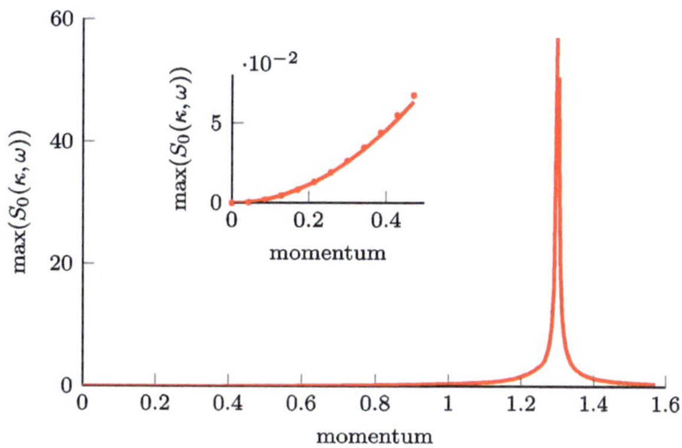

Fig. 4.19 The maximum of the two-particle contribution to the spectral function $S_0(\kappa, \omega)$ for different momentum slices. The *full line* is a guide to the eye. In the inset we show a close-up of the small momentum region, the *full line* is quadratic fit. Calculations were done at $\gamma = 2$ with bond dimension $D = 32$

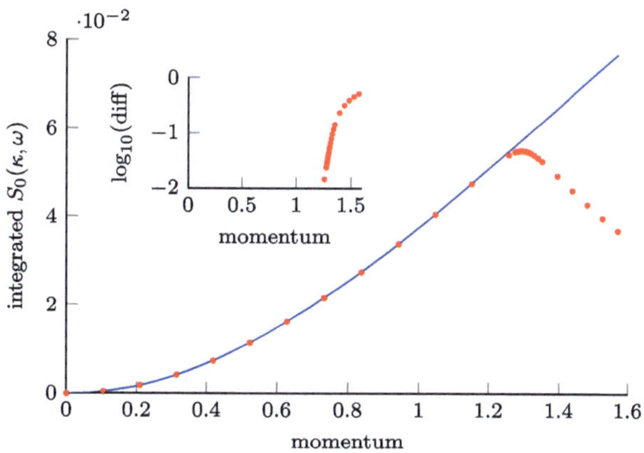

Fig. 4.20 The integrated spectral function $\int d\omega/2\pi S(\kappa, \omega)$ as a function of the momentum κ (*red dots*) compared with the momentum space correlation function $s_0(\kappa)$ (*blue line*). In the inset we plot the (log10 of the) difference between the two; values below 10^{-2} are not shown. Calculations were done at $\gamma = 2$ with bond dimension $D = 32$

at the point where the bound state reaches stability: from that point on the stable bound state contributes a δ peak to the spectral function.

We have also plotted the integrated spectral function in Fig. 4.20. Before the formation of the bound state, we see that the sum rules are completely satisfied (up to numerical errors), which shows that the one- and two-particle sectors indeed capture the full spectral function, at least in this momentum range (see also Ref. [84]). Again, we clearly see the $\propto \kappa^2$ dependence at small momenta. After the bound state has formed, however, the two-magnon part loses increasing spectral weight to the bound state.

Magnetization Process

Let us now turn on the magnetic field. For $SU(2)$ invariant systems, this perturbation does not affect the singlet ground state and induces a Zeeman splitting of the elementary magnon excitation. When the magnetic field reaches the value of the gap, one of the components of the triplet forms a pseudo-condensate (no real condensate can form in one dimension); the system undergoes a continuous phase transition from a commensurate phase with zero magnetization to an incommensurate phase with non-zero magnetization [87].

Analogously as the treatment for the spin-1 Heisenberg chain [Sect. 4.1], the magnon condensation can be treated with the approximate Bethe ansatz. In Fig. 4.21 the result for the magnetization curve of the ladder is plotted. Again, the very accurate results are clearly visible in a very broad regime, up to magnetizations of $m = 0.2$ (one out of five rungs contains a magnon).

The equal-time correlation functions of the magnetized ladder can also be inferred from the mapping to a Luttinger liquid. Indeed, since we have seen above that the S^x_π operator essentially creates a magnon out of the vacuum at momentum π and the S^z_0 operator creates a two-magnon state at momentum 0, we can translate the expressions for the Bose gas correlators [Eq. (3.45)] to the magnetized ladder as

$$\langle (S^x_\pi)_{i'} (S^x_\pi)_i \rangle = A_x \frac{(-1)^{i-i'}}{|i-i'|^{1/2K}} - B_x (-1)^{i-i'} \frac{\cos(2\pi m (i-i'))}{|i-i'|^{2K+1/2K}} \tag{4.6}$$

$$\langle (S^z_0)_{i'} (S^z_0)_i \rangle = m^2 - \frac{K}{2\pi^2 |i-i'|^2} + A_z \frac{\cos(2\pi m (i-i'))}{|i-i'|^{2K}}, \tag{4.7}$$

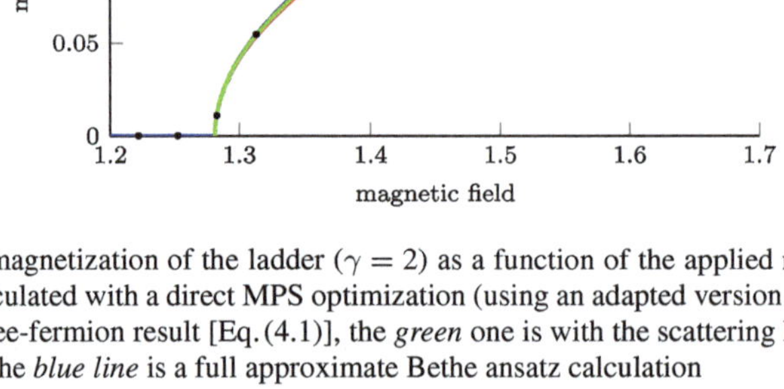

Fig. 4.21 The magnetization of the ladder ($\gamma = 2$) as a function of the applied magnetic field h. The *dots* are calculated with a direct MPS optimization (using an adapted version of Ref. [88]), the *red line* is the free-fermion result [Eq. (4.1)], the *green* one is with the scattering length correction [Eq. (4.2)], and the *blue line* is a full approximate Bethe ansatz calculation

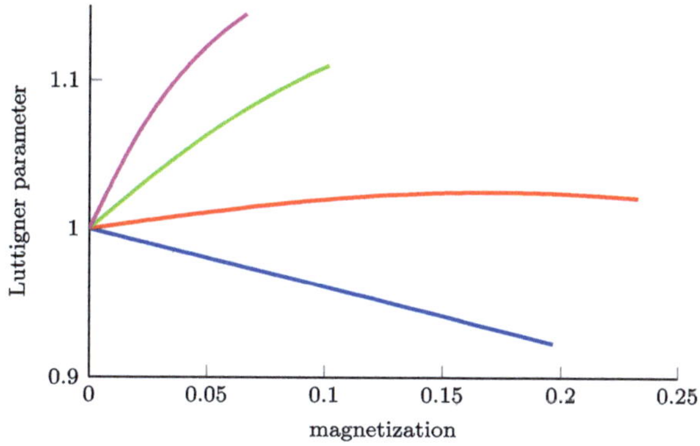

Fig. 4.22 The LL parameter as a function of the magnetization for $\gamma = 5$ (*blue*), $\gamma = 2$ (*red*), $\gamma = 1$ (*green*) and $\gamma = 1/2$ (*magenta*)

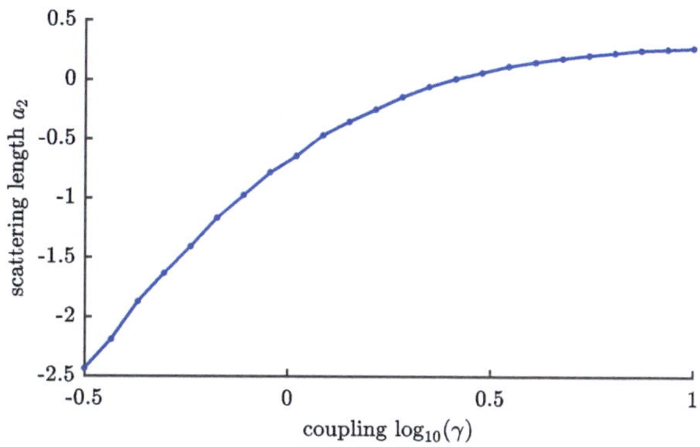

Fig. 4.23 The scattering length for different values of the interchain coupling $\log_{10}(\gamma)$

in accordance with Ref. [89]. The power-law decay of these correlation functions is controlled by the LL parameter K. In Fig. 4.22 we have plotted K as a function of the magnetization m for the ladder at different values of γ. At very low magnetization $m \to 0$ the LL parameter reaches the universal value of 1, but it appears that, beyond this limiting value, $K(m)$ changes qualitatively as we vary γ. The same behaviour was observed in Ref. [89] by fitting the analytic form of the correlation functions (4.6) and (4.7) with DMRG calculations.

We have seen that the first deviation from the universal free-fermion value of the Luttinger parameter K is determined by the scattering length as

$$K(m) = 1 - 2am.$$

In Fig. 4.23 we have plotted the scattering length as a function of the interchain coupling γ. Based on the above equation, the change of the sign of a confirms the varying qualitative behaviour of $K(m)$ as observed in Fig. 4.22 and in Ref. [89]. This change of sign can be identified as the difference between so-called attractive and repulsive Luttinger liquids [79].

Finally, we can study the magnetization process at finite temperatures using the thermodynamic Bethe ansatz. In Fig. 4.24 we have plotted the magnetization curve for

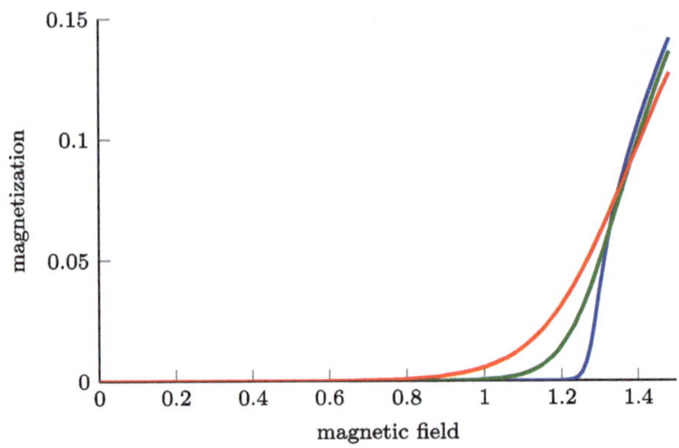

Fig. 4.24 The magnetization as a function of the magnetic field h for three values of the temperature: $T = .01\Delta$ (*blue*), $T = .045\Delta$ (*green*) and $T = .08\Delta$ (*red*)

different temperatures, showing that the zero-temperature square-root dependence around the phase transition is smoothed out at finite temperature. Note that we have included the other components of the magnon triplet – they are thermally excited as well – in a decoupled fashion. In a more correct analysis we would have to solve the fully coupled Bethe equations for the three components, but this falls outside the scope of this dissertation.

4.3 The Frustrated and Dimerized Spin Chain

An excellent model to study the signature of quasiparticle interactions in the spectral function is the frustrated and dimerized spin-1/2 Heisenberg chain, given by the Hamiltonian (see Fig. 4.25)

$$H = \sum_n \left(J_1(1 + \delta(-1)^n)(\vec{S}_n \cdot \vec{S}_{n+1}) + J_2(\vec{S}_n \cdot \vec{S}_{n+2}) \right).$$

It has been shown [90] that spin-Peierls compounds as $CuGe_2$ and NaV_2O_5 can be described with this model, where the dimerization term arises from a three-dimensional coupling of the spin chains. The physics of spinon confinement due to this three-dimensional coupling has recently attracted a lot of experimental and theoretical attention [91–93].

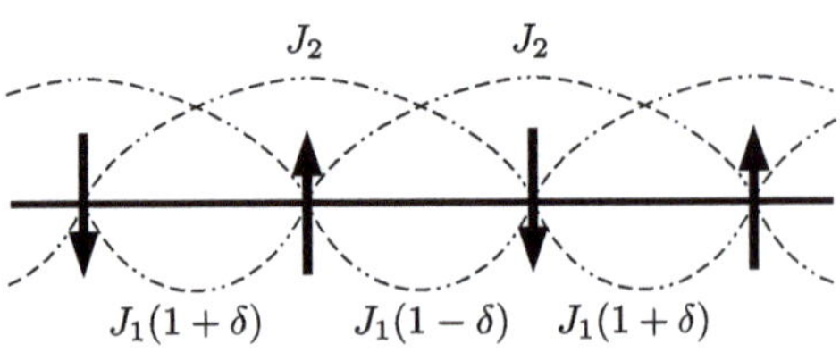

Fig. 4.25 The frustrated and dimerized chain

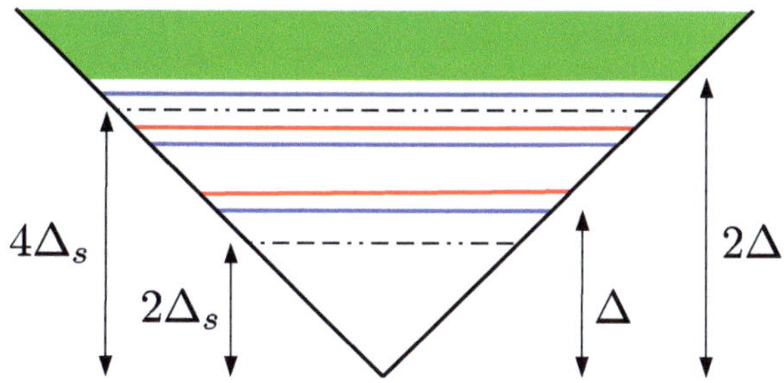

Fig. 4.26 A sketch of the bound-state spectrum; different excitations can be seen as bound state solutions of a linear potential. The strength of the potential ($\propto \delta$) can be tuned more or less independently from the mass Δ_s of the confined solitons. A number of triplet (*blue*) and singlet (*red*) $s\bar{s}$ bound states appear above the two-soliton energy $2\Delta_s$; these appear at energies $2\Delta_s + E_{b,i}$, with $E_{b,i}$ the (positive) binding energy of the i'th solution of the linear confining $s\bar{s}$ potential. Above the edge of the $s\bar{s}$-$s\bar{s}$ continuum (*green*) these solutions are no longer stable against decay into scattering states of two $s\bar{s}$ states. Note that the stability of a bound state depends on its momentum (see Fig. 4.27)

Without explicit dimerization ($\delta = 0$) it is known [94, 95] that a gap opens at $J_2/J_1 \approx 0.2412$ [96], accompanied by a spontaneous lattice dimerization and a twofold-degenerate ground state. The elementary excitations in this regime can be pictured as dressed defects in the ground state dimerization pattern [97] and behave as topological solitons s and anti-solitons $\bar{s}$ interpolating between the two ground states. The solitons have spin 1/2. No bound states occur [98], so the physical spectrum starts with a soliton/anti-soliton scattering continuum at $2\Delta_s$ (with Δ_s the soliton gap).

The dimerization δ favours one of the two ground states and confines the solitons into bound states with a linear potential between a $s\bar{s}$ pair [98–100]. This implies that the $s\bar{s}$ continuum is split up into a stack of discrete triplet and singlet bound states (see Fig. 4.26). If Δ is the energy of the lowest-lying triplet, a two-triplet continuum will start at 2Δ, such that $s\bar{s}$ bound states with an energy above this threshold will not be stable. This is the effect of string breaking: if the energy cost of having the wrong ground state between the $s\bar{s}$ is too high, the bound state will decay to a $s\bar{s}$-$s\bar{s}$ pair. Not too far up in the continuum, however, we expect that unstable, yet long-lived, bound states will leave its signature in the spectral function.

We will study this $s\bar{s}$-$s\bar{s}$ continuum through a computation of the spectral function for small dimerization δ such that the underlying soliton physics can be observed. We will look at the momentum-frequency resolved spin-spin dynamical correlation function as observed in inelastic neutron scattering, defined as

$$S(q, \omega) = \int \mathrm{d}t\, \mathrm{e}^{i\omega t} \sum_n \mathrm{e}^{iqn} \langle S_n^-(t) S_0^+(0) \rangle$$

where $S_n^{+/-}(t)$ are the ladder operators at site n in the Heisenberg picture and the expectation value $\langle \dots \rangle$ is with respect to the ground state.

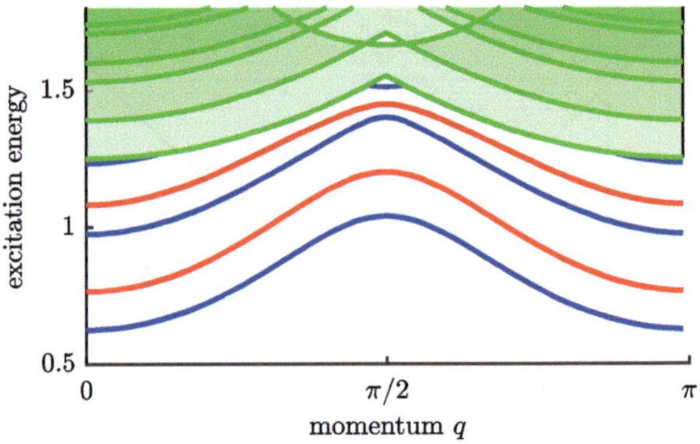

Fig. 4.27 The elementary excitation spectrum for $J_1 = 1$, $J_2 = 1/2$ and $\delta = 0.08$. Because the ground state has a two-site unit cell, the Brillouin zone is confined to momenta $q \in [0, \pi)$ (in units of inverse lattice spacing a^{-1}). One can observe two stable triplet (*blue*) and singlet (*red*) $s\bar{s}$ bound states over the full momentum range, while a third triplet is stable only around the minima and maxima. Above these one-particle lines, there are different two-particle scattering continua (*green*). These continua are obtained by simply adding the momenta and energies of the one-particle dispersions. A third triplet is stable only in limited regions

In Fig. 4.27 the elementary excitation spectrum is plotted for $J_2 = 1/2$ and $\delta = 0.08$, a value of the dimerization for which two triplet and singlet $s\bar{s}$ bound states are stable over the full Brillouin zone. One third triplet excitation is stable close the dispersion's minimum, whereas it merges into the continuum and loses stability for larger momenta. At the maximum of the dispersion, the triplet emerges again from the two-particle continuum. This bound state was also observed in systems with larger dimerization [35], and is not straightforwardly interpreted through the underlying soliton physics (see Fig. 4.31). In fact, in Ref. [35] the same bound state was observed with the use of a perturbative continuous unitary transformation, starting from the isolated dimer limit ($\delta \to \infty$). In Fig. 4.31 we have plotted excitation spectra for different values of δ, showing that the bound state does not vanish if we tune up δ – in contrast to the bound state around momentum 0. Therefore, it seems plausible that the physical origin of this bound state is not connected to the underlying soliton physics (which is only valid for small δ).

Let us first look at the stability of the third triplet bound state at the minimum of the dispersion as a function of δ. In Fig. 4.28 we have plotted momentum slices of $S(q, \omega)$ at $q = \pi$ inside the two-particle continuum. For $\delta = 0.09$ most of the spectral weight is in the δ peak of the bound state. As δ is increased, the bound state comes closer to the continuum at $\omega_{\text{edge}} = 2\Delta(\delta)$, and the two-particle continuum gains spectral weight and becomes sharply peaked. Just as the bound state enters the continuum, the spectral function diverges; by plotting the maximum of the spectral function as a function of δ [inset of Fig. 4.28], the exact value for δ can be pinpointed for which the third triplet $s\bar{s}$ becomes unstable. For larger δ, the maximum of the spectral function travels through the continuum as a signature of an unstable yet long-lived $s\bar{s}$ state (Fig. 4.29).

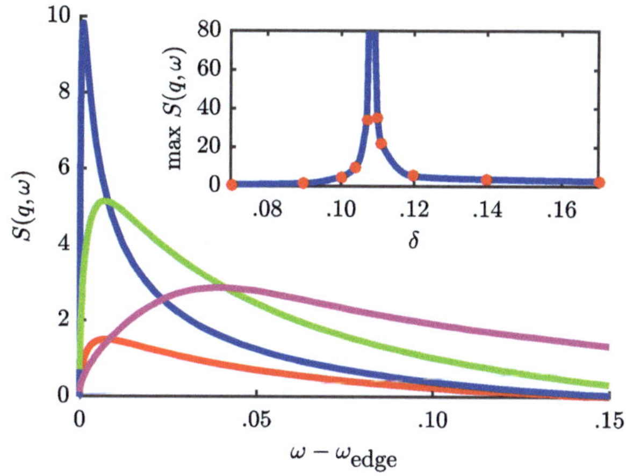

Fig. 4.28 The spectral function $S(q, \omega)$ at momentum $q = \pi$ just above the first triplet-triplet threshold for $J_1 = 1$, $J_2 = 1/2$, and different values of the dimerization: $\delta = 0.09$ (*red*), $\delta = 0.104$ (*blue*), $\delta = 0.12$ (*green*) and $\delta = 0.17$ (*magenta*); the energies ω have been shifted by twice the triplet gap $2\Delta(\delta)$. For $\delta = 0.09$ the third triplet bound state (at the minimum of the dispersion relation) falls below the continuum, but, as the gap becomes smaller, the peak above the threshold becomes larger. As the bound state enters the continuum, the spectral function diverges. For larger δ the bound state decays into two $s\bar{s}$ states and becomes a resonance in the spectral function. This resonance travels to higher energies as δ is further increased. *Inset* The maximum of the spectral function (above ω_{edge}) as a function of δ showing a divergence at $\delta \approx 0.11$ exactly where the third bound state becomes unstable. The *blue line* is a guide to the eye, the largest data points fall outside the plotting region

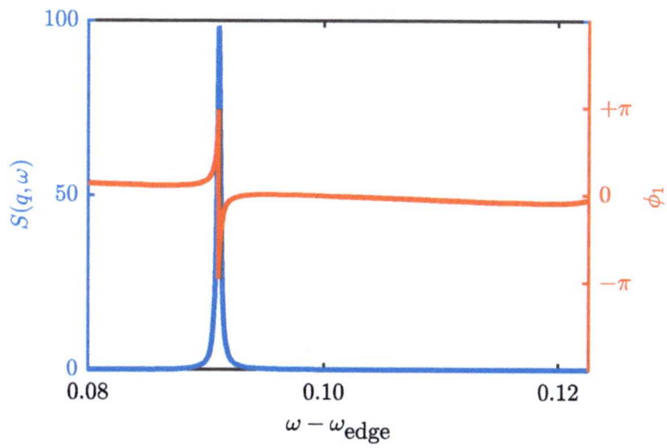

Fig. 4.29 The spectral function $S(q, \omega)$ (*blue*) and the scattering phase in the $S = 1$ sector $\phi_1(q, \omega)$ (*red*) as a function of the rescaled energy $\omega - \omega_{\text{edge}}$ and for fixed momentum $q = 0.56\pi$, where ω_{edge} is the lower edge of the two-particle continuum. We can observe that the scattering phase rotates rapidly at the point where the resonance is located, signalling a strong interaction between the particles at this point. Off the resonance, the dispersion of the scattering phase behaves smoothly. At $\omega - \omega_{\text{edge}} \approx 0.123$ one enters the folding region (with a diverging density of states); beyond that point the S matrix is a 18×18 matrix, which can no longer be written in terms of three phases

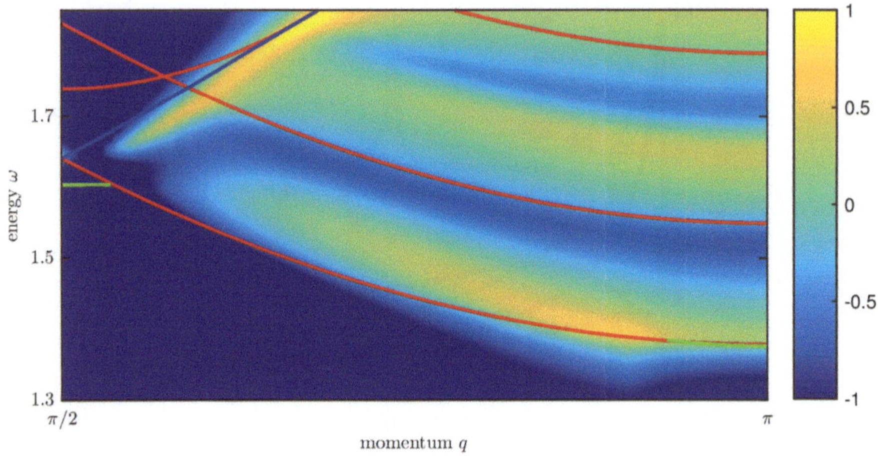

Fig. 4.30 The logarithm of the two-particle spectral function $\log_{10} S(q, \omega)$ in the second half of the Brillouin zone for $J_1 = 1$, $J_2 = 1/2$ and $\delta = 0.1$. Only the two-particle contributions are plotted, the sharp δ-functions of the one-particle states are not shown. A small imaginary frequency $\epsilon = 0.01$ is added for aesthetic reasons (note that we have perfect resolution in q and ω as our methods work in the thermodynamic limit, see the other plots) and to mimic experimental resolution. We have also plotted the edges of the different two-particle continua in *red*. The lower edge of the first band consists of states with momenta $\kappa_1 = \pi/2 + q/2$ and $\kappa_2 = -\pi/2 + q/2$. The *blue line* indicates the states with equal individual momenta $\kappa_{1,2} = q/2$, leading to a diverging density of states on that line. The *green line* follows the dispersion of the third $s\bar{s}$ bound state where it is stable. We can see that the dispersion of the higher two-particle continua gives a banded structure to the spectral function

Figure 4.27 illustrated that the bound state loses stability for larger momenta. In Fig. 4.30 the two-particle spectral function is plotted in the second half of the Brillouin zone (which carries most of its weight) for $J_2/J_1 = 1/2$ and $\delta = 0.1$, for which the third bound state is stable at the minimum of the two-particle continuum. We can see now that, as the momentum moves away from the minimum, the bound state enters the continuum as a sharp resonance and survives throughout a large portion of the Brillouin zone.

Further up in frequency, the two-particle spectral function shows a non-trivial structure, because of the different overlapping two-particle continua. In the absence of special symmetries, there are non-zero scattering amplitudes linking the different two-particle sectors. As a result, the eigenstates will mix up these sectors and the spectral function obtains its characteristic banded structure.

Another strong resonance appears when the bound state at the maximum of the dispersion relation enters the continuum. Surprisingly, the bound state does not connect immediately to the resonance at lower momenta, as another resonance seems to run away with all the spectral weight. This other resonance can be explained as a combined effect from (i) the attractive interaction between the triplet bound states, and (ii) a divergence of the density of states within the continuum. The latter is a consequence of the folding inside the two-triplet continuum: there are regions inside the scattering continuum for which there are two different combinations of one-triplet states that give rise to the same total momentum and energy. The boundary lines of this folding region exhibit, just like the edges of the two-particle continuum, a square root divergence in the density of states. Figure 4.30 shows, however, that the resonance does not coincide with this line. The reason is that the particles have an attractive interaction and, consequently, a negative binding energy which brings the resonance down in energy. One can see that this binding energy goes to zero as it travels towards the upper edge of the continuum, until it merges with a strong signal at the upper edge.

In order to corroborate this picture, the particle interactions can be further characterized by studying the two-triplet S matrix. In the case of triplet-triplet scattering, this S matrix is a 9×9 unitary matrix, which can be diagonalized by going to the total-spin basis. Moreover, because of $SU(2)$ invariance, the S matrix will be constant within every subspace of total spin, so that the information in the S matrix reduces to three scattering phases for every value of the total spin. As only the eigenstates with total spin $S = 1$ contribute to the spectral function, the scattering phase in this sector is plotted in Fig. 4.29, showing a drastic rotation of the scattering phase right at the point where the spectral function has its resonance. This confirms that the resonance is indeed a consequence of a strong attractive triplet-triplet interaction.

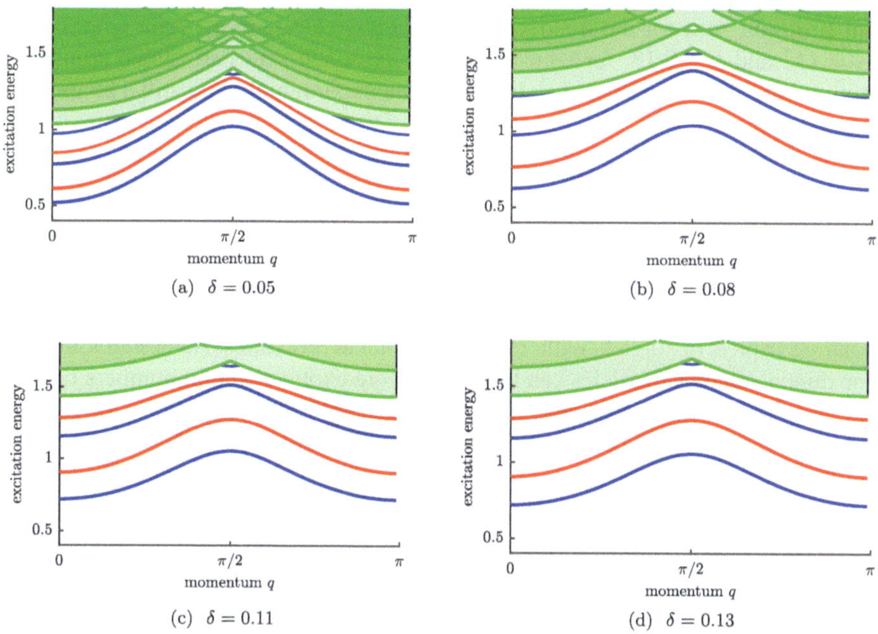

Fig. 4.31 The excitation spectrum for $J_1 = 1$, $J_2 = 1/2$ and four different values of δ. The triplet lines are in *blue*, the singlet lines are *red*

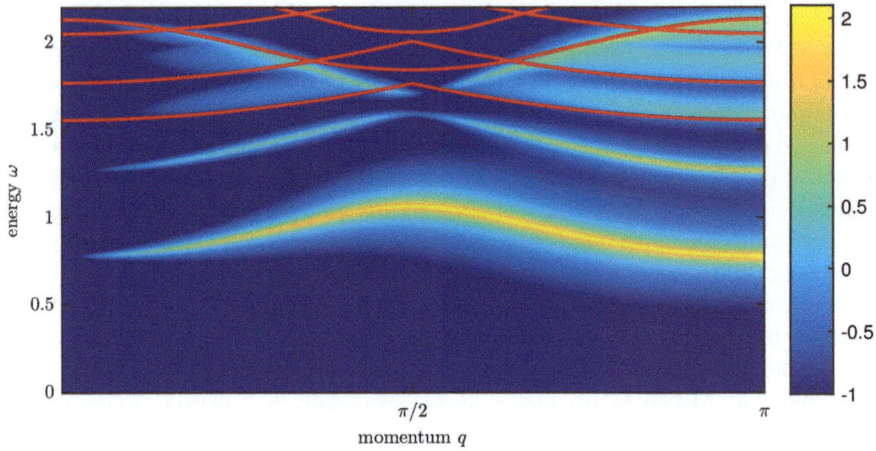

Fig. 4.32 The logarithm of the spectral function $\log_{10} S(q, \omega)$ in the full Brillouin zone for $J_1 = 1$, $J_2 = 1/2$ and $\delta = 0.13$. A small imaginary frequency $\epsilon = 0.01$ is added, such that the one-particle δ peaks can be plotted as Lorentzians. We have also plotted the edges of the different two-particle continua in *red*

For completeness' sake, in Fig. 4.32 we have plotted the spectral function in the full Brillouin zone for the frustrated and dimerized Heisenberg chain with $J_1 = 1$, $J_2 = 1/2$ and $\delta = 0.13$, for which only two $s\bar{s}$ bound states are stable around the minimum of the two-particle band. Around the maximum there is still a bound state (see Fig. 4.31).

4.4 The Gapped XXZ Chain

All the previous applications were focused on systems where the elementary excitations are triplets or singlets without any topological properties. As is well known, in one dimension quasiparticles can have a lot more exotic quantum numbers because of e.g. spin fractionalization and spin-charge separation. We will study the spin-1/2 XXZ chain, which can be solved exactly and provides a non-trivial benchmark for our methods.

The model is defined by the Hamiltonian

$$H = \sum_n \left(S_n^x S_{n+1}^x + S_n^y S_{n+1}^y + \Delta S_n^z S_{n+1}^z \right),$$

and is gapped for $\Delta > 1$. The system is integrable and the spectrum can be computed exactly with the Bethe ansatz [101, 102]. In this regime there is a two-fold degenerate ground state with a finite Néel order parameter, and the elementary excitations are spinons [103]. For a long time, the elementary excitations were believed to be spin-1 magnons [104] (for the isotropic case); only after a careful analysis, it was realized that the spin-1 excited states were in fact scattering states with spin-1/2 [103, 105]. Targeting the individual spinon states can only be done by changing the boundary conditions – in this case, going to an odd number of sites. Using Korepin's method [106], the spinon-spinon S matrix was computed for the isotropic case [103, 105] and for the gapless regime $0 < \Delta < 1$ [107].

Within our MPS framework we can easily capture one-spinon states in the thermodynamic limit by applying the topological excitation ansatz and, subsequently, compute the two-spinon S matrix. In its present form, however, the XXZ chain does not have a translation-invariant ground state. Therefore, we apply the following unitary transformation U to the model

$$U = \prod_{m \text{odd}} e^{i\pi\sigma^x},$$

a spin flip on every second site. This maps the original Hamiltonian to

$$H' = \sum_n \left(S_n^x S_{n+1}^x - S_n^y S_{n+1}^y - \Delta S_n^z S_{n+1}^z \right),$$

which now does have a translation-invariant ground state. The Brillouin zone – which ranged from 0 to π in the original model because of the broken translation invariance – is now again doubled to cover the full $[0, 2\pi)$ range. Whereas the original model contained a doublet of spinons in a reduced range, the transformed model has a single kink state in the double range. When computing observables, all operators have to be first transformed to this basis in order to obtain the physically correct results.

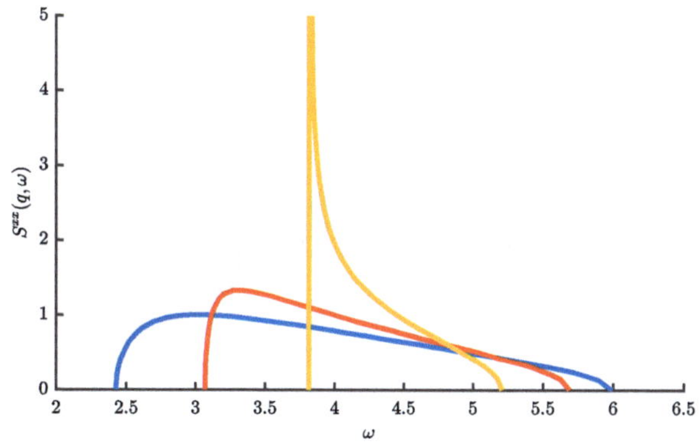

Fig. 4.33 The spectral function $S^{xx}(q, \omega)$ of the XXZ chain with $\Delta = 4$; momentum cuts at $q = \pi/8$ (*blue*), $q = 2\pi/8$ (*red*) and $q = 3\pi/8$ (*orange*). The *orange curve* shows a divergence at the band's edge. Note that the plotting range cuts off the *orange curve*; because our method works directly in the thermodynamic limit, we can reproduce this divergence to arbitrary precision. A comparison with Fig. 5 in Ref. [109] shows a very strong similarity

The first contributions to the spectral function

$$S^{xx}(q,\omega) = \int \mathrm{d}t \sum_n \mathrm{e}^{i\omega t} \mathrm{e}^{iqn} \langle S_n^x(t) S_0^x(0) \rangle$$

come from two-spinon states, which can be computed exactly [108, 109] using Bethe-ansatz techniques. In Fig. 4.33 we have plotted three momentum cuts for $\Delta = 4$, which can be compared to the plots in Ref. [109]. The similarity seems very good, which shows the accuracy of our method (Fig. 4.34).

4.5 The Anisotropic Heisenberg Spin-1 Chain

In the previous sections we have looked at $SU(2)$ invariant spin chains that undergo a quantum phase transition between a gapped spin-liquid phase and gapless Luttinger liquid phase at a critical value of the applied magnetic field. Because of $SU(2)$ invariance the excitations are described as a $S = 1$ triplet of magnons that undergo a Zeeman shift due to the magnetic field; one of the components condenses into the ground state at a critical field equal to the magnon gap.

In many $S = 1$ compounds such as NENP [110, 111], NDMAP [112–114] and NDMAZ [115] there are significant crystal-field anisotropies, which split up the magnon triplet in zero field, leading to a richer phase diagram. The phase transition due to a magnetic field with crystal-field anisotropies is, instead of the C/IC transition for the Heisenberg chain, an Ising transition [116], and can be modelled with a bosonic Landau–Ginzburg theory [71, 117] or a theory of coupled Majorana fermions [70].

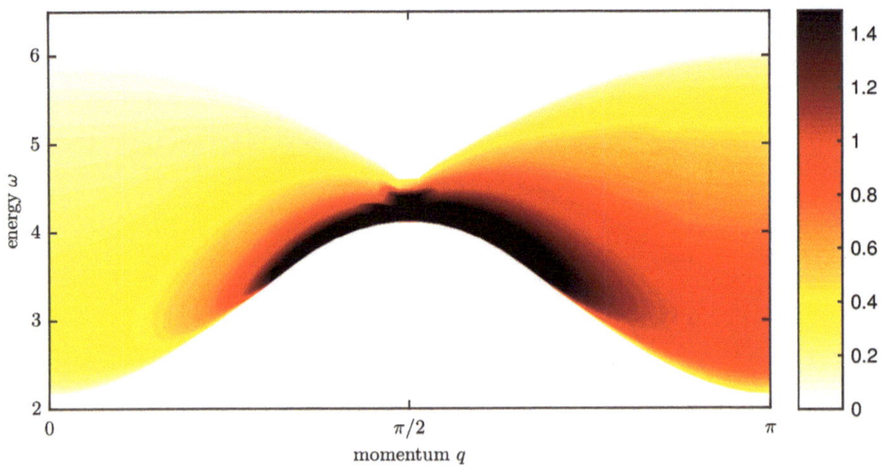

Fig. 4.34 The two-spinon part of the spectral function $S^{xx}(q,\omega)$ of the XXZ chain with $\Delta = 4$; momentum cuts at $q = \pi/8$ (*blue*), $q = 2\pi/8$ (*red*) and $q = 3\pi/8$ (*orange*). A comparison with Ref. [109] shows a very strong similarity

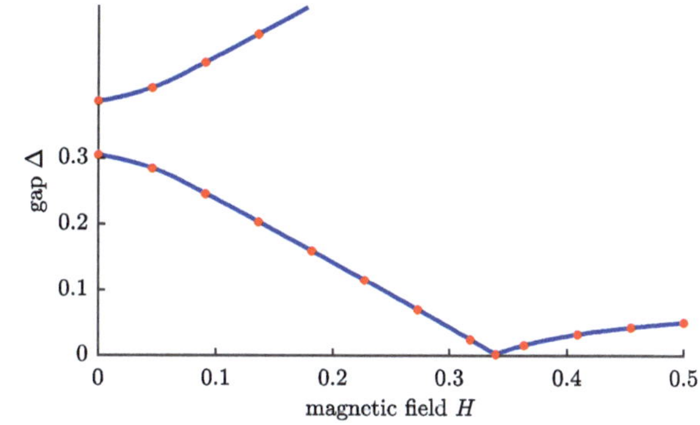

Fig. 4.35 The gap of the anisotropic spin-1 Heisenberg chain with $D = 0.1$ and $E = 0.02$ as a function of magnetic field

The Hamiltonian of these compounds in a magnetic field is

$$H = J\sum_n \vec{S}_n \cdot \vec{S}_{n+1} - h\sum_n S_n^z + D\sum_n \left(S_n^z\right)^2 + E\sum_n \left(\left(S_n^x\right)^2 - \left(S_n^y\right)^2\right).$$

We will look at systems where $0 < E < D$, for which the excitation spectrum consists of three modes with different gaps.[7] The magnetic field does not commute with the zero-field Hamiltonian, such that the field is no longer a simple Zeeman shift of the magnon triplet. In Fig. 4.35 the field dependence of the gap is plotted. If $E \ll D$ the third mode is a lot higher in energy, so that only the two low-lying excitations are taken into account.

We can now try to write down an effective-field theory using the basis of free states with non-trivial S matrix through the use of the Faddeev–Zamolodchikov (FZ) operators [see Sect. 3.9]. The two relevant excitation modes have dispersions $\omega_\alpha(\kappa)$ and $\omega_\beta(\kappa)$ and are created out of the ground state by $Z_\alpha^\dagger(\kappa)$ and $Z_\beta^\dagger(\kappa)$. The low-energy part of the zero-field Hamiltonian can then be written as

$$H_0 = \int_{-\pi}^{\pi} \frac{\mathrm{d}\kappa}{2\pi} \left(\omega_\alpha(\kappa) Z_\alpha^\dagger(\kappa) Z_\alpha(\kappa) + \omega_\beta(\kappa) Z_\beta^\dagger(\kappa) Z_\beta(\kappa)\right).$$

Writing down this "effective Hamiltonian" is only correct if the approximate Bethe ansatz [Sect. 3.7] holds. We now apply a magnetic field H, a perturbation that can projected onto the basis of one- and two-particle states. In this approximation the Hamiltonian remains quadratic in the FZ operators

$$\begin{aligned} H = &\int_{-\pi}^{\pi} \frac{\mathrm{d}\kappa}{2\pi} \left(\omega_\alpha(\kappa) Z_\alpha^\dagger(\kappa) Z_\alpha(\kappa) + \omega_\beta(\kappa) Z_\beta^\dagger(\kappa) Z_\beta(\kappa)\right) \\ &+ h\int_{-\pi}^{\pi} \frac{\mathrm{d}\kappa}{2\pi} \left(M_p^{\alpha\beta} Z_\alpha^\dagger(\kappa) Z_\beta(\kappa) + M_n^{\alpha\beta}(\kappa) Z_\alpha(-\kappa) Z_\beta(\kappa) + M_n^{\alpha\beta *}(\kappa) Z_\beta^\dagger(\kappa) Z_\alpha^\dagger(-\kappa)\right) \end{aligned}$$

---

[7] For $E = 0$ the two lowest-lying modes are still degenerate and the magnetic field commutes with the Hamiltonian; the analysis of the previous section still applies in this case.

where

$$M_p^{\alpha\beta}(\kappa) = \langle \Phi_\alpha(\kappa)|M|\Phi_\beta(\kappa)\rangle$$
$$M_n^{\alpha\beta}(\kappa) = \langle \Psi(A)|M|\Upsilon_{\beta\alpha}(\kappa, -\kappa)\rangle$$

are the particle preserving and non-preserving parts of the perturbation,[8] and we have defined

$$M = \frac{1}{|\mathcal{L}|}\sum_n S_n^z.$$

For general S matrices, a canonical Bogoliubov rotation of the FZ operators proves not to be possible and a more sophisticated strategy will have to be developed for diagonalizing this effective Hamiltonian.

By approximating the FZ operators as fermionic, however, solving this Hamiltonian amounts to performing a Bogoliubov transformation. Indeed, if only the low-lying modes are occupied, it makes sense to approximate the S matrix by a free-fermionic one, so the effective theory is written as (at zero field)

$$H_0 = \int_{-\pi}^{\pi} \frac{\mathrm{d}\kappa}{2\pi}\left(\omega_\alpha(\kappa)\alpha_\kappa^\dagger\alpha_\kappa + \omega_\beta(\kappa)\beta_\kappa^\dagger\beta_\kappa\right)$$

with fermionic creation and annihilation operators

$$[\alpha_\kappa, \alpha_{\kappa'}^\dagger]_+ = [\beta_\kappa, \beta_{\kappa'}^\dagger]_+ = 2\pi\delta(\kappa-\kappa')$$
$$[\alpha_\kappa, \alpha_{\kappa'}]_+ = [\alpha_\kappa, \beta_{\kappa'}]_+ = [\alpha_\kappa, \beta_{\kappa'}^\dagger]_+ = [\alpha_\kappa^\dagger, \beta_{\kappa'}]_+ = [\beta_\kappa, \beta_{\kappa'}]_+ = 0.$$

The Hamiltonian is rewritten as

$$H_0 = \int_0^{+\infty} \frac{\mathrm{d}\kappa}{2\pi}\gamma_a^\dagger(\kappa)H_{ab}^0(\kappa)\gamma_b(\kappa) + \text{cst.}$$

with

$$\gamma_a(\kappa) = \begin{pmatrix}\alpha_\kappa \\ \alpha_{-\kappa}^\dagger \\ \beta_\kappa \\ \beta_{-\kappa}^\dagger\end{pmatrix} \qquad H_{ab}^0(\kappa) = \begin{pmatrix}\omega_\alpha(\kappa) & & & \\ & -\omega_\alpha(\kappa) & & \\ & & \omega_\beta(\kappa) & \\ & & & -\omega_\beta(\kappa)\end{pmatrix}$$

and we have taken the momentum integration to infinity (as only low-lying momenta will be occupied this is ok). The perturbation can be expressed in the basis of one- and two-particle states, so the Hamiltonian remains quadratic in the fermionic operators α_κ and β_κ

[8]The factor $|\mathcal{L}|$ represents the number of sites, which is diverging in the thermodynamic limit.

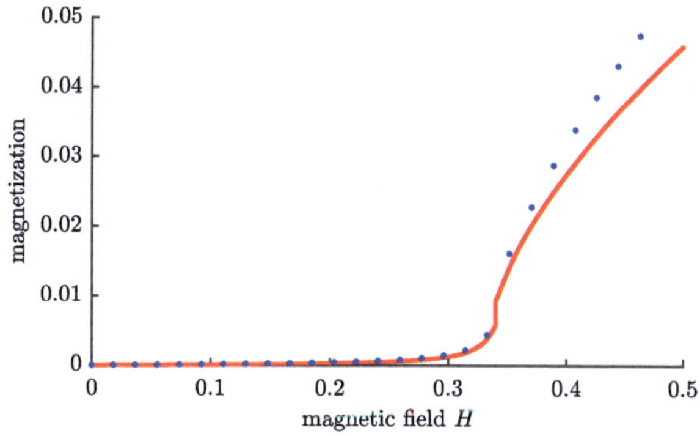

Fig. 4.36 The magnetization as a function of the magnetic field for $D = 0.1$ and $E = 0.02$. The *blue dots* are computed with a direct MPS calculation, the *red line* is using the approximate Bethe ansatz

$$H = \int_0^{+\infty} \frac{d\kappa}{2\pi} \gamma_a^\dagger(\kappa) H_{ab}(\kappa) \gamma_b(\kappa) + \text{cst.} \qquad \rightarrow \qquad H_{ab}(\kappa) = H_{ab}^0(\kappa) + S_{ab}(\kappa)$$

The matrix $H_{ab}(\kappa)$ is diagonalized with a Bogoliubov rotation $\tilde{\gamma}(\kappa) = U(\kappa)\gamma(\kappa)$ so that $\tilde{H}_{ab}(\kappa) = U(\kappa)H_{ab}(\kappa)U^\dagger(\kappa)$ diagonal. The unitarity of $U(\kappa)$ ensures that the rotated operators $\tilde{\alpha}_\kappa$ and $\tilde{\beta}_\kappa$ satisfy the same fermionic commutation relations as the original ones. The ground state $|0\rangle$ is annihilated by the rotated operators, $\tilde{\alpha}_\kappa|0\rangle = \tilde{\beta}_\kappa|0\rangle = 0$. The magnetization of the ground state under a magnetic field is then computed as

$$\begin{aligned} m = \langle 0|M|0\rangle &= \frac{1}{|\mathcal{L}|} \int_0^{+\infty} \frac{d\kappa}{2\pi} \langle 0| \left(\tilde{S}_{22}\tilde{\alpha}_{-\kappa}\tilde{\alpha}_{-\kappa}^\dagger + \tilde{S}_{44}\tilde{\beta}_{-\kappa}\tilde{\beta}_{-\kappa}^\dagger \right) |0\rangle \\ &= \int_0^{+\infty} \frac{d\kappa}{2\pi} \left(\tilde{S}_{22} + \tilde{S}_{44} \right) \end{aligned}$$

with $\tilde{S}_{ab}(\kappa) = U(\kappa)S_{ab}(\kappa)U^\dagger(\kappa)$ and $|\mathcal{L}|$ the (diverging) number of sites.

In Fig. 4.36 the result is plotted for a specific choice of parameters. Clearly the approximate Bethe ansatz captures the qualitative behaviour of the Ising transition, and follows the direct MPS calculations quantitatively to some extent. Note that these results are preliminary; e.g. only a few momentum points were taken in the evaluation of the integral in the above expression for the magnetization. Better results should be possible in a more thorough simulation.

4.6 Future Applications

In the last two sections of Chap. 3 we have already listed a few extensions of our framework that can lead to interesting new developments. In this section, we rather highlight some of the more straightforward applications of the existing framework.

Dynamical Scattering Theory

In Sect. 3.6.4 we have shown explicitly how the collision of wave packets can be simulated using the stationary two-particle scattering states and that it is the derivative of the scattering phase that determines the displacement of the wave packets after the collision process. Similar conclusions have been reached for the scattering of magnons on a ferromagnetic background in Refs. [118, 119].

An interesting confirmation for our methods would be to prepare an initial state of two wave packets and time-evolve it by direct MPS methods for real-time evolution, and afterwards comparing with the results from stationary scattering theory. Interestingly, this setup would involve a scattering process of effective particles against a strongly-correlated background state, and involves non-trivial physics. In particular, we can investigate the effect of integrability and its breaking on the scattering processes.

Spectral Functions at Finite Temperature

The form-factor approach for evaluating spectral functions at temperatures below the excitation gap ($e^{-\beta\Delta} \ll 1$) [see Sect. 3.8.4] has been applied to the integrable Ising model [120] and the non-linear sigma model [121], as well as non-integrable ladders [122] and alternating spin chains [123]. In these works, it has been shown that the thermal broadening of the one-particle δ peak acquires an asymmetry, an effect that is contradicted by the semi-classical approximation of Ref. [124].

It would be interesting to investigate this problem for models that are non-integrable and not treated in perturbation theory from a dispersionless limit. For example, the thermal broadening and shift of the Haldane gap in the spin-1 Heisenberg chain can be computed and compared to the experimental results in Ref. [125].

The Two-Particle S Matrix and Integrability Breaking

Both of the previous physical settings are intimately connected with the special properties of integrability and the ways in which it can be broken. In particular, both the absence of thermalization and the ballistic transport at finite temperatures are explained by the local conserved quantities in integrable systems, which are, in turn, intimately connected with the factorizability of the S matrix. Therefore, it seems that we should be able to trace back the effects of integrability breaking to the factorizability of the S matrix.

As a first measure we can investigate the violations of the Yang–Baxter equation: This is a purely a property of the two-particle S matrix, which we can straightforwardly compute, but provides a consistency condition for the factorizability of three-particle scattering processes. The XXZ model in a transverse magnetic field would be the perfect test model, because it allows for a smooth application of an integrability-breaking term without closing the excitation gap.

Long-Range Interactions

It is well-known that Lieb–Robinson bounds break down when the interactions in the Hamiltonian are no longer local, as e.g. in the long-range Heisenberg model

$$H = \sum_{ij} \frac{\vec{S}_i \cdot \vec{S}_j}{|i - j|^\alpha}.$$

For dipolar interactions, a power of $\alpha = 3$ is observed [126], whereas other powers can be obtained in e.g. ion traps [8, 17]. The spread of correlations through these systems is no longer confined by the Lieb–Robinson light cone [9], so we expect that the quasiparticle velocities diverge in these systems. This effect should be reflected as a cusp in the quasiparticle dispersion relations [126].

Although we have no theoretical foundations for the applicability of the excitation ansatz for long-ranged interactions, we still expect that the particle picture should hold for this case as well. On the side of the implementation, the long-ranged Hamiltonian confronts us with yet another infinite sum, but it should be possible to sum all terms efficiently [127]. An investigation of the quasiparticle properties can shed new light on the physics of long-ranged quantum spin models.

References

1. P. Calabrese, J. Cardy, Evolution of entanglement entropy in one-dimensional systems. J. Stat. Mech. Theory Exp. **2005**, P04010 (2005). doi:10.1088/1742-5468/2005/04/P04010
2. P. Calabrese, J. Cardy, Quantum quenches in extended systems. J. Stat. Mech. Theory Exp. **2007**, P06008 (2007). doi:10.1088/1742-5468/2007/06/P06008
3. T. Kinoshita, T. Wenger, D.S. Weiss, A quantum Newton's cradle. Nature **440**, 900 (2006), http://www.nature.com/nature/journal/v440/n7086/suppinfo/nature04693_S1.html
4. M. Rigol, V. Dunjko, M. Olshanii, Thermalization and its mechanism for generic isolated quantum systems. Nature **452**, 854 (2008). doi:10.1038/nature06838
5. T. Giamarchi, C. Rüegg, O. Tchernyshyov, Bose–Einstein condensation in magnetic insulators. Nat. Phys. **4**, 198 (2008). doi:10.1038/nphys893
6. M. Klanjsek, H. Mayaffre, C. Berthier, M. Horvatić, B. Chiari, O. Piovesana, P. Bouillot, C. Kollath, E. Orignac, R. Citro, T. Giamarchi, Controlling Luttinger liquid physics in spin ladders under a magnetic field. Phys. Rev. Lett. **101**, 137207 (2008). doi:10.1103/PhysRevLett.101.137207
7. C. Rüegg, K. Kiefer, B. Thielemann, D. McMorrow, V. Zapf, B. Normand, M. Zvonarev, P. Bouillot, C. Kollath, T. Giamarchi, S. Capponi, D. Poilblanc, D. Biner, K. Krämer, Thermodynamics of the spin luttinger liquid in a model ladder material. Phys. Rev. Lett. **101**, 247202 (2008). doi:10.1103/PhysRevLett.101.247202
8. P. Jurcevic, B.P. Lanyon, P. Hauke, C. Hempel, P. Zoller, R. Blatt, C.F. Roos, Quasiparticle engineering and entanglement propagation in a quantum many-body system. Nature **511**, 202 (2014). doi:10.1038/nature13461
9. P. Hauke, L. Tagliacozzo, Spread of correlations in long-range interacting quantum systems. Phys. Rev. Lett. **111**, 207202 (2013). doi:10.1103/PhysRevLett.111.207202
10. M.B. Stone, Y. Chen, D.H. Reich, C. Broholm, G. Xu, J.R.D. Copley, J.C. Cook, Magnons and continua in a magnetized and dimerized spin-1/2 chain. Phys. Rev. B **90**, 094419 (2014). doi:10.1103/PhysRevB.90.094419
11. K.W. Plumb, K. Hwang, Y. Qiu, L.W. Harriger, G.E. Granroth, A.I. Kolesnikov, G.J. Shu, F.C. Chou, C. Rüegg, Y.B. Kim, Y.-J. Kim, Quasiparticle-continuum level repulsion in a quantum magnet. Nat. Phys. **12**, 224 (2015). doi:10.1038/nphys3566
12. D.A. Tennant, B. Lake, A.J.A. James, F.H.L. Essler, S. Notbohm, H.-J. Mikeska, J. Fielden, P. Kögerler, P.C. Canfield, M.T.F. Telling, Anomalous dynamical line shapes in a quantum

magnet at finite temperature. Phys. Rev. B **85**, 014402 (2012). doi:10.1103/PhysRevB.85.014402
13. D. Schmidiger, P. Bouillot, S. Muhlbauer, S. Gvasaliya, C. Kollath, T. Giamarchi, A. Zheludev, Spectral and thermodynamic properties of a strong-leg quantum spin ladder. Phys. Rev. Lett. **108**, 167201 (2011). doi:10.1103/PhysRevLett.108.167201
14. M. Cheneau, P. Barmettler, D. Poletti, M. Endres, P. Schauss, T. Fukuhara, C. Gross, I. Bloch, C. Kollath, S. Kuhr, Light-cone-like spreading of correlations in a quantum many-body system. Nature **481**, 484 (2012). doi:10.1038/nature10748
15. T. Fukuhara, A. Kantian, M. Endres, M. Cheneau, P. Schauß, S. Hild, D. Bellem, U. Schollwöck, T. Giamarchi, C. Gross, I. Bloch, S. Kuhr, Quantum dynamics of a mobile spin impurity. Nat. Phys. **9**, 235 (2013). doi:10.1038/nphys2561
16. T. Fukuhara, P. Schauß, M. Endres, S. Hild, M. Cheneau, I. Bloch, C. Gross, Microscopic observation of magnon bound states and their dynamics. Nature **502**, 76 (2013). doi:10.1038/nature12541
17. P. Jurcevic, P. Hauke, C. Maier, C. Hempel, B.P. Lanyon, R. Blatt, C.F. Roos, Spectroscopy of interacting quasiparticles in trapped ions. Phys. Rev. Lett. **115**, 100501 (2015). doi:10.1103/PhysRevLett.115.100501
18. C.A. Hayward, D. Poilblanc, L.P. Lévy, Magnetization of $Cu_2(C_5H_{12}N_2)_2Cl_4$: a Heisenberg spin-ladder system. Phys. Rev. B **54**, R12649 (1996). doi:10.1103/PhysRevB.54.R12649
19. D. Poilblanc, J. Riera, C.A. Hayward, C. Berthier, M. Horvatić, Spin dynamics of the spin-Peierls compound $CuGeO_3$ under a magnetic field. Phys. Rev. B **55**, R11941 (1997). doi:10.1103/PhysRevB.55.R11941
20. D. Augier, D. Poilblanc, S. Haas, A. Delia, E. Dagotto, Dynamical properties of the spin-Peierls compound α'-NaV_2O_5. Phys. Rev. B **56**, R5732 (1997). doi:10.1103/PhysRevB.56.R5732
21. N. Laflorencie, D. Poilblanc, Simulations of pure and doped low-dimensional spin-1/2 gapped systems, in *Quantum Magnetism*, eds. by U. Schollwöck, J. Richter, D.J.J. Farnell, R.F. Bishop (Springer, Berlin, 2004), pp. 227–252. doi:10.1007/BFb0119595
22. S. Capponi, D. Poilblanc, Thermodynamic properties of the coupled dimer system $Cu_2(C_5H_{12}N_2)_2Cl_4$. Phys. Rev. B **75**, 092406 (2007). doi:10.1103/PhysRevB.75.092406
23. S.R. White, I. Affleck, Spectral function for the $S = 1$ Heisenberg antiferromagetic chain. Phys. Rev. B **77**, 134437 (2008). doi:10.1103/PhysRevB.77.134437
24. P.E. Dargel, A. Wöllert, A. Honecker, I.P. McCulloch, U. Schollwöck, T. Pruschke, Lanczos algorithm with matrix product states for dynamical correlation functions. Phys. Rev. B **85**, 205119 (2012). doi:10.1103/PhysRevB.85.205119
25. A. Holzner, A. Weichselbaum, I.P. McCulloch, U. Schollwöck, J. von Delft, Chebyshev matrix product state approach for spectral functions. Phys. Rev. B **83**, 195115 (2011). doi:10.1103/PhysRevB.83.195115
26. M.B. Hastings, Solving gapped Hamiltonians locally. Phys. Rev. B **73**, 085115 (2006). doi:10.1103/PhysRevB.73.085115
27. L. Masanes, Area law for the entropy of low-energy states. Phys. Rev. A **80**, 6 (2009). doi:10.1103/PhysRevA.80.052104
28. J. Eisert, M. Cramer, M.B. Plenio, Colloquium: area laws for the entanglement entropy. Rev. Mod. Phys. **82**, 277 (2010). doi:10.1103/RevModPhys.82.277
29. J.-S. Caux, R. Hagemans, J.M. Maillet, Computation of dynamical correlation functions of Heisenberg chains: the gapless anisotropic regime. J. Stat. Mech. Theory Exp. **2005**, P09003 (2005). doi:10.1088/1742-5468/2005/09/P09003
30. J.-S. Caux, J.M. Maillet, Computation of dynamical correlation functions of Heisenberg chains in a magnetic field. Phys. Rev. Lett. **95**, 077201 (2005). doi:10.1103/PhysRevLett.95.077201
31. J.-S. Caux, Correlation functions of integrable models: a description of the ABACUS algorithm. J. Math. Phys. **50**, 095214 (2009). doi:10.1063/1.3216474
32. S. Trebst, H. Monien, C. Hamer, Z. Weihong, R. Singh, Strong-coupling expansions for multiparticle excitations: continuum and bound states. Phys. Rev. Lett. **85**, 4373 (2000). doi:10.1103/PhysRevLett.85.4373

33. K.P. Schmidt, G.S. Uhrig, Excitations in one-dimensional S = 1/2 quantum antiferromagnets. Phys. Rev. Lett. **90**, 227204 (2003). doi:10.1103/PhysRevLett.90.227204
34. C. Knetter, K.P. Schmidt, M. Grüninger, G.S. Uhrig, Fractional and integer excitations in quantum antiferromagnetic spin-1/2 ladders. Phys. Rev. Lett. **87**, 167204 (2001). doi:10.1103/PhysRevLett.87.167204
35. K.P. Schmidt, C. Knetter, G.S. Uhrig, Spectral properties of the dimerized and frustrated S = 1/2 chain. Phys. Rev. B **69**, 104417 (2004). doi:10.1103/PhysRevB.69.104417
36. J. Haegeman, B. Pirvu, D.J. Weir, J.I. Cirac, T.J. Osborne, H. Verschelde, F. Verstraete, Variational matrix product ansatz for dispersion relations. Phys. Rev. B **85**, 100408 (2012). doi:10.1103/PhysRevB.85.100408
37. J. Haegeman, T.J. Osborne, F. Verstraete, Post-matrix product state methods: to tangent space and beyond. Phys. Rev. B **88**, 075133 (2013). doi:10.1103/PhysRevB.88.075133
38. V. Zauner, D. Draxler, L. Vanderstraeten, M. Degroote, J. Haegeman, M.M. Rams, V. Stojevic, N. Schuch, F. Verstraete, Transfer matrices and excitations with matrix product states. New J. Phys. **17**, 053002 (2015). doi:10.1088/1367-2630/17/5/053002
39. A. Milsted, J. Haegeman, T.J. Osborne, Matrix product states and variational methods applied to critical quantum field theory. Phys. Rev. D **88**, 085030 (2013). doi:10.1103/PhysRevD.88.085030
40. D. Draxler, J. Haegeman, T.J. Osborne, V. Stojevic, L. Vanderstraeten, F. Verstraete, Particles, holes, and solitons: a matrix product state approach. Phys. Rev. Lett. **111**, 020402 (2013). doi:10.1103/PhysRevLett.111.020402
41. B. Buyens, J. Haegeman, K. Van Acoleyen, H. Verschelde, F. Verstraete, Matrix product states for gauge field theories. Phys. Rev. Lett. **113**, 091601 (2014). doi:10.1103/PhysRevLett.113.091601
42. B. Buyens, K. Van Acoleyen, J. Haegeman, F. Verstraete, Matrix product states for Hamiltonian lattice gauge theories (2014), arXiv:1411.0020
43. B. Buyens, J. Haegeman, H. Verschelde, F. Verstraete, K. Van Acoleyen, Confinement and string breaking for QED $_2$ in the Hamiltonian picture (2015), arXiv:1509.00246
44. F.D.M. Haldane, Nonlinear field theory of large-spin Heisenberg antiferromagnets: semiclassically quantized solitons of the one-dimensional easy-axis Néel state. Phys. Rev. Lett. **50**, 1153 (1983b). doi:10.1103/PhysRevLett.50.1153
45. F.D.M. Haldane, Continuum dynamics of the 1-D Heisenberg antiferromagnet: identification with the O(3) nonlinear sigma model. Phys. Lett. A **93**, 464 (1983). doi:10.1016/0375-9601(83)90631-X
46. E.H. Lieb, T. Schultz, D. Mattis, Two soluble models of an antiferromagnetic chain. Ann. Phys. **16**, 407 (1961). doi:10.1016/0003-4916(61)90115-4
47. I. Affleck, E.H. Lieb, A proof of part of Haldane's conjecture on spin chains. Lett. Math. Phys. **12**, 57 (1986). doi:10.1007/BF00400304
48. I. Affleck, The quantum Hall effects, σ-models at $\Theta = \pi$ and quantum spin chains. Nucl. Phys. B **257**, 397 (1985). doi:10.1016/0550-3213(85)90353-0
49. I. Affleck, Quantum spin chains and the Haldane gap. J. Phys. Condens. Matter **1**, 3047 (1989), http://cat.inist.fr/?aModele=afficheN&cpsidt=6916826
50. A.B. Zamolodchikov, A.B. Zamolodchikov, Factorized S-matrices in two dimensions as the exact solutions of certain relativistic quantum field theory models. Ann. Phys. **120**, 253 (1979). doi:10.1016/0003-4916(79)90391-9
51. J. Parkinson, J. Bonner, Spin chains in a field: crossover from quantum to classical behavior. Phys. Rev. B **32**, 4703 (1985). doi:10.1103/PhysRevB.32.4703
52. M. Nightingale, H. Blöte, Gap of the linear spin-1 Heisenberg antiferromagnet: a Monte Carlo calculation. Phys. Rev. B **33**, 659 (1986). doi:10.1103/PhysRevB.33.659
53. M. Takahashi, Monte Carlo calculation of elementary excitation of spin chains. Phys. Rev. Lett. **62**, 2313 (1989). doi:10.1103/PhysRevLett.62.2313
54. M. Takahashi, Excitation spectra of S = 1 antiferromagnetic chains. Phys. Rev. B **50**, 3045 (1994). doi:10.1103/PhysRevB.50.3045

55. S.R. White, D. Huse, Numerical renormalization-group study of low-lying eigenstates of the antiferromagnetic S = 1 Heisenberg chain. Phys. Rev. B **48**, 3844 (1993). doi:10.1103/PhysRevB.48.3844
56. S.R. White, Density-matrix algorithms for quantum renormalization groups. Phys. Rev. B **48**, 10345 (1993). doi:10.1103/PhysRevB.48.10345
57. S.R. White, Density matrix formulation for quantum renormalization groups. Phys. Rev. Lett. **69**, 2863 (1992). doi:10.1103/PhysRevLett.69.2863
58. E. Sorensen, I. Affleck, Large-scale numerical evidence for Bose condensation in the S = 1 antiferromagnetic chain in a strong field. Phys. Rev. Lett. **71**, 1633 (1993). doi:10.1103/PhysRevLett.71.1633
59. H. Ueda, K. Kusakabe, Determination of boundary scattering, magnon-magnon scattering, and the Haldane gap in Heisenberg spin chains. Phys. Rev. B **84**, 054446 (2011). doi:10.1103/PhysRevB.84.054446
60. E. Sorensen, I. Affleck, Equal-time correlations in Haldane-gap antiferromagnets. Phys. Rev. B **49**, 15771 (1994). doi:10.1103/PhysRevB.49.15771
61. E. Sorensen, I. Affleck, S(k) for Haldane-gap antiferromagnets: large-scale numerical results versus field theory and experiment. Phys. Rev. B **49**, 13235 (1994). doi:10.1103/PhysRevB.49.13235
62. G. Källén, On the definition of the renormalization in quantum electrodynamics. Helv. Phys. Acta **25**, 417 (1952)
63. H. Lehmann, Über Eigenschaften von Ausbreitungsfunktionen und Renormierungskonstanten quantisierter Felder. Il Nuovo Cimento **11**, 342 (1954). doi:10.1007/BF02783624
64. M.E. Peskin, D.V. Schroeder, *An Introduction to Quantum Field Theory* (Westview Press, Colorado, 1995)
65. I. Affleck, R. Weston, Theory of near-zero-wave-vector neutron scattering in Haldane-gap antiferromagnets. Phys. Rev. B **45**, 4667 (1992). doi:10.1103/PhysRevB.45.4667
66. P. Hohenberg, W. Brinkman, Sum rules for the frequency spectrum of linear magnetic chains. Phys. Rev. B **10**, 128 (1974). doi:10.1103/PhysRevB.10.128
67. R.M. Konik, P. Fendley, Haldane-gapped spin chains as Luttinger liquids: correlation functions at finite field. Phys. Rev. B **66**, 144416 (2002). doi:10.1103/PhysRevB.66.144416
68. I. Affleck, Luttinger liquid parameter for the spin-1 Heisenberg chain in a magnetic field. Phys. Rev. B **72**, 132414 (2005). doi:10.1103/PhysRevB.72.132414
69. G. Fáth, Luttinger liquid behavior in spin chains with a magnetic field. Phys. Rev. B **68**, 134445 (2003). doi:10.1103/PhysRevB.68.134445
70. A.M. Tsvelik, Field-theory treatment of the Heisenberg spin-1 chain. Phys. Rev. B **42**, 10499 (1990). doi:10.1103/PhysRevB.42.10499
71. I. Affleck, Bose condensation in quasi-one-dimensional antiferromagnets in strong fields. Phys. Rev. B **43**, 3215 (1991). doi:10.1103/PhysRevB.43.3215
72. J. Lou, S. Qin, T.-K. Ng, Z. Su, I. Affleck, Finite-size spectrum, magnon interactions, and magnetization of S = 1 Heisenberg spin chains. Phys. Rev. B **62**, 3786 (2000). doi:10.1103/PhysRevB.62.3786
73. T. Masuda, A. Zheludev, H. Manaka, L.-P. Regnault, J.-H. Chung, Y. Qiu, Dynamics of composite haldane spin chains in IPA-$CuCl_3$. Phys. Rev. Lett. **96**, 047210 (2006). doi:10.1103/PhysRevLett.96.047210
74. S. Notbohm, P. Ribeiro, B. Lake, D. Tennant, K.P. Schmidt, G.S. Uhrig, C. Hess, R. Klingeler, G. Behr, B. Büchner, M. Reehuis, R. Bewley, C. Frost, P. Manuel, R. Eccleston, One- and Two-Triplon spectra of a cuprate ladder. Phys. Rev. Lett. **98**, 027403 (2007). doi:10.1103/PhysRevLett.98.027403
75. A. Shapiro, C.P. Landee, M.M. Turnbull, J. Jornet, M. Deumal, J.J. Novoa, M.A. Robb, W. Lewis, Synthesis, structure, and magnetic properties of an antiferromagnetic spin-ladder complex: bis(2,3-dimethylpyridinium) tetrabromocuprate. J. Am. Chem. Soc. **129**, 952 (2007). doi:10.1021/ja066330m
76. D. Schmidiger, P. Bouillot, T. Guidi, R. Bewley, C. Kollath, T. Giamarchi, A. Zheludev, Spectrum of a magnetized strong-leg quantum spin ladder. Phys. Rev. Lett. **111**, 107202 (2013a). doi:10.1103/PhysRevLett.111.107202

77. P. Bouillot, C. Kollath, A.M. Läuchli, M. Zvonarev, B. Thielemann, C. Rüegg, E. Orignac, R. Citro, M. Klanjsek, C. Berthier, M. Horvatić, T. Giamarchi, Statics and dynamics of weakly coupled antiferromagnetic spin-1/2 ladders in a magnetic field. Phys. Rev. B **83**, 054407 (2011). doi:10.1103/PhysRevB.83.054407
78. D. Schmidiger, S. Mühlbauer, A. Zheludev, P. Bouillot, T. Giamarchi, C. Kollath, G. Ehlers, A.M. Tsvelik, Symmetric and asymmetric excitations of a strong-leg quantum spin ladder. Phys. Rev. B **88**, 094411 (2013). doi:10.1103/PhysRevB.88.094411
79. M. Jeong, D. Schmidiger, H. Mayaffre, M. Klanjšek, C. Berthier, W. Knafo, G. Ballon, B. Vignolle, S. Krämer, A. Zheludev, M. Horvatić, Dichotomy between attractive and repulsive Tomonaga–Luttinger liquids in spin ladders, (2016), arXiv:1604.05252
80. D. Shelton, A. Nersesyan, A.M. Tsvelik, Antiferromagnetic spin ladders: crossover between spin S = 1/2 and S = 1 chains. Phys. Rev. B **53**, 8521 (1996). doi:10.1103/PhysRevB.53.8521
81. K.P. Schmidt, C. Knetter, G.S. Uhrig, Raman response in antiferromagnetic two-leg S = 1/2 Heisenberg ladders. Europhys. Lett. **56**, 877 (2001). doi:10.1209/epl/i2001-00601-y
82. C. Knetter, K.P. Schmidt, G.S. Uhrig, High order perturbation theory for spectral densities of multi-particle excitations: S = 1/2 two-leg Heisenberg ladder. The Eur. Phys. J. B Condens. Matter **36**, 525 (2003). doi:10.1140/epjb/e2004-00008-2
83. K. Coester, S. Clever, F. Herbst, S. Capponi, K.P. Schmidt, A generalized perspective on non-perturbative linked cluster expansions. Europhys. Lett. **110**, 20006 (2015). doi:10.1209/0295-5075/110/20006
84. K.P. Schmidt, G.S. Uhrig, Spectral properties of magnetic excitations in cuprate two-leg ladder systems. Mod. Phys. Lett. B **19**, 1179 (2005). doi:10.1142/S0217984905009237
85. S. Larochelle, M. Greven, Susceptibilities and spin gaps of weakly coupled spin ladders. Phys. Rev. B **69**, 092408 (2004). doi:10.1103/PhysRevB.69.092408
86. T. Fischer, S. Duffe, G.S. Uhrig, Adapted continuous unitary transformation to treat systems with quasi-particles of finite lifetime. New J. Phys. **12**, 033048 (2010). doi:10.1088/1367-2630/12/3/033048
87. H. Schulz, Critical behavior of commensurate-incommensurate phase transitions in two dimensions. Phys. Rev. B **22**, 5274 (1980). doi:10.1103/PhysRevB.22.5274
88. J. Haegeman, J.I. Cirac, T.J. Osborne, I. Pizorn, H. Verschelde, F. Verstraete, Time-dependent variational principle for quantum lattices. Phys. Rev. Lett. **107**, 070601 (2011). doi:10.1103/PhysRevLett.107.070601
89. T. Hikihara, A. Furusaki, Spin correlations in the two-leg antiferromagnetic ladder in a magnetic field. Phys. Rev. B **63**, 134438 (2001). doi:10.1103/PhysRevB.63.134438
90. D. Augier, D. Poilblanc, Dynamical properties of low-dimensional $CuGeO_3$ and NaV_2O_5 spin-Peierls systems. The Eur. Phys. J. B **1**, 19 (1998). doi:10.1007/s100510050148
91. B. Lake, A.M. Tsvelik, S. Notbohm, D. Alan Tennant, T.G. Perring, M. Reehuis, C. Sekar, G. Krabbes, B. Büchner, Confinement of fractional quantum number particles in a condensed-matter system. Nat. Phys. **6**, 50 (2009). doi:10.1038/nphys1462
92. R. Coldea, D.A. Tennant, E.M. Wheeler, E. Wawrzynska, D. Prabhakaran, M. Telling, K. Habicht, P. Smeibidl, K. Kiefer, Quantum criticality in an Ising chain: experimental evidence for emergent E8 symmetry. Science **327**, 177 (2010). doi:10.1126/science.1180085
93. Z. Wang, J. Wu, S. Xu, W. Yang, C. Wu, A.K. Bera, A.T.M.N. Islam, B. Lake, D. Kamenskyi, P. Gogoi, H. Engelkamp, A. Loidl, J. Deisenhofer, Deconfining spinon excitations of an XXZ quantum antiferromagnet in a transverse magnetic field (2015), arXiv:1512.01753
94. F.D.M. Haldane, Spontaneous dimerization in the S = 1/2 Heisenberg antiferromagnetic chain with competing interactions. Phys. Rev. B **25**, 4925 (1982). doi:10.1103/PhysRevB.25.4925
95. F.D.M. Haldane, Erratum: spontaneous dimerization in the S = 1/2 Heisenberg antiferromagnetic chain with competing interactions. Phys. Rev. B **26**, 5257 (1982). doi:10.1103/PhysRevB.26.5257
96. K. Okamoto, K. Nomura, Fluid-dimer critical point in S = 1/2 antiferromagnetic Heisenberg chain with next nearest neighbor interactions. Phys. Lett. A **169**, 433 (1992). doi:10.1016/0375-9601(92)90823-5

97. B.S. Shastry, B. Sutherland, Excitation spectrum of a dimerized next-neighbor antiferromagnetic chain. Phys. Rev. Lett. **47**, 964 (1981). doi:10.1103/PhysRevLett.47.964
98. E. Sørensen, I. Affleck, D. Augier, D. Poilblanc, Soliton approach to spin-Peierls antiferromagnets: large-scale numerical results. Phys. Rev. B **58**, R14701 (1998). doi:10.1103/PhysRevB.58.R14701
99. I. Affleck, Soliton confinement and the excitation spectrum of spin-Peierls antiferromagnets, in *Dynamical Properties of Unconventional Magnetic Systems*, eds. by A. Skjeltorp, D. Sherrington. NATO ASI Series, vol. 349 (Springer, Netherlands, 1998), pp. 123–131. doi:10.1007/978-94-011-4988-4_6
100. D. Augier, E. Sørensen, J. Riera, D. Poilblanc, Soliton bound states in the Raman spectrum of pure and doped spin-Peierls chains. Phys. Rev. B **60**, 1075 (1999). doi:10.1103/PhysRevB.60.1075
101. H. Bethe, Zur Theorie der Metalle. I. Eigenwerte und Eigenfunktionen der linearen Atomkette. Zeitschrift für Physik **71**, 205 (1931). doi:10.1007/BF01341708
102. R. Orbach, Linear antiferromagnetic chain with anisotropic coupling. Phys. Rev. **112**, 309 (1958). doi:10.1103/PhysRev.112.309
103. L.D. Faddeev, L.A. Takhtajan, What is the spin of a spin wave? Phys. Lett. A **85**, 375 (1981). doi:10.1016/0375-9601(81)90335-2
104. J. des Cloizeaux, J.J. Pearson, Spin-wave spectrum of the antiferromagnetic linear chain. Phys. Rev. **128**, 2131 (1962). doi:10.1103/PhysRev.128.2131
105. L.D. Faddeev, L.A. Takhtadzhyan, Spectrum and scattering of excitations in the one-dimensional isotropic Heisenberg model. J. Math. Sci. (1984), http://www.springerlink.com/index/M41546840870l1G1.pdf
106. V.E. Korepin, Direct calculation of the S matrix in the massive thirring model. Theor. Math. Phys. **41**, 953 (1979). doi:10.1007/BF01028501
107. A. Doikou, R.I. Nepomechie, Discrete symmetries and S-matrix of the XXZ chain. J. Phys. A Math. General **31**, L621 (1998). doi:10.1088/0305-4470/31/37/001
108. A. Bougourzi, M. Karbach, G. Müller, Exact two-spinon dynamic structure factor of the one-dimensional s = 1/2 Heisenberg-Ising antiferromagnet. Phys. Rev. B **57**, 11429 (1998). doi:10.1103/PhysRevB.57.11429
109. J.-S. Caux, J. Mossel, I.P. Castillo, The two-spinon transverse structure factor of the gapped Heisenberg antiferromagnetic chain. J. Stat. Mech. Theory Exp. **2008**, P08006 (2008). doi:10.1088/1742-5468/2008/08/P08006
110. K. Katsumata, H. Hori, T. Takeuchi, M. Date, A. Yamagishi, J.P. Renard, Magnetization process of an S = 1 linear-chain Heisenberg antiferromagnet. Phys. Rev. Lett. **63**, 86 (1989). doi:10.1103/PhysRevLett.63.86
111. Y. Ajiro, T. Goto, H. Kikuchi, T. Sakakibara, T. Inami, High-field magnetization of a quasi-one-dimensional S = 1 antiferromagnet $Ni(C_2H_8N_2)_2NO_2(ClO_4)$: observation of the Haldane gap. Phys. Rev. Lett. **63**, 1424 (1989). doi:10.1103/PhysRevLett.63.1424
112. Y. Chen, Z. Honda, A. Zheludev, C. Broholm, K. Katsumata, S.M. Shapiro, Field-induced three- and two-dimensional freezing in a quantum spin liquid. Phys. Rev. Lett. **86**, 1618 (2001). doi:10.1103/PhysRevLett.86.1618
113. A. Zheludev, Z. Honda, Y. Chen, C.L. Broholm, K. Katsumata, S.M. Shapiro, Quasielastic neutron scattering in the high-field phase of a Haldane antiferromagnet. Phys. Rev. Lett. **88**, 077206 (2002). doi:10.1103/PhysRevLett.88.077206
114. M. Hagiwara, Z. Honda, K. Katsumata, A.K. Kolezhuk, H.-J. Mikeska, Zeeman levels with exotic field dependence in the high field phase of an S = 1 Heisenberg antiferromagnetic chain. Phys. Rev. Lett. **91**, 177601 (2003). doi:10.1103/PhysRevLett.91.177601
115. A. Zheludev, Z. Honda, K. Katsumata, R. Feyerherm, K. Prokes, Field-induced commensurate long-range order in the Haldane-gap system $Ni(C_5H_{14}N_2)_2N_3(ClO_4)$. Europhys. Lett. **55**, 868 (2001). doi:10.1209/epl/i2001-00362-7
116. F.H.L. Essler, I. Affleck, Haldane-gap chains in a magnetic field. J. Stat. Mech. Theory Exp. **2004**, P12006 (2004). doi:10.1088/1742-5468/2004/12/P12006

117. I. Affleck, Theory of Haldane-gap antiferromagnets in applied fields. Phys. Rev. B **41**, 6697 (1990). doi:10.1103/PhysRevB.41.6697
118. M. Ganahl, M. Haque, H.G. Evertz, Quantum bowling: particle-hole transmutation in one-dimensional strongly interacting lattice models (2013), arXiv:1302.2667
119. R. Vlijm, M. Ganahl, D. Fioretto, M. Brockmann, M. Haque, H.G. Evertz, J.-S. Caux, Quasi-soliton scattering in quantum spin chains. Phys. Rev. B **92**, 214427 (2015). doi:10.1103/PhysRevB.92.214427
120. F.H.L. Essler, R.M. Konik, Finite-temperature lineshapes in gapped quantum spin chains. Phys. Rev. B **78**, 100403 (2008). doi:10.1103/PhysRevB.78.100403
121. F.H.L. Essler, R.M. Konik, Finite-temperature dynamical correlations in massive integrable quantum field theories. J. Stat. Mech. Theory Exp. **2009**, P09018 (2009). doi:10.1088/1742-5468/2009/09/P09018
122. W.D. Goetze, U. Karahasanovic, F.H.L. Essler, Low-temperature dynamical structure factor of the two-leg spin-1/2 Heisenberg ladder. Phys. Rev. B **82**, 104417 (2010). doi:10.1103/PhysRevB.82.104417
123. A. James, F.H.L. Essler, R.M. Konik, Finite-temperature dynamical structure factor of alternating Heisenberg chains. Phys. Rev. B **78**, 094411 (2008). doi:10.1103/PhysRevB.78.094411
124. K. Damle, S. Sachdev, Spin dynamics and transport in gapped one-dimensional Heisenberg antiferromagnets at nonzero temperatures. Phys. Rev. B **57**, 8307 (1998). doi:10.1103/PhysRevB.57.8307
125. G. Xu, C. Broholm, Y.-A. Soh, G. Aeppli, J.F. Ditusa, Y. Chen, M. Kenzelmann, C.D. Frost, T. Ito, K. Oka, H. Takagi, Mesoscopic phase coherence in a quantum spin fluid. Science **317**, 1049 (2007). doi:10.1126/science.1143831
126. D. Peter, S. Müller, S. Wessel, H.P. Büchler, Anomalous behavior of spin systems with dipolar interactions. Phys. Rev. Lett. **109**, 025303 (2012). doi:10.1103/PhysRevLett.109.025303
127. J.C. Halimeh, V. Zauner-Stauber, Enriching the dynamical phase diagram of spin chains with long-range interactions (2016), arXiv:1610.02019

Chapter 5
Towards a Particle Theory in Two Dimensions

In one-dimensional quantum spin systems the ground-state correlations are particularly strong, to the effect that the elementary excitations are collective modes that do not have a direct connection with the microscopic degrees of freedom. Nonetheless, in the last two chapters we have shown that it is still possible to develop an effective particle description of the low-energy physics. At this point, two questions naturally pop up: (i) can we develop a similar formalism for two-dimensional quantum spin systems, and (ii) is such a formalism needed in two dimensions?

The latter question was answered affirmatively in Sect. 2.3. It was argued that, despite the fact that a lot of two-dimensional quantum systems develop long-range order at zero temperature and allow for a more straightforward computation of the excitation spectra, there is still a large class of quantum spin systems that exhibit strong quantum correlations and defy perturbative or mean-field treatments. The excitations in these systems are again collective and display exotic phenomena such as fractionalization and anyonic statistics.

In this chapter, we will work towards an affirmative answer to the first question. The style of this chapter will be different from the two previous ones, as we are not yet in a position to present a nicely polished framework for computing one-particle excitations and their properties for generic two-dimensional systems. Instead, we will focus more on ground-state algorithms, where optimizing the numerical efficiency and benchmarking will be more of an issue than developing a consistent framework. Only in the last sections we will work towards capturing the quasiparticle excitations.

The two-dimensional versions of MPS go under the name of projected entangled-pair states (PEPS) [1, 2], which have emerged as a viable candidate for capturing the ground states of strongly-correlated quantum lattice models in two dimensions. It is by explicitly modelling the distribution of entanglement in low-energy states of local Hamiltonians, that PEPS parametrize the "physical corner of Hilbert space". Indeed, PEPS have a built-in area law for the entanglement entropy [3], they provide a natural characterization of topological order [4–8], and they can realize bulk-boundary

L. Vanderstraeten, *Tensor Network States and Effective Particles for Low-Dimensional Quantum Spin Systems*, Springer Theses,
DOI 10.1007/978-3-319-64191-1_5

correspondences explicitly [9–11]. Moreover, PEPS can be formulated directly in the thermodynamic limit [12], a feature that allows to focus on bulk physics without any finite-size or boundary effects.

An efficient optimization of the parameters in a PEPS has proven to be more challenging. According to the variational principle, finding the best approximation to the ground state for a given Hamiltonian H reduces to the minimization of the energy expectation value. For infinite PEPS this amounts to a highly non-linear optimization problem for which the evaluation of e.g. the gradient of the energy functional is a hard problem. For that reason, the state-of-the-art PEPS algorithms have taken recourse to imaginary-time evolution [12–14]: a trial PEPS state is evolved with the operator $\mathrm{e}^{-\tau H}$, which should result in a ground-state projection for very long times τ. This imaginary-time evolution is integrated by applying small time steps $\delta\tau$ with a Trotter-Suzuki decomposition and, after each time step, truncating the PEPS bond dimension in an approximate way. This truncation can be done by a purely local singular-value decomposition – the so-called simple-update [13] algorithm – or by taking the full PEPS wave function into account – the full-update [12] and fast full-update [14] algorithms.

These imaginary-time algorithms have allowed very accurate simulations of frustrated spin systems [15–25] and strongly-correlated electrons [25–28], but it remains unclear whether they succeed in finding the optimal state in a given variational class of PEPS. Although computationally very cheap, ignoring the environment in the simple-update scheme is often a bad approximation for systems with large correlations. The full-update scheme takes the full wave function into account for the truncation, but requires the inversion of the effective environment which is potentially ill-conditioned. This problem was solved by regularizing the environment appropriately and fixing the gauge of the PEPS tensor [14, 29]. Nonetheless, the truncation procedure in the full-update scheme is not guaranteed to provide the *globally* optimal truncated tensors in the sense that the global overlap of the truncated and the original PEPS is maximized. Indeed, the truncated tensor is optimized *locally* and afterwards put in at every site in the lattice to give an updated (global) PEPS wave function.

Similar issues have been at the center of attention in the context of MPS, where a number of different strategies have been around for optimizing ground-state approximations directly in the thermodynamic limit – see Sect. 2.6.4. We have seen that the so-called tangent-space methods have reinterpreted this problem of finding an optimal MPS by (i) identifying the class of matrix product states as a non-linear manifold embedded in physical Hilbert space, and (ii) formulating a minimization problem of the global energy functional on this manifold. From this perspective, imaginary-time evolution can be interpreted as a steepest-decent method, but more advanced optimization methods such as conjugate-gradient or quasi-Newton algorithms can find an optimal matrix product state much more efficiently. Besides better ground-state optimization algorithms, the tangent-space concept also leads to better algorithms for simulating the low-energy dynamics on top of an MPS ground state. The time-dependent variational principle dictates how to simulate real-time evolution within the MPS manifold: at every time step, the time-evolved state is projected onto the tangent space. More importantly, the elementary excitations are accurately captured by the tangent-space excitation ansatz; the last two chapters provide ample proof of the versatility of the ansatz.

In this chapter we show how to lift these tangent-space methods to two dimensions, by introducing a contraction scheme based on the concept of a "corner environment" [Sect. 5.2]. Using these environments, in the next section [Sect. 5.3] we present a PEPS algorithm that optimizes the global energy functional using a conjugate-gradient optimization method. In contrast to other methods, this algorithm has a clear convergence criterion, which can guarantee that an optimal state has been reached. Moreover, it allows to impose physical symmetries on the PEPS more easily. On the fly, the contraction scheme also allows to compute the energy variance of the variational ground state – an unbiased measure of the accuracy of the variational ansatz *and* a tool for better energy extrapolations – as well as general two-point correlation functions and static structure factors. We apply [Sect. 5.4] this new optimization scheme to the transverse-field Ising model and two Heisenberg models. In the last two sections [Sects. 5.5 and 5.6], we make the next step towards an effective particle description by introducing the ansatz for elementary excitations on a PEPS ground state and applying it to find the excitation spectrum of two frustration-free models.

5.1 A PEPS and Its Tangent Space

Consider an infinite square lattice with every site hosting a quantum degree of freedom with dimension d. For this quantum spin system, a PEPS can be introduced formally as

$$|\Psi(A)\rangle = \sum_{\{s\}} \mathcal{C}_2(A) |\{s\}\rangle \tag{5.1}$$

where $\mathcal{C}_2(\ldots)$ is the contraction of an infinite tensor network. This contraction is most easily represented graphically as

$\mathcal{C}_2(A) =$ 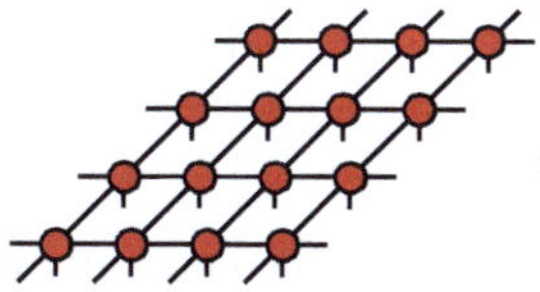 ,

with the red circle always representing the same five-legged tensor A

$A^s_{u,r,d,l} =$ 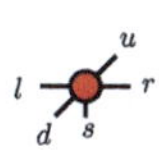

.

In order to obtain a physical state, a tensor A is associated with every site in the lattice and all virtual indices (u, r, d, l) are contracted in the network. The physical indices s are left open, such that a coefficient is obtained for every spin configuration in the superposition in Eq. (5.1). The graphical representation is then obtained by connecting links that are contracted and leaving the physical links open.

The virtual degrees of freedom in the PEPS carry the quantum correlations and mimic the entanglement structure of low-energy states. The dimension of the virtual indices is called the bond dimension D and can be tuned in order to enlarge the variational class; as such, it acts as a refinement parameter for the variational PEPS ansatz.

Similar to the one-dimensional case, a PEPS is invariant under gauge transformations of the form

$$A^s_{l,d,r,u} \rightarrow \sum_{l',d',r',u'} X_{l,l'} Y_{d,d'} A^s_{l',d',r',u'} (X^{-1})_{r',r} (Y^{-1})_{u',u}$$

with X and Y invertible $D \times D$ matrices. Yet this gauge freedom does not allow to straightforwardly define a canonical form for a PEPS in the sense that it is not possible to define orthonormal bases for all connections in the PEPS simultaneously. This will be reflected in the next sections by the fact that the contraction of a PEPS is necessarily a non-trivial task.

Finally, in every point $|\Psi(A)\rangle$ on the PEPS manifold we can define a tangent space spanned by the tangent vectors

$$|\Phi(B;A)\rangle = \sum_{j\in\{l,d,r,u,s\}} B^j \frac{\partial}{\partial A_j} |\Psi(A)\rangle = \sum_{i\in\mathcal{L}}$$

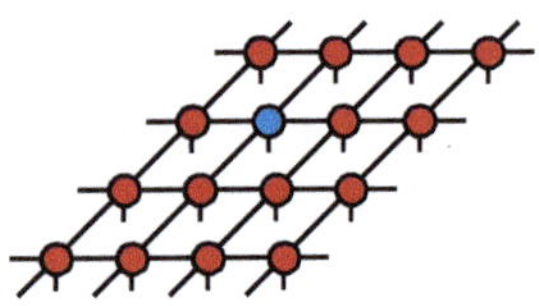

(5.2)

where in the sum the tensor B is represented as a blue tensor at site i and the index j runs over all tensor entries.

5.2 Transfer Matrices and Approximating Effective Environments

The norm of an infinite PEPS can be pictorially represented as

$$\langle\Psi(A)|\Psi(A)\rangle =$$

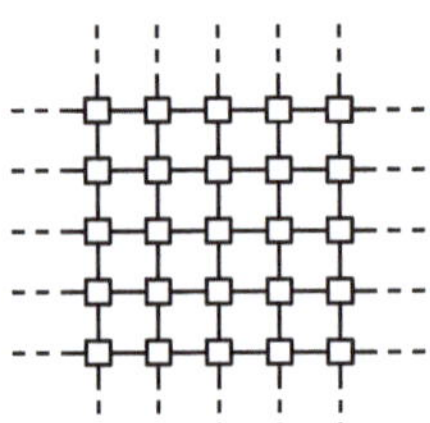

,

where every block represents the tensor a obtained by contracting the tensor A with its conjugate $\bar{A}$ over the physical index, i.e.

$a =$ 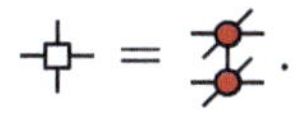 .

Throughout the rest of this chapter, the virtual indices of the ket and bra level are grouped into one index, so that these "top view" representations of double-layer tensor contractions are simplified.

The norm of the PEPS is thus obtained by the contraction of an infinite tensor network and can, in general, only be done approximately. Different numerical methods have been developed to contract these infinite networks efficiently, which allows the evaluation of the norm of a PEPS, as well as expectation values and correlation functions.

5.2.1 *The Linear Transfer Matrix*

The first and most straightforward strategy is based on the linear transfer matrix $\mathcal{T}$, graphically represented as

$\mathcal{T} =$ 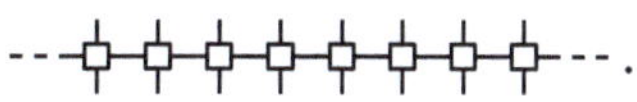 .

This object carries all the correlations in the PEPS from one row in the network to the next. One can of course define a similar transfer matrix in the vertical direction, and even diagonal transfer matrices can be considered. Naturally the transfer matrix is interpreted as an operator from the top to the bottom indices, so that the full contraction of the two-dimensional network reduces to successively multiplying copies of $\mathcal{T}$. In the thermodynamic limit, the norm of a PEPS is thus given by

$$\langle \Psi(A) | \Psi(A) \rangle = \lim_{N \to \infty} \mathcal{T}^N = \lambda^N$$

with λ the leading eigenvalue of the transfer matrix. The associated leading eigenvector or fixed point contains all the information on the correlations of a half-infinite part of the lattice.

An exact representation of the fixed point is only possible in a number of special cases and approximate methods have to be devised in general. Given the versatility of MPS for approximating the ground state of local gapped Hamiltonians [Sect. 2.6], one expects that this class of states might provide a good variational ansatz for the case of gapped transfer matrices as well. Whereas MPS approximations for fixed points go way back [30], a variety of efficient tensor-network methods have been developed [1, 31] recently. Here we use an algorithm [32] in the spirit of Ref. [33], that treats the linear and corner transfer matrix [Sect. 5.2.2] on a similar footing.

The fixed-point equation can be stated graphically as

(5.3)

For this to equation to hold, a relation of the form

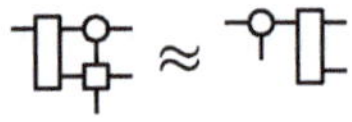

should hold to a very high precision. Indeed, if this tensor (rectangle) exists, it maps the action of the transfer matrix back to the same MPS fixed point. The virtual dimension of the MPS fixed point will be denoted as χ and can be tuned to improve the accuracy of the PEPS contraction.

Given that the fixed point equation can be solved efficiently, the PEPS can now be normalized to one by rescaling the A tensor such that the largest eigenvalue λ of the transfer matrix equals unity. With the MPS fixed point, the expectation value of a local operator at an arbitrary site i,

$$\langle O_i \rangle = \frac{\langle \Psi(A) | O_i | \Psi(A) \rangle}{\langle \Psi(A) | \Psi(A) \rangle},$$

can be easily computed. First the upper and lower half of the network are replaced by the fixed points,

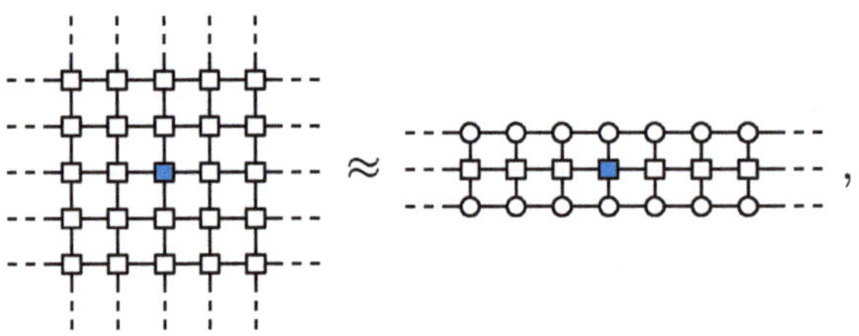

where a coloured block tensor always indicates the presence of a physical operator at that site. The resulting effective one-dimensional network can be evaluated exactly by finding the leading left and right eigenvectors (fixed points) of the channel operator,

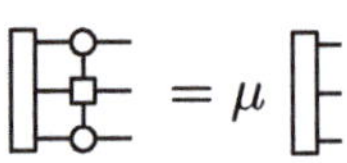

and

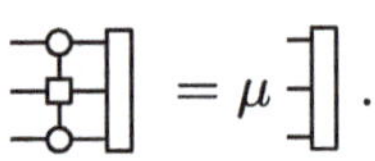

The eigenvalue μ depends on the normalization of the MPS tensors in the upper and lower fixed points of the linear transfer matrix, and its value can be put to one. The fixed points are determined up to a factor, which can be fixed by imposing that the norm of the PEPS

$$\langle \Psi(A) | \Psi(A) \rangle =$$ 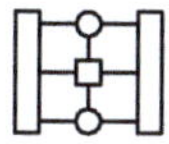

equals unity such that

$$\langle O_i \rangle = \langle \Psi(A) | \, O_i \, | \Psi(A) \rangle =$$ 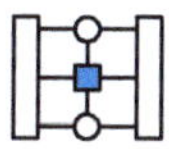 .

5.2.2 *The Corner Transfer Matrix*

Another set of methods for contracting two-dimensional tensor networks relies on the concept of the corner transfer matrix, which was first applied to classical lattice systems [30, 34–36] and recently used extensively in tensor network simulations [27, 37, 38]. The strategy now is to break up the infinite tensor network in different regions, and represent these as tensors with a fixed dimension. Graphically, the set-up is

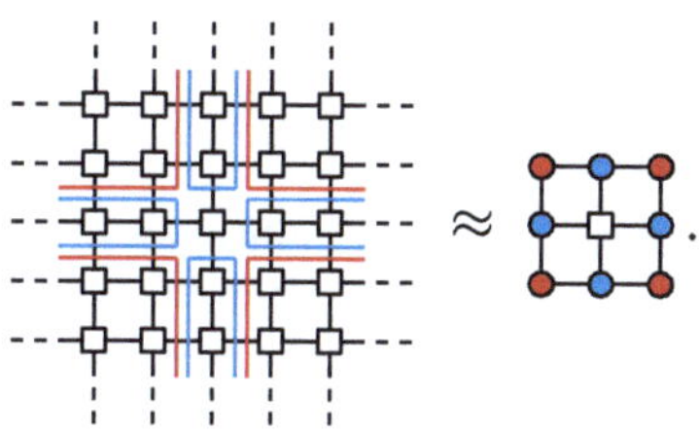

A red tensor is a compression of one of the corners of the network, whereas the blue tensors capture the effect of an infinite row of a tensors. Together, they provide an effective one-site environment for the computation of the norm of the PEPS or local expectation values.

This scheme can now be extended in order to evaluate non-local expectation values such as general two-point correlation functions. Indeed, by not compressing the blue region above, one can construct an environment that looks like

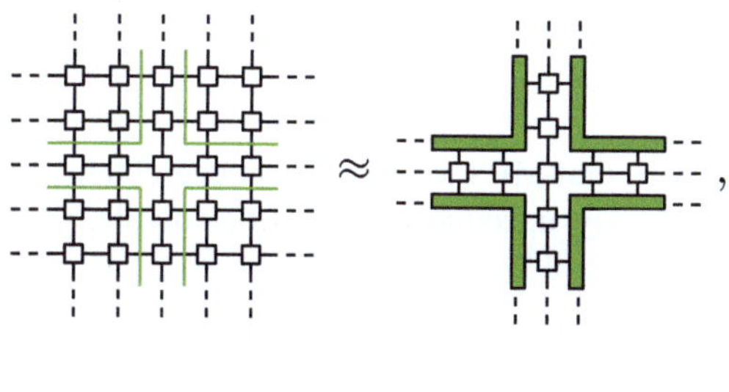

so that one could evaluate operators that have an arbitrary location in the lattice.

Finding this effective "corner environment" can again been done by solving a fixed point equation. Indeed, the green corner-shaped environment should be the result of an infinite number of iterations of an equally corner-shaped transfer matrix; the fixed point equation is

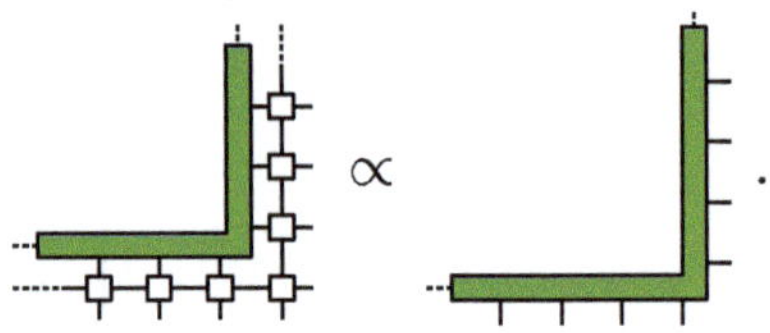

Very far from the corner this equation reduces to the one for the linear transfer matrix. This implies that, asymptotically, the fixed point can be well approximated by an MPS. Let us therefore make the ansatz that the full fixed point can be approximated as an MPS, where we put an extra tensor on the virtual level to account for the corner. With this ansatz, the fixed point equation is given by

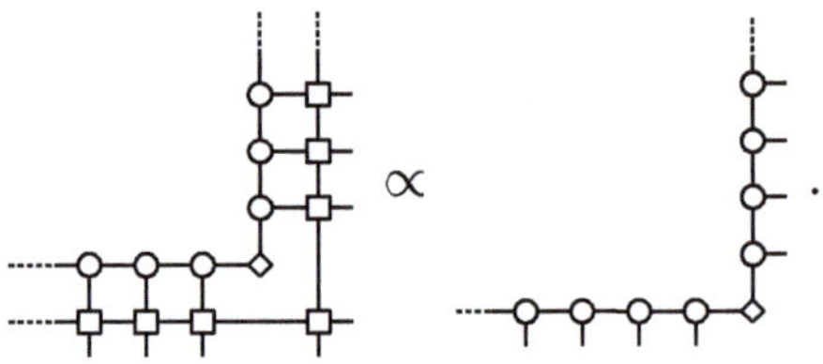

(5.4)

Note, firstly, that the corner transfer matrix introduces two new sites at every application[1] and therefore does not have a fixed point in the strict sense. Nevertheless, repeated application of C is likewise expected to result in a state with a converged (i.e. translation-invariant) structure, up to the corner itself. We therefore model the "fixed point" using the tensors of the fixed point of the linear transfer matrix and insert a new corner tensor. Applying the corner transfer matrix C once and using the tensor from above gives rise to

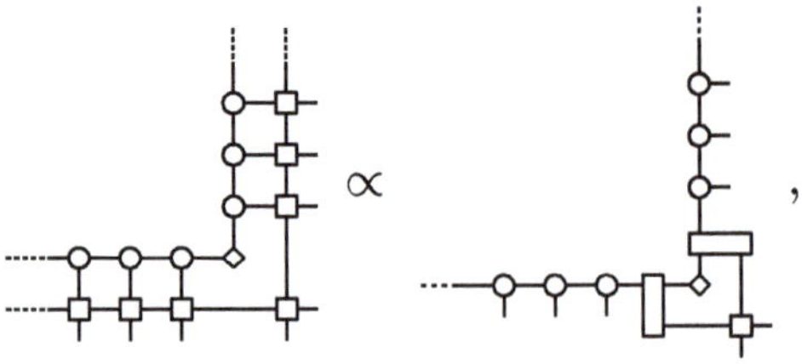

which shows that the original MPS tensors are indeed obtained after application of C, except on the two newly introduced sites. In principle, two new MPS tensors

[1]This algorithm is similar to infinite-size DMRG, see Sect. 2.6.4.

connected by a new corner matrix could appear. However, since the unique MPS tensor of the fixed point of $\mathcal{T}$ seems to capture the correct structure on the further sites, and these two can be assumed to have originated from previous applications of $\mathcal{C}$, we can make the ansatz that these tensors should also be put on the two new sites. With this ansatz, we obtain a linear fixed point equation for the corner matrix itself, which corresponds to a simple eigenvalue equation. With the corner matrix as only variational parameters in the fixed point equation for $\mathcal{C}$, we essentially have a linear subspace as ansatz and can therefore easily measure the error obtained by projecting onto this subspace.

5.2.3 Channel Environments

Once we have found (i) the fixed points of the linear transfer matrix in all directions [Eq. (5.3)], and (ii) the four corner tensors [Eq. (5.4)], we can contract the network corresponding to the norm, a local expectation value or a correlation function of the PEPS. Let us assume that the tensor A is normalized such that the largest eigenvalue of the linear transfer matrix is unity. Computing the norm $\langle\Psi(A)|\Psi(A)\rangle$ with a channel environment then reduces to the contraction of

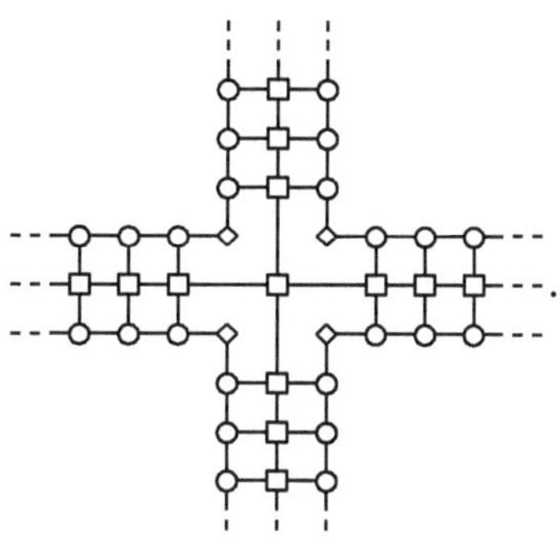

An infinitely long channel can be contracted by computing the fixed point ρ_L of the "channel operator". Therefore the eigenvector corresponding to the largest eigenvalue should be found, i.e.

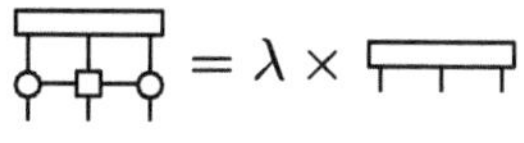

for the top channel. The boundary MPS tensors have to be rescaled such that the largest eigenvalue λ is put to one. Similarly, the fixed point in the other direction ρ_R is defined as

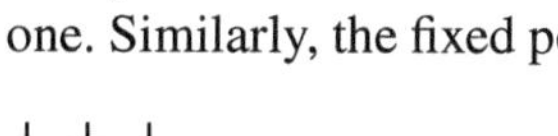

The inner product of the left and right fixed point is put to one. For further use, we note that, by subtracting the projector on the leading eigenvector, an operator is constructed that has spectral radius strictly smaller than one

$\rho\Big($ 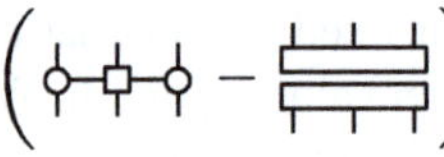 $\Big) < 1.$

The norm of the PEPS is then reduced to

$\langle \Psi(A) | \Psi(A) \rangle =$ 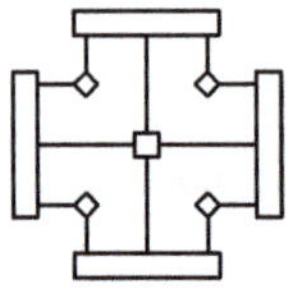 ,

which can be put to one by rescaling the corner tensors by the appropriate scalar. With these conventions, the norm of the state is well-defined and expectation values can be safely computed. For a local one-site operator O we have

$\langle \Psi(A) | O | \Psi(A) \rangle =$ 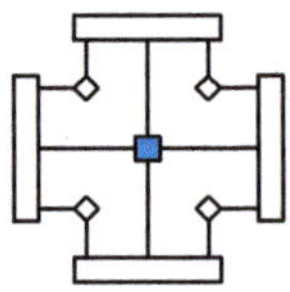 ,

and, similarly, the expectation value of a two-site operator is

$\langle \Psi(A) | O | \Psi(A) \rangle =$ 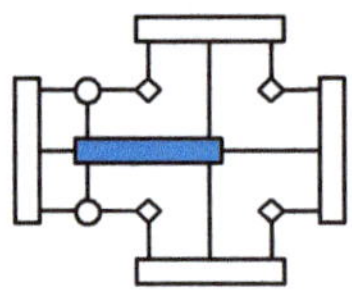 ,

where the two-site operator can of course be oriented in the other channels as well.

The real power of the channel environment is now that arbitrary two-point correlation functions can be computed straightforwardly. Indeed, the expectation value of two operators at generic locations in the lattice is computed as

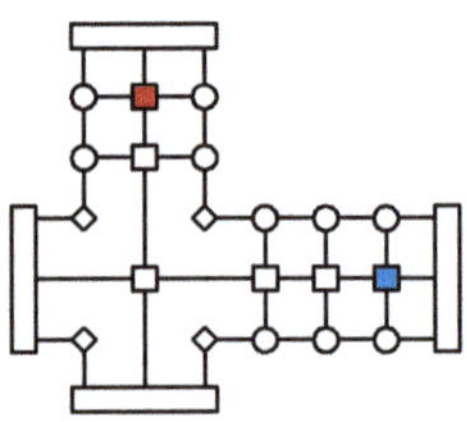 .

In fact, even three-point correlation functions can be evaluated by orienting the corners in the right way, as in e.g. the contraction

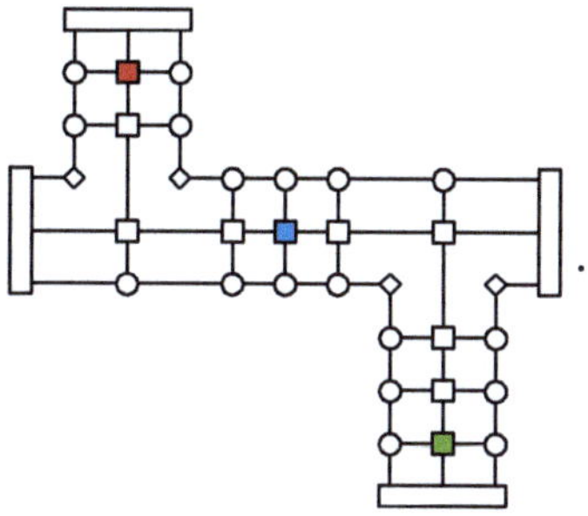

5.2.4 Static Structure Factor

As an example of the power of the corner environment, we will explicitly show how to compute a static correlation function directly in momentum space, i.e. the static structure factor $s(\vec{q})$

$$s(\vec{q}) = \frac{1}{|\mathcal{L}|} \sum_{i,j\in\mathcal{L}} \mathrm{e}^{i\vec{q}\cdot(\vec{n}_i - \vec{n}_j)} \langle \Psi(A)| O_i^\dagger O_j |\Psi(A)\rangle_c$$

where only the connected part is taken up in the correlator, or, equivalently, the operators have been redefined such that their ground state expectation value is zero.

The momentum superposition of all relative positions of the operators can be evaluated explicitly by moving the operators independently through the channels and summing all contributions. This infinite number of contributions can be resummed by noting that one obtains a geometric series inside the channels. Summing all different contributions from an operator moving in the top channel can be done by introducing a new momentum-resolved operator that captures the momentum superposition[2]

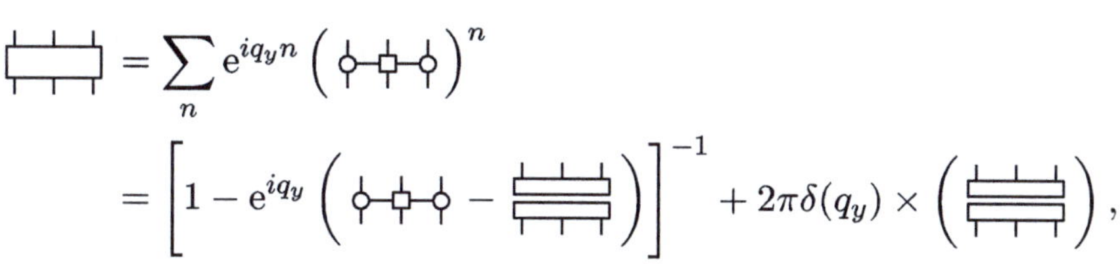

(5.5)

[2]The way in which these geometric sums are explicitly computed is very similar to the procedures that we have applied for the case of MPS in Chap. 3, including the inversion of a regularized version of the transfer matrix.

where we have separated the projector onto the largest eigenvector. As we will see, the diverging δ contribution will always drop out, such that the inverse is well-defined. The momentum superposition inside the channel can be represented as

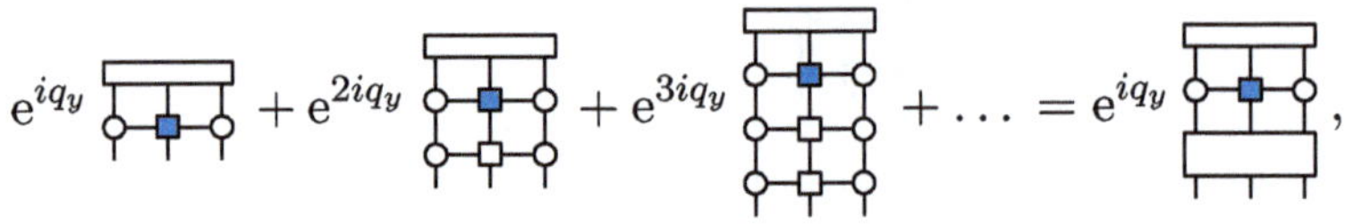

where the component along the channel fixed point is indeed always zero – this component would correspond to the disconnected part of the correlation function. The geometric series converges for every value of the momentum and the inverse can be taken without a problem.

By independently letting the two operators travel through the channels, all relative positions can be taken into account. In addition, we also need the contribution where the two operators act on the same site. The full expression is given by

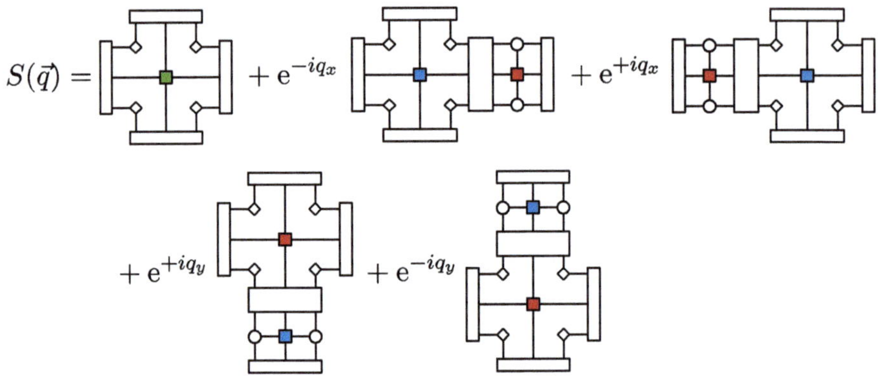

(5.6)

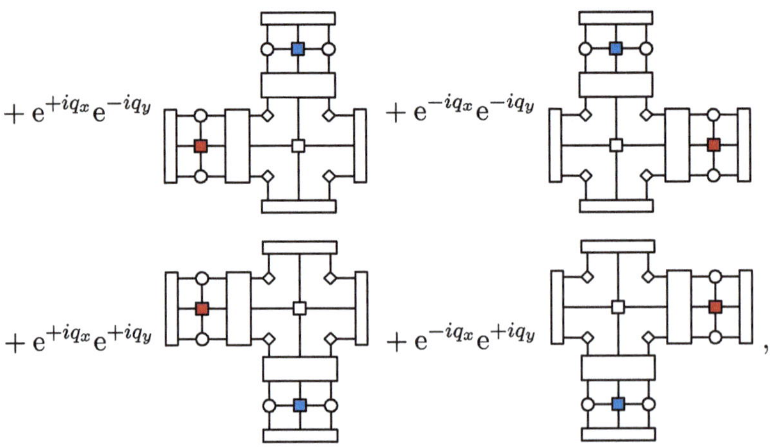

(5.7)

where the green tensor represents the action of the two operators at the same site, and the blue and red tensors represent actions of the operators on the ket and bra level.

5.3 Ground-State Optimization

The PEPS ansatz defines a variational class of states that should approximate the ground state of two-dimensional quantum lattice systems in the thermodynamic limit. The system is described by its Hamiltonian, which we assume to consist of nearest-neighbour interactions, i.e.

$$H = \sum_{\langle ij \rangle} h_{ij},$$

and its lattice structure, for which we will confine ourselves to the square lattice. The following can be straightforwardly extended to different lattices, larger unit cells or longer-range Hamiltonians.

As dictated by the variational principle, finding the best approximation to the ground state of H now amounts to solving the highly non-linear minimization problem

$$\min_A \frac{\langle \Psi(A) | H | \Psi(A) \rangle}{\langle \Psi(A) | \Psi(A) \rangle}. \tag{5.8}$$

As we have seen, the evaluation of this energy functional for a certain tensor A is already non-trivial, but can be done efficiently using a variety of numerical methods. Yet the evaluation of the energy is not enough, as efficient numerical optimization algorithms also rely on the evaluation of the gradient or higher-order derivatives of the energy functional. For a translation-invariant PEPS, the gradient is a highly non-trivial object; it requires the evaluation of the change in energy from a variation in the tensor A, for which the effect of local *and* non-local contributions should be added. In this respect, it is quite similar to a zero-momentum structure factor, and can be evaluated using the channel environment that we introduced in Sect. 5.2. In Sect. 5.3.1 we run through the different diagrams for the gradient's explicit evaluation, and show that it can be computed efficiently.

The easiest algorithm using the gradient is the steepest-descent method, where in each iteration one minimizes the energy in the direction of the gradient. One iteration i corresponds to an update of the A tensor as

$$A_{i+1} \rightarrow A_i + \alpha \tilde{A}_i$$

with $\tilde{A}_i = -g_i$ (g_i is the gradient at iteration i). The value of $\alpha > 0$ is determined with a line-search algorithm. The performance can be greatly enhanced by implementing a non-linear conjugate-gradient method, where the search direction is a linear combination of the gradient and the direction of the previous iteration

$$\tilde{A}_i = -g_i + \beta_i \tilde{A}_{i-1}.$$

For each non-linear optimization problem, the parameter β_i can be chosen from a set of different prescriptions [39–41]. Here we have exclusively used the Fletcher–Reeves scheme [42], according to which

$$\beta_i = \frac{\|g_i\|^2}{\|g_{i-1}\|^2}.$$

Crucially, these algorithms have a clear convergence criterion: when the norm of the gradient is sufficiently small, the energy cannot be further optimized and an optimal solution has been found.

Note that these direct optimization methods allow to control the number of variational parameters and to impose certain symmetries on the PEPS tensor A: the iterative search can be easily confined to a certain subspace of the PEPS variational class by e.g. projecting the gradient onto this subspace in each iteration. Moreover, this direct optimization strategy allows to start from a random input tensor A and systematically converge to an optimal solution – all the results in Sect. 5.4 were obtained by starting from a random initial tensor.

5.3.1 Computing the Gradient

The objective function f that we want to minimize [see Eq. (5.8)] is a real function of the complex-valued A, or, equivalently, the independent variables A and $\bar{A}$. The gradient g is then obtained by differentiating $f(\bar{A}, A)$ with respect to $\bar{A}$

$$\begin{aligned} g &= 2 \times \frac{\partial f(\bar{A}, A)}{\partial \bar{A}} \\ &= 2 \times \frac{\partial_{\bar{A}} \langle \Psi(\bar{A}) | H | \Psi(A) \rangle}{\langle \Psi(\bar{A}) | \Psi(A) \rangle} - 2 \times \frac{\langle \Psi(\bar{A}) | H | \Psi(A) \rangle}{\langle \Psi(\bar{A}) | \Psi(A) \rangle^2} \partial_{\bar{A}} \langle \Psi(\bar{A}) | \Psi(A) \rangle, \end{aligned}$$

where we have clearly indicated A and $\bar{A}$ as independent variables. In the implementation we will always make sure that the PEPS is properly normalized, such that the numerators drop out. By subtracting from every term in the Hamiltonian its expectation value, the full Hamiltonian can be redefined as

$$H \rightarrow H - \langle \Psi(\bar{A})|H|\Psi(A)\rangle, \tag{5.9}$$

such that the gradient takes on the simple form

$$g = 2 \times \partial_{\bar{A}} \langle \Psi(\bar{A})|H|\Psi(A)\rangle.$$

The gradient is thus obtained by differentiating the energy expectation value $\langle \Psi(\bar{A})|H|\Psi(A)\rangle$ with respect to every $\bar{A}$ tensor in the bra level and taking the sum of all contributions. Every term in this infinite sum is obtained by omitting one $\bar{A}$ tensor and leaving the indices open. The full infinite summation is then obtained by letting the Hamiltonian operator and this open spot in the network travel through the channels separately, just as in the case of the structure factor in Sect. 5.2.4.

Let us first define a new tensor that captures the infinite sum of Hamiltonian operators acting inside a channel,

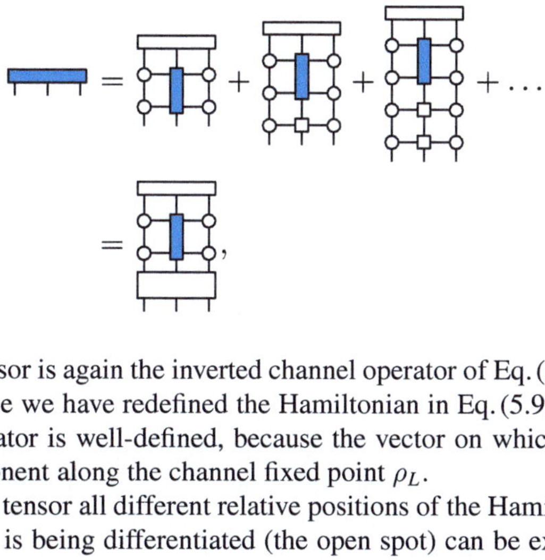

where the big tensor is again the inverted channel operator of Eq. (5.5) with momentum zero. Because we have redefined the Hamiltonian in Eq. (5.9), the inversion of the channel operator is well-defined, because the vector on which the inverse acts has a zero component along the channel fixed point ρ_L.

With this blue tensor all different relative positions of the Hamiltonian terms and the tensor $\bar{A}$ that is being differentiated (the open spot) can be explicitly summed, similarly as in the expression for the structure factor [Eq. (5.7)]. There are a few more terms because every Hamiltonian term corresponds to a two-site operator and has different orientations.

The full expression is

$g =$ 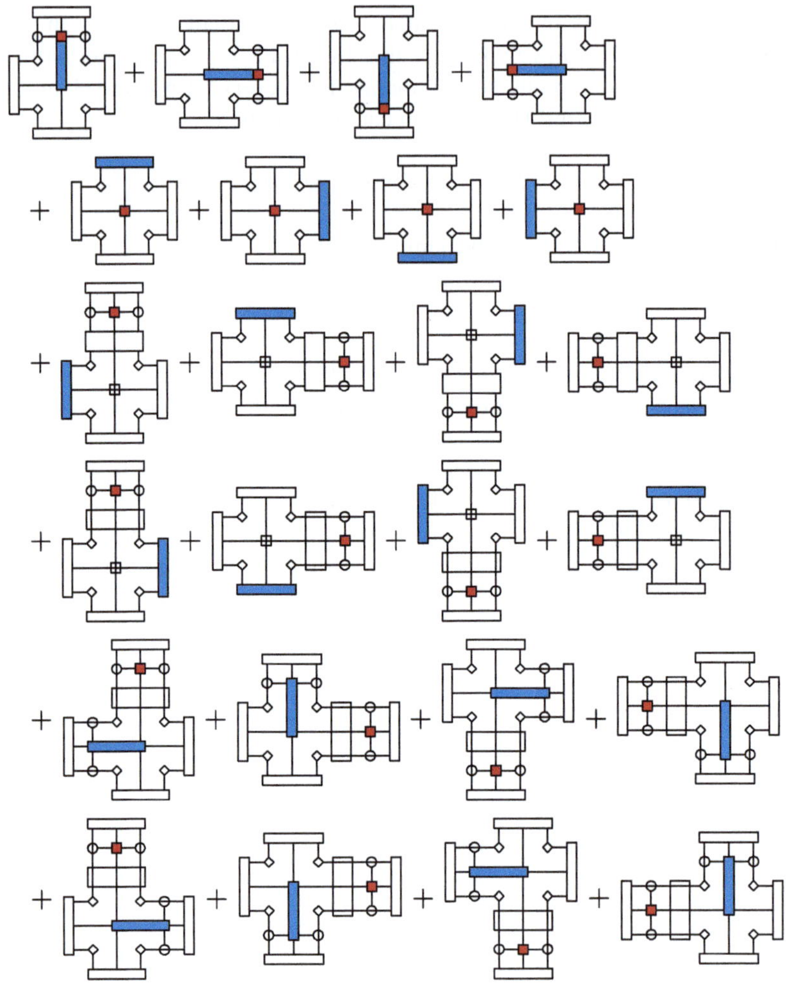,

where the red tensor indicates where the open spot in the bra level of the diagram is. Note that the diagrams on the same line are always related by a rotation; in the case that the PEPS tensor A is rotationally invariant, these diagrams give exactly the same contribution. This implies that the gradient corresponding to a rotationally invariant tensor A is itself rotationally invariant, which gives an easy way for incorporating symmetries in the PEPS optimization.

5.3.2 *The Energy Variance*

Like any variational method, the PEPS ansatz is a priori not guaranteed to provide an accurate parametrization of a ground state. It is expected that increasing the PEPS bond dimension gives a good test for the reliability of the simulation: an extrapolation in D should provide the correct results. One problem is that it is unclear how the energy or order parameter behave as a function of D [25]. A better and completely unbiased extrapolation quantity is the energy variance [43], defined as

$$v = \langle\Psi(A)|\,(H - e)^2\,|\Psi(A)\rangle,$$

with $e = \langle\Psi(A)|H|\Psi(A)\rangle$ the energy expectation value. It measures to what extent a variational wave function approximates the ground state (or more generally, an eigenstate) of the Hamiltonian.

Because the variance can be interpreted as a zero-momentum structure factor of the Hamiltonian operator, the computation of the energy variance is again similar. In addition to the green tensor above, we will also need the following geometric series

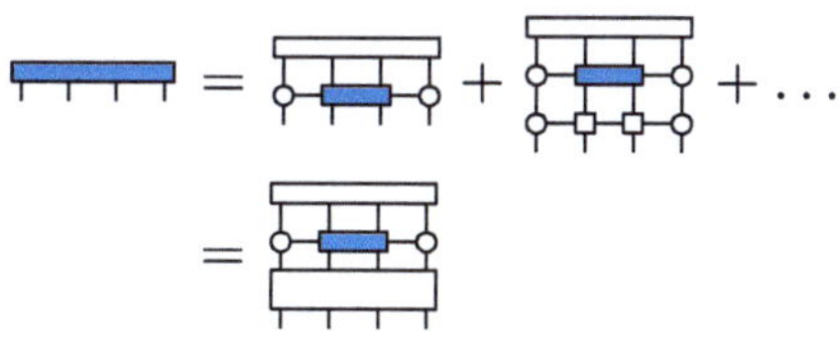

where

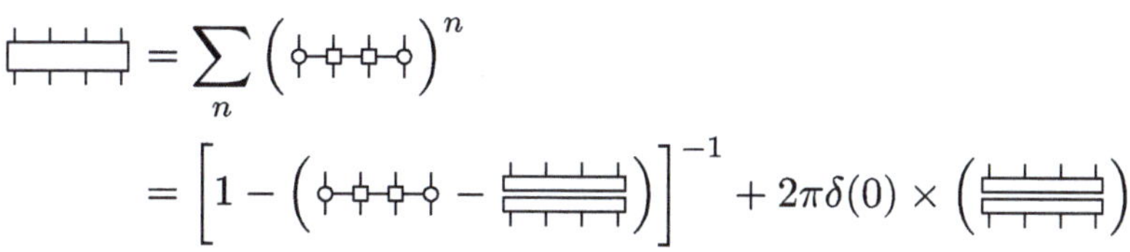

with the fixed points of the two-site channels properly normalised. We again renormalize the Hamiltonian as

$$H \rightarrow H - \langle\Psi(A)|H|\Psi(A)\rangle$$

such that disconnected contributions always drop out and the inverse of the operator above is well-defined. The blue tensor has $\chi^2 D^4$ elements, so its computation is by far the most costly step for the variance evaluation. Approximating it by a tensor decomposition might reduce the cost considerably, but for our purposes this has not been necessary.

Let us now associate to each nearest-neighbour term $\langle ij\rangle$ in the Hamiltonian a variance term as

$$v_{\langle ij\rangle} = \langle\Psi(A)|Hh_{\langle ij\rangle}|\Psi(A)\rangle,$$

such that the energy variance per site is given by[3]

$$v = \frac{1}{|\mathcal{L}|} \langle \Psi(A)|H^2|\Psi(A)\rangle = v_{\langle ij\rangle,\mathrm{hor}} + v_{\langle ij\rangle,\mathrm{ver}},$$

the sum of the variances corresponding to the horizontal and vertical nearest-neighbour terms in the Hamiltonian.

The vertical contribution is given by

$v_{\langle ij\rangle,\mathrm{ver}} =$

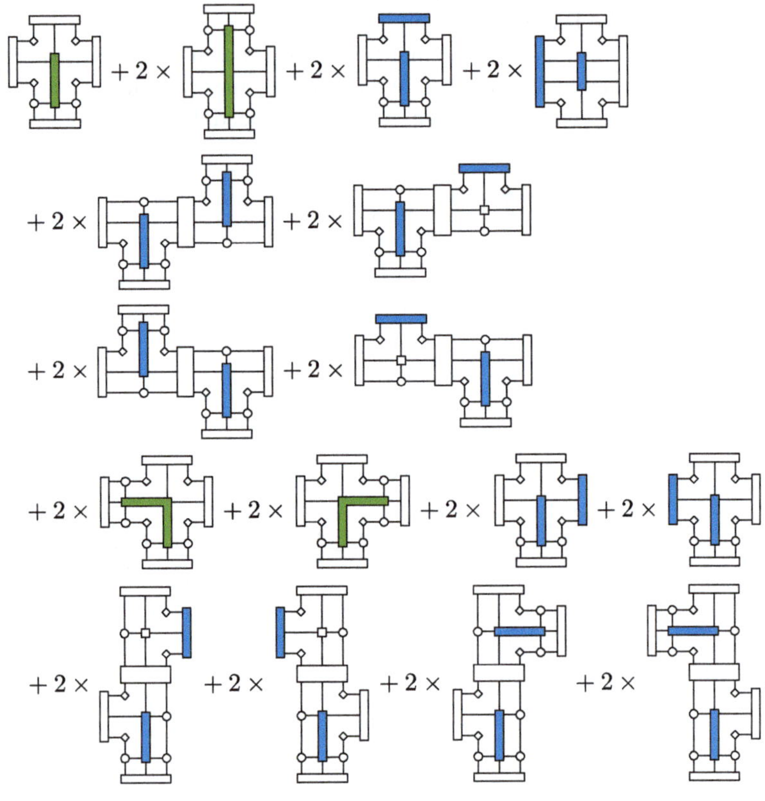

The green tensors represent the double action of the Hamiltonian operator; a two-site tensor if they fully overlap and a three-site tensor if the overlap is on one site only. In this expression, we have explicitly used the rotational invariance of the PEPS tensor A, which can be easily imposed within our framework. Under this symmetry, the horizontal and vertical contributions to the variance are obviously equal, so the

[3]The factor $|\mathcal{L}|$ represents the diverging number of sites in the system, such that v gives a finite variance density in the thermodynamic limit.

above is the complete expression for the variance up to a factor of 4. If A is not rotationally invariant, all the other diagrams can be obtained by rotating the above ones.

5.4 Ground State Optimization: Benchmarks

5.4.1 *Transverse Ising Model*

As a first check, we apply our PEPS algorithm to the transverse Ising model, defined by the Hamiltonian

$$H_{\text{Ising}} = -\sum_{\langle ij \rangle} S_i^z S_j^z + \lambda \sum_i S_i^x .$$

The model exhibits a phase transition at $\lambda_c \approx 3.044$ [44] from a symmetry broken phase to a polarized phase; the order parameter is $m = \langle S^z \rangle$. The model has been extensively studied with the PEPS ansatz [12, 14, 37], and we use the model as a benchmark for our conjugate-gradient method. The results shown in Fig. 5.1 confirm that we capture the magnetization curve accurately. In Fig. 5.2 we have compared our variational search with imaginary-time evolution (full update), showing that we find lower energies and better order parameters, even as the Trotter error goes to zero.

In Fig. 5.3 we provide some details on the convergence of the conjugate-gradient algorithm. In particular, we have found that rather high values of χ were needed to evaluate the gradient accurately close to convergence. Indeed, in the case of a strongly-correlated PEPS, a lot of different terms contribute to the expression for the gradient. Close to convergence the gradient becomes a vector of small magnitude, which can only happen due to the subtle cancellations of a lot of different terms; consequently, finding the gradient accurately is bound to require a large value of

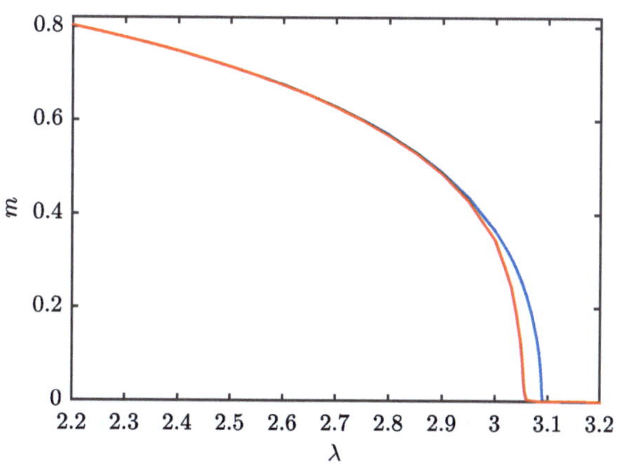

Fig. 5.1 The magnetization curve for the transverse Ising model with bond dimensions $D = 2$ (*blue*) and $D = 3$ (*orange*). We nicely capture the phase transition, although the critical point has been slightly shifted. The critical point can be estimated as the point where the slope of the curve is maximal; we arrive at $\lambda_c \approx 3.09$ ($D = 2$) and $\lambda_c \approx 3.054$ ($D = 3$)

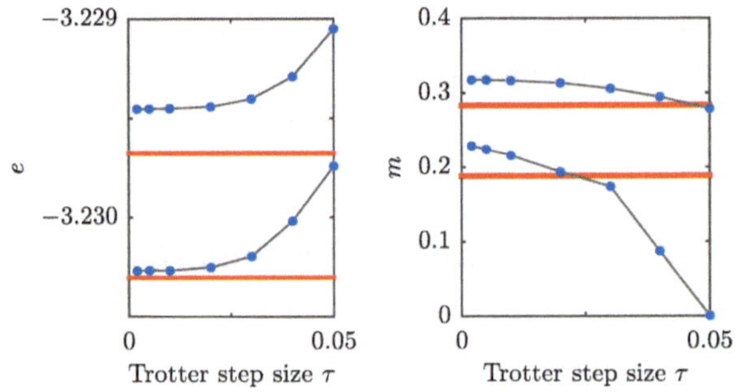

Fig. 5.2 Our variational results compared to the results that are obtained with imaginary-time evolution using the full-update algorithm; the comparison is done for the transverse Ising model at $\lambda = 3.04$ for bond dimensions $D = 2$ and $D = 3$. On the *left* (*right*) we have plotted the convergence for the energy (magnetization) as a function of the Trotter step size of the full-update scheme (*blue points*), and our results (*red line*). The plot clearly shows that, as the Trotter step size goes to zero, the imaginary-time results do not converge to the variational optimum that we obtain. Note that the variational freedom is slightly different: we optimize over a rotationally symmetric PEPS with one-site unit cell, whereas the imaginary-time results break rotational symmetry and work with a two-site unit cell. Although this larger rotationally asymmetric unit-cell might give lower energies, it appears that our optimization still gives better energies and order parameters

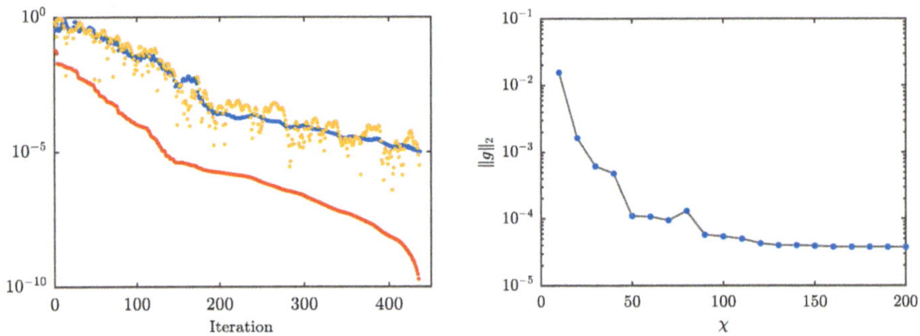

Fig. 5.3 Details on the convergence of the optimization algorithm for the Ising model at $\lambda = 3$. (*Left*) The convergence of the norm of the gradient $\|g\| = \sqrt{g^\dagger g}$ (*blue*), the error in the energy (*red*) and the error in the magnetization (*yellow*), as a function of the iteration. The errors are computed as the relative error with respect to the last iteration. In this $D = 2$ simulation the convergence criterion was $\|g\| \leq 10^{-5}$, a value for which the two plotted observables have clearly converged. (*Right*) The convergence of the norm of the gradient as a function of the bond dimension χ of the corner environment, at a particular iteration of the conjugate-gradient scheme for $D = 3$ (close to convergence). This plot shows that large values of χ are needed to obtain a required tolerance on the norm of the gradient (in this case $\chi \approx 100$)

χ. Note that the large values of χ are only necessary close to convergence, so we grow χ throughout the optimization. We never impose the final value of χ, because it is the correlations in the optimized PEPS that determine the χ needed to reach a certain tolerance on the norm of the gradient.

5.4.2 Heisenberg Models

As a second application, we study two spin-1/2 Heisenberg models on the square lattice, defined by the Hamiltonian

$$H_{\text{Heisenberg}} = \sum_{\langle ij \rangle} S_i^x S_j^x + S_i^y S_j^y + J_z S_i^z S_j^z.$$

The model has been of great theoretical and experimental interest, because of its paradigmatic long-range antiferromagnetic order [45]. In particular, Heisenberg models have proven to be a hard case for the PEPS ansatz [46] because of the large quantum fluctuations around the antiferromagnetic ordering; as such, they provide a proper benchmark for our conjugate-gradient method.

In contrast to most PEPS implementations, we prefer to work with a single-site unit cell, so we perform a sublattice rotation in order to capture the staggered magnetic order in the ground state. Moreover, we impose rotational symmetry on the PEPS tensor A, so that our variational ground state is automatically invariant under rotations of the lattice. In Figs. 5.4 and 5.5 we have plotted the energy expectation value and staggered magnetization after convergence as a function of the bond dimension, for the XY model ($J_z = 0$) and the isotropic Heisenberg model ($J_z = 1$). Comparing with results from imaginary-time evolution [14, 46], we see that our variational method reaches considerably lower energies and order parameters at the same bond dimension.

In addition we also compute the variance of these PEPS variational states, in order to get an idea of how well they approximate the true ground state. The result for the isotropic Heisenberg model is plotted in Fig. 5.6. We observe the expected linear behaviour [43] to some extent, and a zero-variance extrapolation based on the two last points ($D = 4, 5$) improves the estimate of the energy by a factor of two. Better

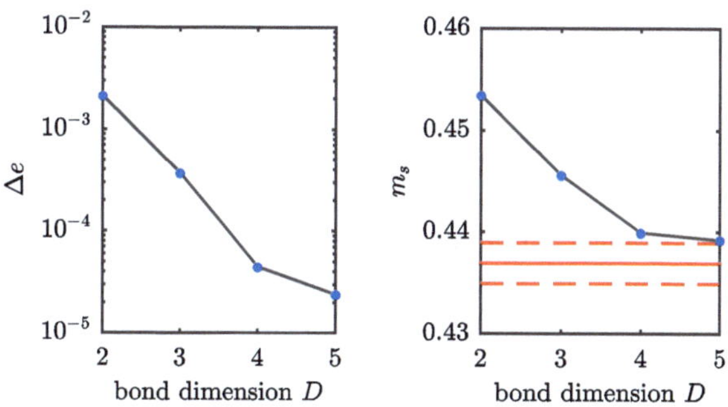

Fig. 5.4 Results for the XY model ($J_z = 0$), compared to the Monte-Carlo (MC) results in Ref. [47]. (*Left*) The relative error of our variational results $\Delta e = |(e_{\text{var}} - e_{\text{MC}})/e_{\text{MC}}|$ as a function of the bond dimension. (*Right*) The staggered magnetization as a function of the bond dimension; the *red line* is the Monte Carlo result with error bars

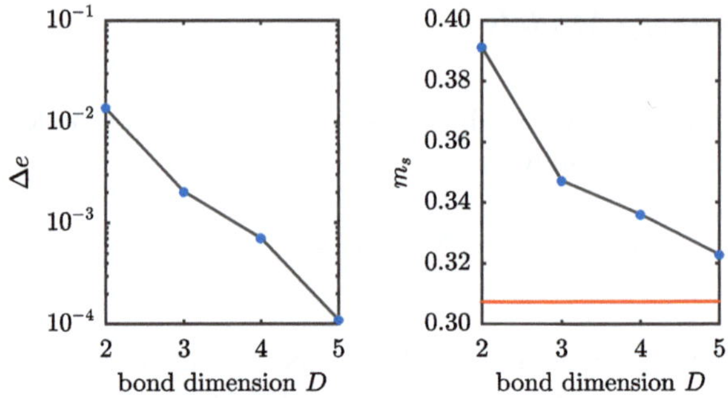

Fig. 5.5 Results for the Heisenberg antiferromagnet ($J_z = 1$), compared to the Monte-Carlo (MC) results in Refs. [48, 49]. (*Left*) The relative error of our variational results, $\Delta e = |(e_{\text{var}} - e_{\text{MC}})/e_{\text{MC}}|$, as a function of the bond dimension. (*Right*) The staggered magnetization as a function of the bond dimension; the *red line* is the Monte-Carlo result for which the error bars are too small to plot

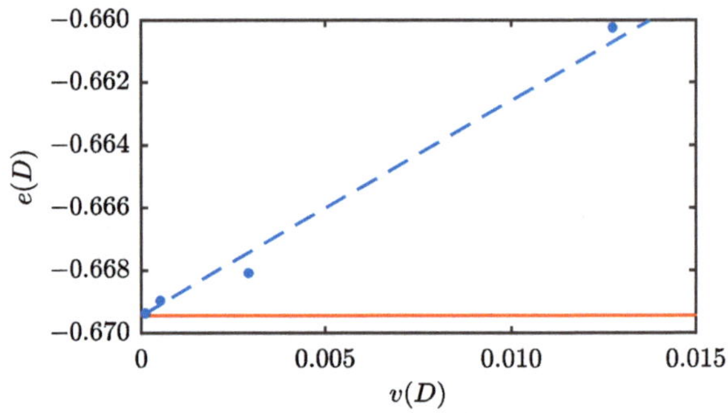

Fig. 5.6 The energy expectation value as a function of the variance per site for the isotropic ($J_z = 1$) Heisenberg antiferromagnet, for four values of the bond dimension ($D = 2 \to 5$). The *red line* represents the Monte Carlo result for the ground state energy [48]. A linear extrapolation with respect to the two best points ($D = 4, 5$) gives an energy with a relative error of $\Delta e \approx 4.7 \times 10^{-5}$. The *striped line* is drawn between the exact MC result and the $D = 5$ point and serves only as a guide to the eye

zero-variance extrapolations should be possible at higher bond dimensions for which the linear behaviour is expected to be stronger.

Another quantity that is within reach of our PEPS framework is the static structure factor, a central quantity for detecting the order in the ground state, and of direct experimental relevance. It is defined as

$$s(\vec{q}) = \frac{1}{|\mathcal{L}|} \sum_{i,j \in \mathcal{L}} \mathrm{e}^{i\vec{q}\cdot(\vec{n}_i - \vec{n}_j)} \langle \vec{S}_i \cdot \vec{S}_j \rangle_c,$$

where only the connected part of the correlation function is taken into account. The disconnected part will give a δ peak at $\vec{q} = (\pi, \pi)$ (the X point), corresponding to the

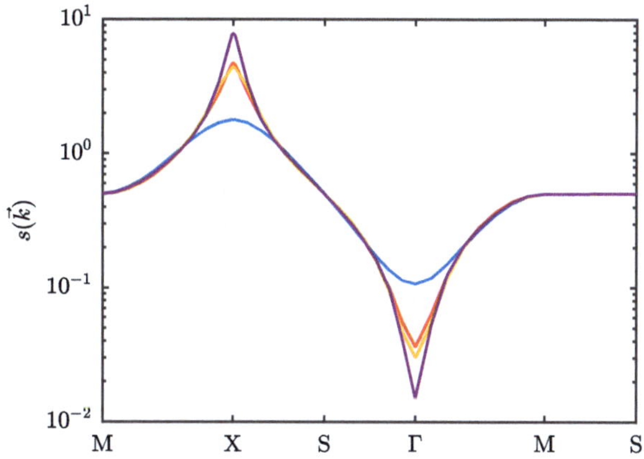

Fig. 5.7 Structure factor $s(\vec{q})$ of the isotropic ($J_z = 1$) Heisenberg antiferromagnet along a path through the Brillouin zone for optimised PEPS states with bond dimensions $D = 2$ (*blue*), $D = 3$ (*red*), $D = 4$ (*orange*) and $D = 5$ (*purple*), in agreement with the results in Refs. [50, 51]. The divergence around the X point and the zero around the Γ point are better reproduced as D increases, although the improvement as a function of D seems not to be smooth

staggered-magnetization order parameter. The strong fluctuations around this point will give an additional $1/q$ divergence, with q the distance from the X point [50]. The structure factor becomes zero at $\vec{q} = (0, 0)$, because the ground state is in a singlet state. In Fig. 5.7 we observe that the regular parts of the structure factor are perfectly reproduced, even at low bond dimensions, whereas the divergences can only be accurately captured by observing the behaviour as a function of the bond dimension.

5.5 Elementary Particle Excitations

The next step towards an effective particle theory in two dimensions is the formulation and implementation of the quasiparticle excitation ansatz. Of course, the MPS ansatz can be straightforwardly translated to the two-dimensional world, but, as we will see in this section, a few of the nice implementation tricks will no longer hold.

5.5.1 The Excitation Ansatz

Suppose we have found a translation-invariant PEPS representation for the ground state of a local Hamiltonian H. The tangent-space ansatz for elementary excitations is given by

$$|\Phi_{\vec{q}}(B)\rangle = \sum_{i\in\mathcal{L}} e^{i\vec{q}\cdot\vec{n}_i}$$ 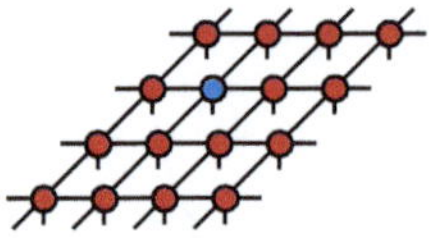 (5.10)

where the red tensors correspond to those of an optimal ground state and only the blue circle represents a new tensor B at site i. This ansatz thus represents the momentum superposition of a local perturbation, which has a finite width (determined by the PEPS bond dimension) by acting via the virtual degrees of freedom.

The blue tensor B, containing all variational freedom in the ansatz, has the same dimensions as the ground state tensor A, yet the number of variational parameters is smaller because of a redundancy in the parametrization. Indeed, through simple insertion one can easily check that the tensors (where the green square is a $D \times D$ matrix)

$B_{0,x} =$ 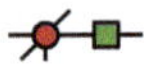 $- e^{iq_x}$ 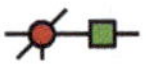

and

$B_{0,y} =$  $- e^{iq_y}$ 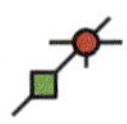

give rise to states with zero norm. Correspondingly, the *effective norm matrix* $\mathsf{N}_{\text{eff}}(\vec{q})$, defined as

$$\langle\Phi_{\vec{q}'}(B')|\Phi_{\vec{q}}(B)\rangle = 4\pi^2\delta(q_x - q_x')\delta(q_y - q_y')\ \boldsymbol{B}'\,\mathsf{N}_{\text{eff}}(\vec{q})\ \boldsymbol{B},$$

will have a number of zero eigenvalues ($\boldsymbol{B}$ is the vector containing all elements of the tensor B). In order for the variational subspace to be well-defined, we will always project out these null modes. In addition, since we want an excitation to be orthogonal to the ground state, we will also project out the component that is proportional to the ground state tensor A.

Once we have defined the variational subspace, we can minimize the energy in order to find the best approximation to the true excitation. Since the subspace is linear, this can be done by solving the Rayleigh–Ritz problem

$$\mathsf{H}_{\text{eff}}(\vec{q})\boldsymbol{B} = \omega\mathsf{N}_{\text{eff}}(\vec{q})\boldsymbol{B}$$

where the *effective Hamiltonian matrix* $\mathsf{H}_{\text{eff}}(\vec{q})$ is similarly defined as (E_0 is the ground-state energy)

$$\langle\Phi_{\vec{q}'}(B')|H - E_0|\Phi_{\vec{q}}(B)\rangle = 4\pi^2\delta(q_x - q_x')\delta(q_y - q_y')\ \boldsymbol{B}'\,\mathsf{H}_{\text{eff}}(\vec{q})\ \boldsymbol{B}.$$

The matrix elements of $\mathsf{N}_{\text{eff}}(\vec{q})$ and $\mathsf{H}_{\text{eff}}(\vec{q})$ contain two- and three-point functions that can be computed with a channel environment.

The smallest eigenvalue ω corresponds to the lowest excitation energy at momentum (κ_x, κ_y). By repeating this procedure for different momenta, we can obtain the dispersion relation for all one-particle excitations in the system. Moreover, with an expression for the wave functions, we can straightforwardly compute their spectral weights.

5.5.2 The Eigenvalue Problem

In contrast to the one-dimensional case in Chap. 3, the effective norm matrix cannot be explicitly brought into diagonal form because here the environment is a fully connected object. Therefore we will have to solve the generalized eigenvalue problem,

$$\mathsf{H}_{\text{eff}}(\vec{q})\boldsymbol{B} = \omega \mathsf{N}_{\text{eff}}(\vec{q})\boldsymbol{B},$$

which can be solved iteratively, but, for which the action of the inverse of the effective norm matrix $\mathsf{N}_{\text{eff}}(\vec{q})$ has to be computed in every iteration. We can again compute the inverse efficiently by solving the linear equation

$$\boldsymbol{C}' = \mathsf{N}_{\text{eff}}(\vec{q})\boldsymbol{C}$$

with an iterative solver. An explicit computation of the effective norm matrix yields the following double (infinite) sum

$$\langle \Phi_{\vec{q}'}(B')|\Phi_{\vec{q}}(B)\rangle = \sum_{i,j\in\mathcal{L}} \mathrm{e}^{i(\vec{q}\cdot\vec{n}_i - \vec{q}'\cdot\vec{n}_j)} \left[B \text{ at site } \vec{n}_i \text{ and } B' \text{at site } \vec{n}_j\right].$$

Because the ground state is translation invariant, we can simplify to a single sum as

$$\begin{aligned}\langle \Phi_{\vec{q}'}(B')|\Phi_{\vec{q}}(B)\rangle = {} & 2\pi\delta(q_x - q_x')2\pi\delta(q_y - q_y') \\ & \times \sum_{m,n'} \mathrm{e}^{iq_x m}\mathrm{e}^{-iq_y n'} \left[B \text{ at site } (m,0) \text{ and } B' \text{ at site } (0,n')\right].\end{aligned}$$

This expression is, of course, very similar to the evaluation of the structure factor in Sect. 5.2.4. In fact, we can just copy the different diagrams,

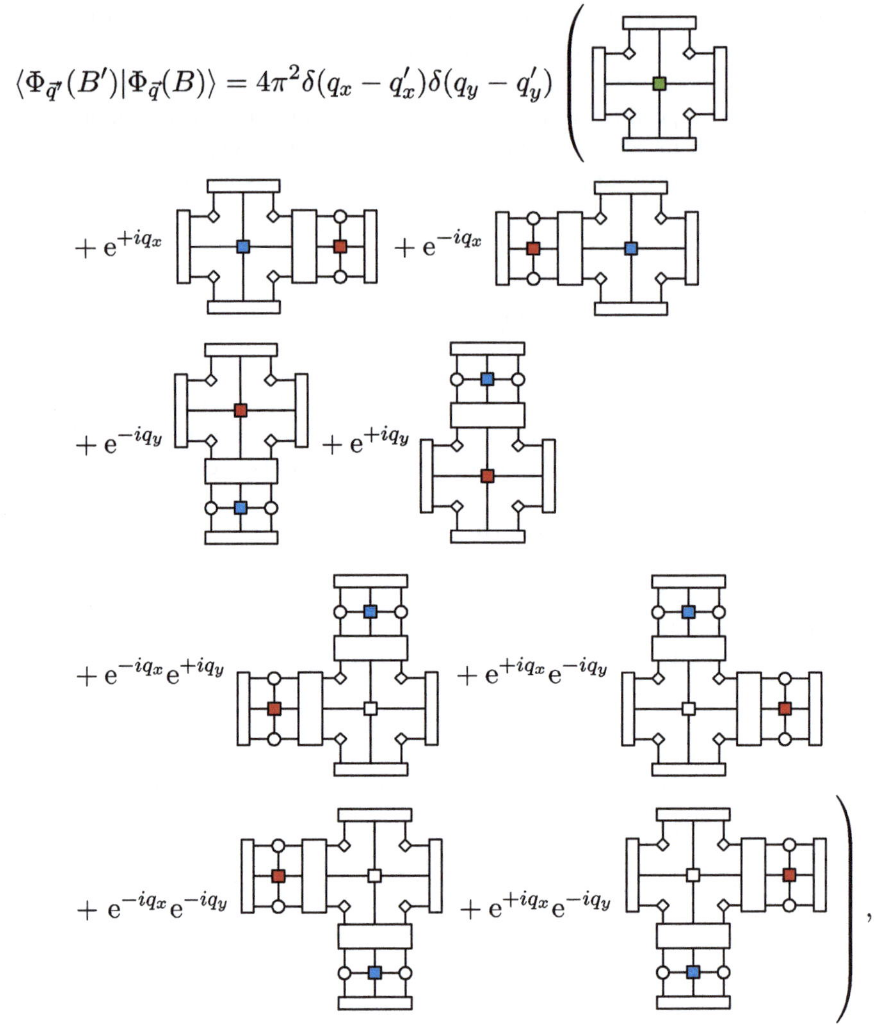

(5.11)

where the green tensor is now obtained by contracting B and B' on the same site, the red tensors indicate the location of the B tensor and the blue tensors the B' tensor; the big channel operators again represent the geometric sum with the proper momentum factors. In an iterative solver, only the action of this expression on a tensor B is needed, which can be computed efficiently. Indeed, all the inverse channel operators can be evaluated iteratively by only computing their action on a vector.

The effective Hamiltonian matrix is essentially a three-point function, which can also be computed with the corner environments. A typical diagram would be

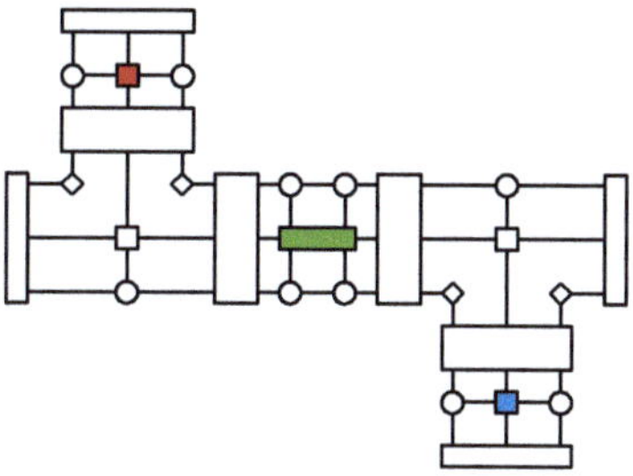

which can be contracted efficiently. The full expression contains a large number of similar terms, and we will not list all terms here, but it should be obvious that an efficient implementation is feasible.

5.6 Excitations in Frustration-Free Systems

The excitation ansatz is completely general in the sense that, given a PEPS approximation for the ground state of a local Hamiltonian, we can compute the excitation spectrum. In this section, however, we will confine ourselves to frustration-free models for which the ground state can be represented exactly as a PEPS with low bond dimension. This implies that the action of the Hamiltonian on the PEPS ground state is zero locally, so that all disconnected terms in the effective Hamiltonian matrix vanish.

5.6.1 Two-Dimensional AKLT Model

As a first application, we study the two-dimensional AKLT model [52, 53] on the square lattice. The one-dimensional AKLT model was first introduced to establish the existence of a gap in a rotationally invariant spin-1 chain, in support of the Haldane conjecture [54, 55]. The construction can be straightforwardly extended to two-dimensional systems, but the existence of a gap – although highly expected – can no longer be established rigorously. On the square lattice the AKLT state can be represented as a PEPS with bond dimension $D = 2$, and is the unique ground state of

$$H_{\text{AKLT}} = \frac{1}{28} \sum_{\langle ij \rangle} h_{ij} + \frac{7}{10} h_{ij}^2 + \frac{7}{45} h_{ij}^3 + \frac{1}{90} h_{ij}^4$$

with $h_{ij} = \vec{S}_i \cdot \vec{S}_j$ the spin-2 Heisenberg interaction.

Let us first study the static correlations of the AKLT state. In Refs. [56, 57] it was argued that these correlations can be expressed as thermal averages of a related classical model in the same number of dimensions, contrary to what one expects for

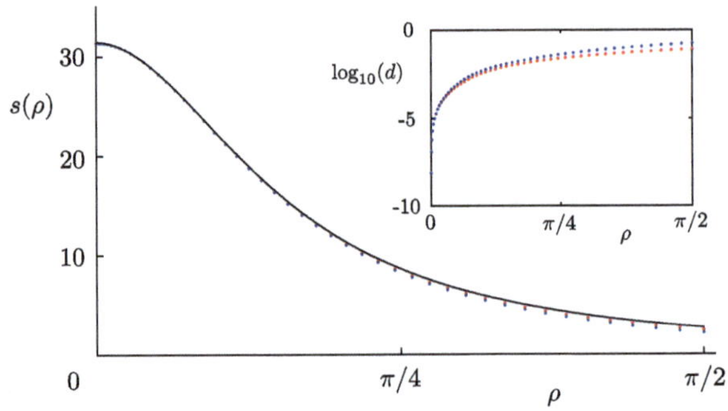

Fig. 5.8 The structure factor $s(\rho)$ where $\rho = ((\kappa_x - \pi)^2 + (\kappa_y - \pi)^2)^{1/2}$ for the two-dimensional AKLT model along the axes $\kappa_y = \pi$ (*red*) and $\kappa_x = \kappa_y$ (*blue*) compared to the form in Eq. (5.12). In the *inset* the ($\log_{10}$ of the) deviations are plotted. The *plot* shows that the classical Ornstein–Zernike form is accurate for a large portion of the Brillouin zone and that the structure factor is nicely isotropic around $\vec{\kappa}^* = (\pi, \pi)$

generic quantum ground states. The real-space correlations then decay according to the 2D Ornstein–Zernike form

$$s(\vec{n}) \propto \mathrm{e}^{-|\vec{n}|/\xi} \frac{\mathrm{e}^{i\vec{q}^*\cdot\vec{n}}}{\sqrt{|\vec{n}|}} \qquad (|\vec{n}| \gg 1)$$

with $\vec{q}^* = (\pi, \pi)$ the oscillation period. In momentum space, the structure factor is thus given by

$$s(\vec{q}) = \sum_{i\in\mathcal{L}} \mathrm{e}^{-i\vec{q}\cdot\vec{n}_i} s(\vec{n}_i) \propto \frac{1}{1 + \xi^2|\vec{q} - \vec{q}^*|^2} \tag{5.12}$$

for momenta close to $\vec{q}^*$ where $s(\vec{q})$ reaches its maximum. We can confirm this result with our methods. Firstly, we can calculate the correlation length by computing the gap of the linear transfer matrix $\mathcal{T}$ using the methods of Ref. [58]. We obtain the value $\xi_{\text{AKLT}} = -1/\log|\lambda_{\mathcal{T}}^{(2)}| = 2.06491$, in reasonable agreement with the value of $\xi^{-1} \approx 0.52$ in Ref. [59]. In Fig. 5.8 we compare the computation of the structure factor with our methods with the form of Eq. (5.12) and observe very good agreement in a large portion of the Brillouin zone.

Based on the behaviour of the structure factor, the elementary excitations are expected to be (threefold-degenerate) magnons. In a first step, we can target the magnons with the single-mode approximation (SMA), which is known to reproduce the one-particle dispersion relation qualitatively in the one-dimensional case [56]. In the present context the SMA wave function is given by ($\alpha = x, y, z$)

$$|\Phi_{\vec{q}}^{\alpha}\rangle_{\text{SMA}} = \sum_{i\in\mathcal{L}} \mathrm{e}^{i\vec{q}\cdot\vec{n}_i} S_i^{\alpha} |\Psi_{\text{AKLT}}\rangle.$$

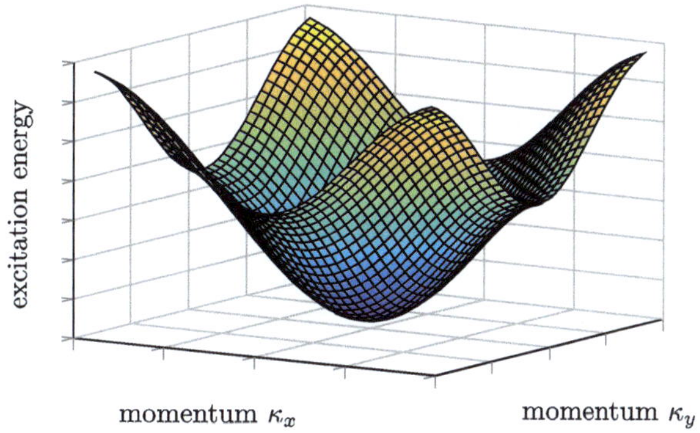

Fig. 5.9 Single mode approximation for the dispersion relation of the two-dimensional AKLT model on a square lattice. The minimum of the triplet dispersion relation is at momentum $\kappa_{x,y} = \pi$

The norm of these states $\langle \Phi^{\alpha}_{\vec{q}} | \Phi^{\alpha}_{\vec{q}} \rangle$ is equal to the static structure factor, while the energy expectation value reduces to an easy PEPS contraction. The dispersion relation is shown in Fig. 5.9; the spectrum consists of an elementary triplet with its minimum at $\vec{q} = (\pi, \pi)$ and a gap $\Delta_{\text{SMA}} = 0.0199$.

Next we can determine the excitations variationally with the excitation ansatz [Eq. (5.10)]. To improve on the SMA result, we enlarge our variational subspace by perturbing the ground state on a larger region.[4] By introducing a B tensor on a block of two by two sites, we are able to estimate the gap at $\Delta_{\text{var}} = 0.0147$,[5] in excellent agreement with the value $2\Delta = 0.03$ in Ref. [60], obtained through a computation of the magnetization curve. In addition, we can compute the characteristic velocity through the second derivative of the dispersion relation in its minimum and we obtain $v_{\text{var}} = 0.04115$. Finally, in Table 5.1 we have listed the spectral weight of the elementary magnon around its minimum.

By assuming an effective Lorentz invariance around the minimum, we can – via the Källén–Lehmann representation of the correlation functions $s(\vec{q}) \propto (\Delta^2 + v^2|\vec{q} - \vec{q}^*|^2)^{-1/2}$ – estimate the correlation length variationally as

$$\xi_{\text{var}} = \sqrt{\frac{v_{\text{var}}^2}{2\Delta_{\text{var}}^2}} = 1.975$$

in reasonable agreement with the above value. Still better agreement should be possible by growing the block size in the ansatz, though this soon becomes computationally intractable; for block size 2×2 the number of variational parameters is already $\mathcal{O}(D^8 d^4)$.

[4]This procedure of systematically growing the block size was explained in Sect. 3.3.6 for the one-dimensional case.

[5]This value is quite small because of the prefactor of $\frac{1}{28}$ making the AKLT Hamiltonian a projector.

Table 5.1 Spectral weight $w = \sum_\alpha |\langle\Phi_\alpha|S^z|\Psi(A)\rangle|^2$ of the elementary magnon triplet as a function of momentum $q_x = q_y = \pi + \rho/\sqrt{2}$, and the percentage of the integrated spectral function that is saturated in the one-particle sector. We observe that the magnon contains nearly all spectral weight in the minimum, and that this percentage drops when going away from this point

ρ	Spectral weight w	% of sum rule
0.0000π	31.222	99.57
0.0682π	26.058	99.35
0.1364π	17.198	98.84
0.2045π	10.672	96.59
0.2273π	8.9968	94.26
0.2500π	7.3452	88.75

5.6.2 *The Filtered Toric Code Model*

As a second example we study the anyonic excitations in the toric code [61], the easiest example of a model exhibiting topological order and anyonic excitations (fluxes and charges). At the toric code fixed point, the anyons have a flat dispersion relation since all Hamiltonian terms commute. To generate non-trivial dynamics, we perturb the toric code state corresponding to a non-unitary operator

$$f_i = \exp\left(\frac{1}{4}(\beta_x \sigma_i^x + \beta_z \sigma_i^z)\right).$$

acting on every site i. The associated Hamiltonian is

$$H_{\mathrm{FTC}} = \sum_s \tilde{H}_s + \sum_p \tilde{H}_p,$$

where the filtered star operators are given by

$$\tilde{H}_s = \left(\prod_{i\in s} f_i^{-1}\right)\left(1 - \prod_i \sigma_i^x\right)\left(\prod_{i\in s} f_i^{-1},\right)$$

and similar for the plaquette operator $\tilde{H}_p$ with σ_i^z for $i \in p$. In first order, the filtering operation is equivalent to applying a magnetic field to the toric code. In fact, it has been shown in Refs. [58, 62] that the phase diagram is qualitatively similar.

As we have discussed in Sect. 2.6.5, a PEPS can provide a natural description of topological phases, where the topological order of the global state is reflected by a symmetry of the local PEPS tensor A on the virtual level. A variational ansatz for the complete set of elementary excitations in the different anyon sectors is obtained by attaching a half-infinite virtual matrix product operator string to the local B tensor in the excitation ansatz [Eq. (5.10)]. The topological sector is encoded in the type of

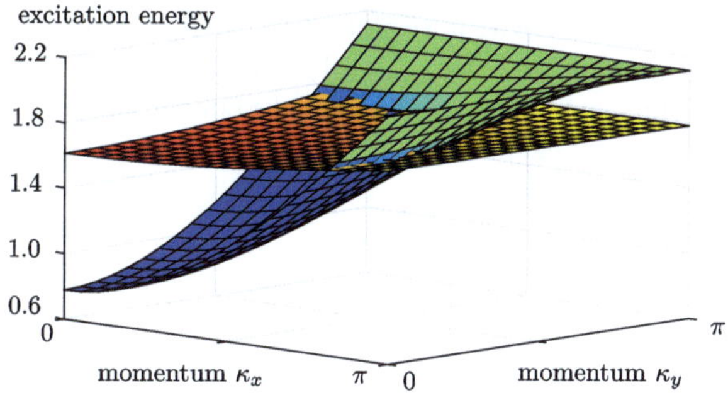

Fig. 5.10 The dispersion relation of the toric code in the flux (*blue-green*) and charge (*red-yellow*) sector for filtering parameters $\beta_x = 0.05$ and $\beta_z = 0.35$. For the former, we have used the topological ansatz with a string of σ^z operators at the virtual level, whereas the latter was obtained with an excitation that carries no string but transforms according to the odd representation of the virtual $\mathbb{Z}_2$ symmetry (note that we did not impose this symmetry on the B-tensor in Eq. (5.10) explicitly)

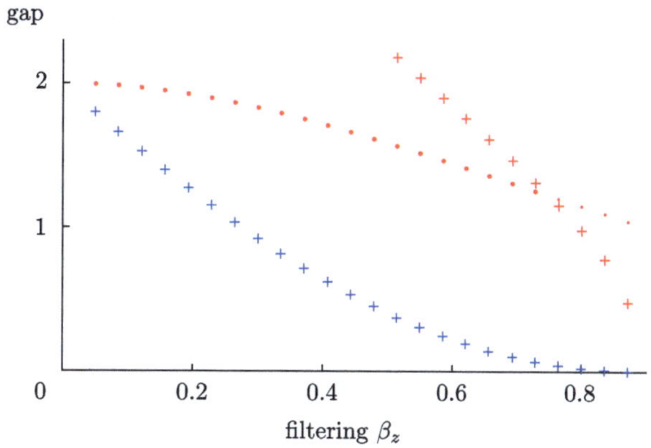

Fig. 5.11 The lowest-lying excitation energies in the charge and flux sector in function of β_z and with $\beta_x = 0$, for which the model can be shown to have a phase transition at $\beta_z = \log(\sqrt{2}+1) \approx 0.88$ [63]. The *color* represents trivial (*red*) or non-trivial (*blue*) flux of the excitations, whereas the symbol represents even (+) or odd (·) charge number. We can observe that the gap in the flux sector closes, signaling the condensation of the flux excitations. Although our variational ansatz is not suited for describing multi-particle states, we can see the two-flux state (even charge in the trivial flux sector) coming down in energy close to the phase transition. The charge excitation remains gapped

string and the virtual symmetry representation of the local tensor [8]. In the case of the toric code, the flux of the excitations is encoded in the presence or absence of a string of σ^z operators, whereas the charge is encoded in the symmetry representation of the B tensor [4]. These topological characteristics are unaltered by the filtering.

In Fig. 5.10 we have plotted the elementary excitation spectrum in both the charge and flux sector for a specific value of the filtering parameters. We can see that the filtering in one direction mostly affects one of the two excitations. This is reflected more clearly in Fig. 5.11, where we have plotted the gap in both sectors along the $\beta_x = 0$ axis. At the critical point $\beta_z = \log(\sqrt{2}+1)$, one can observe that the gap to the one-flux and many-flux states closes, indicating the condensation of flux excitations. The charge sector remains gapped but will cease to exist beyond the transition, for reasons explained in Ref. [58].

5.7 Outlook

In the beginning of this chapter we have already indicated that in this dissertation we only provide the first steps towards an effective particle framework for generic spin systems in two dimensions. Nonetheless, we are confident that the contraction scheme, ground-state optimization algorithm and excitation ansatz will lead the way to a full implementation of tangent-space methods for PEPS. In this last section, we list some further steps that should be taken towards that goal.

Better Optimization Algorithms

From the perspective of numerical optimization, a conjugate-gradient search is only a first step to more advanced schemes such as Newton or quasi-Newton methods. The Hessian of the energy function is crucial in these optimization schemes, the evaluation of which is straightforward with our effective environment. Alternatively, the non-trivial geometric structure of the PEPS manifold can be taken into account in the optimization.

The optimization of PEPS can also greatly benefit from imposing symmetries on the PEPS tensor. In the simulations above, we have always imposed rotation invariance, but other symmetries might reduce the computational cost significantly. Also, it should prove worthwile to reduce the number of variational parameters by trying to figure out better effective parametrizations. Also, imposing a certain gauge fixing on the PEPS tensor might render the optimisation more efficient.

These different improvements should in the end lead to a black-box code, possibly complemented with other PEPS approaches, that can give valuable insight into the ground-state properties of generic two-dimensional quantum spin systems.

Real-Time Evolution for PEPS

Throughout this dissertation, we have touched multiple times on the time-dependent variational principle [Sect. 3.5], which gives a prescription for simulating time evolution within a variational manifold. In particular, the time evolution of the PEPS tensor A is derived by projecting the Schrödinger equation on the tangent space at every time step, i.e.

$$i\frac{\partial}{\partial t}|\Psi(A)\rangle = H|\Psi(A)\rangle \qquad \rightarrow \qquad i\frac{\partial}{\partial t}|\Psi(A(t)\rangle = \mathsf{P}_{\mathbb{T}}H|\Psi(A(t))\rangle,$$

where the right-hand side is a tangent vector. The corner environment that we introduced above, gives us all the contractions that we need for integrating this differential equation. This implies that we can straightforwardly implement real-time evolution for generic two-dimensional systems.

Quasiparticle Excitations for Generic Models

Finally, we should implement the excitation ansatz for generic, i.e. non-frustration-free, Hamiltonians. This would give us the key for studying the collective low-energy modes of two-dimensional quantum phases, such as quantum spin liquids and topologically-ordered systems. In particular, by using the representation of topological symmetries on the virtual level of the PEPS, we can capture fractionalized quantum numbers and anyonic statistics.

Of course, this would only be the first step for developing an effective particle theory, as we did for one-dimensional systems in the previous chapters. Indeed, a formulation of the scattering problem in two dimensions would lead to the definition of the S matrix, which we expect to have drastically different properties than its one-dimensional counterpart. Moreover, we expect that an accurate resolution of the spectral functions will lay bare the quasiparticle interactions and the physics of bound-state formation.

References

1. F. Verstraete, J.I. Cirac, Renormalization algorithms for quantum-many body systems in two and higher dimensions (2004), arXiv:cond-mat/0407066
2. R. Orús, A practical introduction to tensor networks: matrix product states and projected entangled pair states. Ann. Phys. **349**, 117 (2014). doi:10.1016/j.aop.2014.06.013
3. J. Eisert, M. Cramer, M.B. Plenio, Colloquium: area laws for the entanglement entropy. Rev. Mod. Phys. **82**, 277 (2010). doi:10.1103/RevModPhys.82.277
4. N. Schuch, J.I. Cirac, D. Pérez-García, PEPS as ground states: degeneracy and topology. Ann. Phys. **325**, 2153 (2010). doi:10.1016/j.aop.2010.05.008
5. T. Wahl, H.-H. Tu, N. Schuch, J.I. Cirac, Projected entangled-pair states can describe chiral topological states. Phys. Rev. Lett. **111**, 236805 (2013). doi:10.1103/PhysRevLett.111.236805
6. M.B. Şahinoğlu, D. Williamson, N. Bultinck, M. Mariën, J. Haegeman, N. Schuch, F. Verstraete, Characterizing topological order with matrix product operators (2014), arXiv:1409.2150
7. O. Buerschaper, Twisted injectivity in projected entangled pair states and the classification of quantum phases. Ann. Phys. **351**, 447 (2014). doi:10.1016/j.aop.2014.09.007
8. N. Bultinck, M. Mariën, D.J. Williamson, M.B. Şahinoğlu, J. Haegeman, F. Verstraete, Anyons and matrix product operator algebras (2015), arXiv:1511.08090
9. J.I. Cirac, D. Poilblanc, N. Schuch, F. Verstraete, Entanglement spectrum and boundary theories with projected entangled-pair states. Phys. Rev. B **83**, 245134 (2011). doi:10.1103/PhysRevB.83.245134
10. N. Schuch, D. Poilblanc, J.I. Cirac, D. Pérez-García, Topological order in the projected entangled-pair states formalism: transfer operator and boundary Hamiltonians. Phys. Rev. Lett. **111**, 090501 (2013). doi:10.1103/PhysRevLett.111.090501
11. S. Yang, L. Lehman, D. Poilblanc, K. Van Acoleyen, F. Verstraete, J.I. Cirac, N. Schuch, Edge theories in projected entangled pair state models. Phys. Rev. Lett. **112**, 036402 (2014). doi:10.1103/PhysRevLett.112.036402
12. J. Jordan, R. Orús, G. Vidal, F. Verstraete, J.I. Cirac, Classical simulation of infinite-size quantum lattice systems in two spatial dimensions. Phys. Rev. Lett. **101**, 250602 (2008). doi:10.1103/PhysRevLett.101.250602

13. H. Jiang, Z. Weng, T. Xiang, Accurate determination of tensor network state of quantum lattice models in two dimensions. Phys. Rev. Lett. **101**, 090603 (2008). doi:10.1103/PhysRevLett.101.090603
14. H.N. Phien, J.A. Bengua, H.D. Tuan, P. Corboz, R. Orús, Infinite projected entangled pair states algorithm improved: fast full update and gauge fixing. Phys. Rev. B **92**, 035142 (2015). doi:10.1103/PhysRevB.92.035142
15. L. Wang, S.G. Chung, Entanglement perturbation theory for infinite quasi-1D quantum systems (2011), arXiv:1108.2532
16. P. Corboz, M. Lajkó, A.M. Läuchli, K. Penc, F. Mila, Spin-orbital quantum liquid on the honeycomb lattice. Phys. Rev. X **2**, 041013 (2012). doi:10.1103/PhysRevX.2.041013
17. H.H. Zhao, C. Xu, Q.N. Chen, Z.C. Wei, M.P. Qin, G.M. Zhang, T. Xiang, Plaquette order and deconfined quantum critical point in the spin-1 bilinear-biquadratic Heisenberg model on the honeycomb lattice. Phys. Rev. B **85**, 134416 (2012). doi:10.1103/PhysRevB.85.134416
18. Z.-C. Gu, H.-C. Jiang, D.N. Sheng, H. Yao, L. Balents, X.-G. Wen, Time-reversal symmetry breaking superconducting ground state in the doped Mott insulator on the honeycomb lattice. Phys. Rev. B **88**, 155112 (2013). doi:10.1103/PhysRevB.88.155112
19. P. Corboz, F. Mila, Tensor network study of the Shastry-Sutherland model in zero magnetic field. Phys. Rev. B **87**, 115144 (2013). doi:10.1103/PhysRevB.87.115144
20. J.O. Iregui, P. Corboz, M. Troyer, Probing the stability of the spin-liquid phases in the Kitaev-Heisenberg model using tensor network algorithms. Phys. Rev. B **90**, 195102 (2014). doi:10.1103/PhysRevB.90.195102
21. P. Corboz, F. Mila, Crystals of bound states in the magnetization plateaus of the Shastry-Sutherland model. Phys. Rev. Lett. **112**, 147203 (2014). doi:10.1103/PhysRevLett.112.147203
22. T. Picot, D. Poilblanc, Nematic and supernematic phases in kagome quantum antiferromagnets under the influence of a magnetic field. Phys. Rev. B **91**, 064415 (2015). doi:10.1103/PhysRevB.91.064415
23. T. Picot, M. Ziegler, R. Orús, D. Poilblanc, Spin-S kagome quantum antiferromagnets in a field with tensor networks. Phys. Rev. B **93**, 060407 (2016). doi:10.1103/PhysRevB.93.060407
24. P. Nataf, M. Lajkó, P. Corboz, A.M. Läuchli, K. Penc, F. Mila, SU(6) Heisenberg model on the honeycomb lattice: competition between plaquette and chiral order (2016), arXiv:1601.00959
25. P. Corboz, Improved energy extrapolation with infinite projected entangled-pair states applied to the two-dimensional Hubbard model. Phys. Rev. B **93**, 045116 (2016). doi:10.1103/PhysRevB.93.045116
26. P. Corboz, R. Orús, B. Bauer, G. Vidal, Simulation of strongly correlated fermions in two spatial dimensions with fermionic projected entangled-pair states. Phys. Rev. B **81**, 165104 (2010). doi:10.1103/PhysRevB.81.165104
27. P. Corboz, J. Jordan, G. Vidal, Simulation of fermionic lattice models in two dimensions with projected entangled-pair states: next-nearest neighbor Hamiltonians. Phys. Rev. B **82**, 245119 (2010). doi:10.1103/PhysRevB.82.245119
28. P. Corboz, T.M. Rice, M. Troyer, Competing states in the t - J model: uniform d-wave state versus stripe state. Phys. Rev. Lett. **113**, 046402 (2014). doi:10.1103/PhysRevLett.113.046402
29. M. Lubasch, J.I. Cirac, M.-C. Bañuls, Algorithms for finite projected entangled pair states. Phys. Rev. B **90**, 64425 (2014). doi:10.1103/PhysRevB.90.064425
30. R.J. Baxter, Dimers on a rectangular lattice. J. Math. Phys. **9**, 650 (1968). doi:10.1063/1.1664623
31. R. Orús, G. Vidal, Infinite time-evolving block decimation algorithm beyond unitary evolution. Phys. Rev. B **78**, 155117 (2008). doi:10.1103/PhysRevB.78.155117
32. J. Haegeman, in preparation
33. J. Haegeman, C. Lubich, I. Oseledets, B. Vandereycken, F. Verstraete, Unifying time evolution and optimization with matrix product states (2014), arXiv:1408.5056
34. R.J. Baxter, Corner transfer matrices. Phys. A **106**, 18 (1981). doi:10.1016/0378-4371(81)90203-X
35. T. Nishino, K. Okunishi, Corner transfer matrix renormalization group method. J. Phys. Soc. Jpn. **65**, 891 (1996). doi:10.1143/JPSJ.65.891

36. T. Nishino, K. Okunishi, Corner transfer matrix algorithm for classical renormalization group. J. Phys. Soc. Jpn. **66**, 3040 (1997). doi:10.1143/JPSJ.66.3040
37. R. Orús, G. Vidal, Simulation of two-dimensional quantum systems on an infinite lattice revisited: corner transfer matrix for tensor contraction. Phys. Rev. B **80**, 094403 (2009). doi:10.1103/PhysRevB.80.094403
38. R. Orús, Exploring corner transfer matrices and corner tensors for the classical simulation of quantum lattice systems. Phys. Rev. B **85**, 205117 (2012). doi:10.1103/PhysRevB.85.205117
39. J.F. Bonnans, J.C. Gilbert, C. Lemaréchal, C.A. Sagastizábal, *Numerical Optimization* (Springer, Berlin, 2006)
40. J. Nocedal, S. Wright, *Numerical Optimization* (Springer, Berlin, 2006)
41. W.W. Hager, H. Zhang, A survey of nonlinear conjugate gradient methods. Pac. J. Optim. **2**, 35 (2006), http://www.caam.rice.edu/~yzhang/caam554/pdf/cgsurvey.pdf
42. R. Fletcher, C. Reeves, Function minimization by conjugate gradients. Comput. J. **7**, 149 (1964)
43. S. Sorella, Generalized Lanczos algorithm for variational quantum Monte Carlo. Phys. Rev. B **64**, 024512 (2001). doi:10.1103/PhysRevB.64.024512
44. H.W.J. Blöte, Y. Deng, Cluster Monte Carlo simulation of the transverse Ising model. Phys. Rev. E **66**, 066110 (2002). doi:10.1103/PhysRevE.66.066110
45. E. Manousakis, The spin-1/2 Heisenberg antiferromagnet on a square lattice and its application to the cuprous oxides. Rev. Mod. Phys. **63**, 1 (1991). doi:10.1103/RevModPhys.63.1
46. B. Bauer, G. Vidal, M. Troyer, Assessing the accuracy of projected entangled-pair states on infinite lattices. J. Stat. Mech.: Theory Exp. **2009**, P09006 (2009). doi:10.1088/1742-5468/2009/09/P09006
47. A.W. Sandvik, C.J. Hamer, Ground-state parameters, finite-size scaling, and low-temperature properties of the two-dimensional S=1/2 XY model. Phys. Rev. B **60**, 6588 (1999). doi:10.1103/PhysRevB.60.6588
48. A.W. Sandvik, Finite-size scaling of the ground-state parameters of the two-dimensional Heisenberg model. Phys. Rev. B **56**, 11678 (1997). doi:10.1103/PhysRevB.56.11678
49. A.W. Sandvik, in *AIP Conference Proceedings*, vol. 1297 (2010), pp. 135–338. doi:10.1063/1.3518900
50. W. Zheng, J. Oitmaa, C.J. Hamer, Series studies of the spin-1/2 Heisenberg antiferromagnet at T=0: magnon dispersion and structure factors. Phys. Rev. B **71**, 184440 (2005). doi:10.1103/PhysRevB.71.184440
51. A. Lüscher, A.M. Läuchli, Exact diagonalization study of the antiferromagnetic spin-1/2 Heisenberg model on the square lattice in a magnetic field. Phys. Rev. B **79**, 195102 (2009). doi:10.1103/PhysRevB.79.195102
52. I. Affleck, T. Kennedy, E.H. Lieb, H. Tasaki, Rigorous results on valence-bond ground states in antiferromagnets. Phys. Rev. Lett. **59**, 799 (1987). doi:10.1103/PhysRevLett.59.799
53. I. Affleck, T. Kennedy, E.H. Lieb, H. Tasaki, Valence bond ground states in isotropic quantum antiferromagnets. Commun. Math. Phys. **115**, 477 (1988). doi:10.1007/BF01218021
54. F.D.M. Haldane, Nonlinear field theory of large-spin Heisenberg antiferromagnets: semiclassically quantized solitons of the one-dimensional easy-axis Néel state. Phys. Rev. Lett. **50**, 1153 (1983). doi:10.1103/PhysRevLett.50.1153
55. F.D.M. Haldane, Continuum dynamics of the 1-D Heisenberg antiferromagnet: identification with the O(3) nonlinear sigma model. Phys. Lett. A **93**, 464 (1983a). doi:10.1016/0375-9601(83)90631-X
56. D.P. Arovas, A. Auerbach, F.D.M. Haldane, Extended Heisenberg models of antiferromagnetism: analogies to the fractional quantum Hall effect. Phys. Rev. Lett. **60**, 531 (1988). doi:10.1103/PhysRevLett.60.531
57. D.P. Arovas, S.M. Girvin, Exact questions to some interesting answers in many body physics, in *Recent Progress in Many-Body Theories*, ed. by T.L. Ainsworth, C.E. Campbell, B.E. Clements, E. Krotscheck (Springer, Boston, 1992). doi:10.1007/978-1-4615-3466-2
58. J. Haegeman, V. Zauner, N. Schuch, F. Verstraete, Shadows of anyons and the entanglement structure of topological phases. Nat. Commun. **6**, 8284 (2015a). doi:10.1038/ncomms9284

59. Y. Hieida, K. Okunishi, Y. Akutsu, Numerical renormalization approach to two-dimensional quantum antiferromagnets with valence-bond-solid type ground state. New J. Phys. **1**, 7 (1999). doi:10.1088/1367-2630/1/1/007
60. A. Garcia-Saez, V. Murg, T.-C. Wei, Spectral gaps of Affleck-Kennedy-Lieb-Tasaki Hamiltonians using tensor network methods. Phys. Rev. B **88**, 245118 (2013). doi:10.1103/PhysRevB.88.245118
61. A. Kitaev, Fault-tolerant quantum computation by anyons. Ann. Phys. **303**, 2 (2003). doi:10.1016/S0003-4916(02)00018-0
62. J. Haegeman, K. Van Acoleyen, N. Schuch, J.I. Cirac, F. Verstraete, Gauging quantum states: from global to local symmetries in many-body systems. Phys. Rev. X **5**, 011024 (2015c). doi:10.1103/PhysRevX.5.011024
63. C. Castelnovo, C. Chamon, Quantum topological phase transition at the microscopic level. Phys. Rev. B **77**, 054433 (2008). doi:10.1103/PhysRevB.77.054433

Appendix A
Technical Details

A.1 Variance Density of The Ground State

Uniform matrix product states prove to offer a very accurate description of ground states of gapped, translation invariant Hamiltonians in the thermodynamic limit. For simplicity's sake, we will restrict to nearest neighbour interactions, so that $\hat{H} = \sum_n \hat{h}_{n,n+1}$. Having found a variationally optimal tensor A for this Hamiltonian (with variational energy density e_0), we can calculate its variance with respect to the state $|\Psi(A)\rangle$ to get an idea of how well it approximates the true ground state. This variance scales with the system size, however, so we should define a local state error as

$$\epsilon_{\text{GS}} = \frac{1}{|\mathcal{L}|}\Delta H_{\text{GS}} = \frac{1}{|\mathbb{Z}|}\langle\Psi(A)|\hat{H}^2|\Psi(A)\rangle,$$

where $|\mathcal{L}|$ represents the diverging system size and we have redefined the Hamiltonian as $h_{n,n+1} \rightarrow h_{n,n+1} - e_0$. A simple calculation shows that the local state error is equal to

$$\begin{aligned}
\epsilon_{\text{GS}} &= \frac{1}{|\mathcal{L}|}\sum_n\sum_{n'}\langle\Psi(A)|\hat{h}_{n,n+1}\hat{h}_{n',n'+1}|\Psi(A)\rangle \\
&= \sum_{n'}\langle\Psi(A)|\hat{h}_{0,1}\hat{h}_{n',n'+1}|\Psi(A)\rangle \\
&= 2\times(l|H^{AA}_{AA}\sum_{n=0}^{+\infty}E^n H^{AA}_{AA}|r) + (l|H^{AA(A)}_{(A)AA}|r) + (l|H^{(A)AA}_{AA(A)}|r) + (l|J^{AA}_{AA}|r)
\end{aligned}$$

where we have used the following notations

$$H^{AB}_{CD} = \sum_{ss'tt'} A^s B^t \otimes \overline{C}^{s'}\overline{D}^{t'}\langle s't'|\hat{h}|st\rangle$$

L. Vanderstraeten, *Tensor Network States and Effective Particles for Low-Dimensional Quantum Spin Systems*, Springer Theses,
DOI 10.1007/978-3-319-64191-1

$$H_{DE(F)}^{(A)BC} = \sum_{ss'tt'uu'} A^s B^t C^u \otimes \overline{D}^{s'} \overline{E}^{t'} \overline{F}^{u'} \times \sum_v \langle vu'|\hat{h}|tu\rangle\langle s't'|\hat{h}|sv\rangle$$

$$J_{CD}^{AB} = \sum_{ss'tt'} A^s B^t \otimes \overline{C}^{s'} \overline{D}^{t'} \times \sum_{vw} \langle vw|\hat{h}|st\rangle\langle s't'|\hat{h}|vw\rangle$$
$$= \sum_{ss'tt'} A^s B^t \otimes \overline{C}^{s'} \overline{D}^{t'} \times \langle s't'|\hat{h}^2|st\rangle.$$

The contribution from the infinite sum can again be computed by inverting the transfer matrix, such that we obtain

$$\epsilon_{\text{GS}} = 2 \times (l|H_{AA}^{AA}(1-E)^P H_{AA}^{AA}|r) + (l|H_{(A)AA}^{AA(A)}|r) + (l|H_{AA(A)}^{(A)AA}|r) + (l|J_{AA}^{AA}|r).$$

A.2 Variance of the One-Particle Ansatz

If we write the one-particle ansatz as

$$|\Phi_\kappa(B)\rangle = \sum_n e^{i\kappa n}|\chi(n)\rangle \quad \text{with } |\chi(n)\rangle = \sum_{\{s\}} \boldsymbol{v}_L^\dagger \left[\prod_{m<n} A^{s_m}\right] B^{s_n} \left[\prod_{m>n} A^{s_m}\right] \boldsymbol{v_R}|\{s\}\rangle,$$

where B is in the left gauge, such that the site dependent states are orthonormalized as

$$\langle\chi(n')|\chi(n)\rangle = \delta_{nn'}.$$

The variance of the Hamiltonian with respect to this state can be calculated as (we denote $\Delta(\kappa) = \langle\Phi_\kappa(B)|H|\Phi_\kappa(B)\rangle$)

$$\epsilon_{\text{EX}} = \langle\Phi_{\kappa'}(B)|(\hat{H} - \Delta(\kappa))^2|\Phi_\kappa(B)\rangle$$
$$= \sum_n e^{i\kappa n} \sum_{n'} e^{-i\kappa' n'} \langle\chi(n')|\hat{H}^2|\chi(n)\rangle - \Delta(\kappa)^2\langle\Phi_{\kappa'}(B)|\Phi_\kappa(B)\rangle$$
$$= 2\pi\delta(\kappa-\kappa')\left(\sum_{n'=-\infty}^{+\infty} e^{-i\kappa n'}\langle\chi(n')|\hat{H}^2|\chi(0)\rangle - \Delta(\kappa)^2\right).$$

Does this expression make sense? First of all, the sum breaks off for high enough n', i.e. $\langle\chi(n')|\hat{H}^2|\chi(0)\rangle \to 0$ for n' large enough, as we will see later on. The infinite δ-prefactor is also no problem as this is just the norm of the momentum superposition state. The contribution $\langle\chi(0)|\hat{H}^2|\chi(0)\rangle$ is somewhat more problematic however, as it contains an infinite contribution from the ground state error. Therefore, we subtract the (infinite) ground state variance from this expression. We get the following

$$\epsilon_{\text{EX}} = 2\pi\delta(\kappa-\kappa') \left[\langle\chi(0)|\hat{H}^2 - \Delta H_{\text{GS}}|\chi(0)\rangle + \sum_{n'=1}^{+\infty} \left(\mathrm{e}^{-i\kappa n'} \langle\chi(n')|\hat{H}^2|\chi(0)\rangle + c.c. \right) - \Delta(\kappa)^2 \right].$$

In the calculations it will become clear that this is indeed a finite expression.

The first contribution. We will first calculate the contribution $\langle\chi(0)|\hat{H}^2 - \Delta H_{\text{GS}}|\chi(0)\rangle$. We have two infinite sums and one infinite quantity in this contribution, so we have to be precise in our summations. We have

$$\langle\chi(0)|\hat{H}^2 - \Delta H_{\text{GS}}|\chi(0)\rangle = \sum_{n,n'} \left(\langle\chi(0)|\hat{h}_{n,n+1}\hat{h}_{n',n'+1}|\chi(0)\rangle - \langle\Psi[A]|\hat{h}_{n,n+1}\hat{h}_{n',n'+1}|\Psi[A]\rangle\langle\chi(0)|\chi(0)\rangle \right).$$

Every term for n can be calculated individually, making sure that the right number of ground state errors ϵ_{GS} is subtracted

$$\begin{aligned}
&\langle\chi(0)|\hat{H}^2 - \Delta H_{\text{GS}}|\chi(0)\rangle \\
&= \sum_{n=-\infty}^{-3} \Big[(l| \left(H^{AA}_{AA}(1-E)^P H^{AA}_{AA} E^A_A + H^{AA(A)}_{(A)AA} E^A_A + J^{AA}_{AA} E^A_A + H^{(A)AA}_{AA(A)} \right) (E^A_A)^{|n|-3} E^B_B |r) \\
&\quad + (l|H^{AA}_{AA} \sum_{n'=n+2}^{-2} (E^A_A)^{|n|-|n'|-2} H^{AA}_{AA} (E^A_A)^{|n'|-2} E^B_B |r) \\
&\quad + (l|H^{AA}_{AA}(E^A_A)^{|n|-3} \left(H^{AB}_{AB} + E^A_A H^{BA}_{BA} + E^A_A E^B_B (1-E)^P H^{AA}_{AA} \right) |r) - \epsilon_{\text{GS}} (l|E^B_B|r) \Big] \\
&+ (l| \Big(H^{AA}_{AA}(1-E)^P H^{AA}_{AA} E^B_B + H^{(A)AA}_{AA(A)} E^B_B + J^{AA}_{AA} E^B_B + H^{AA(B)}_{(A)AB} \\
&\qquad\qquad + H^{AA}_{AA} H^{BA}_{BA} + H^{AA}_{AA} E^B_B (1-E)^P H^{AA}_{AA} \Big) |r) - \epsilon_{\text{GS}} (l|E^B_B|r) \\
&+ (l| \left(H^{AA}_{AA}(1-E)^P H^{AB}_{AB} + H^{(A)AB}_{AA(B)} + J^{AB}_{AB} + H^{AB(A)}_{(A)BA} + H^{AB}_{AB}(1-E)^P H^{AA}_{AA} \right) |r) \\
&\quad - \epsilon_{\text{GS}} (l|E^B_B|r) \\
&+ (l| \left(H^{AA}_{AA}(1-E)^P H^{BA}_{BA} + H^{(A)BA}_{AB(A)} + J^{BA}_{BA} + H^{BA(A)}_{(B)AA} + H^{BA}_{BA}(1-E)^P H^{AA}_{AA} \right) |r) \\
&\quad - \epsilon_{\text{GS}} (l|E^B_B|r) \\
&+ (l| \Big(H^{AA}_{AA}(1-E)^P E^B_B H^{AA}_{AA} + H^{AB}_{AB} H^{AA}_{AA} + H^{(B)AA}_{BA(A)} + E^B_B J^{AA}_{AA} \\
&\qquad\qquad + E^B_B H^{AA(A)}_{(A)AA} + E^B_B H^{AA}_{AA}(1-E)^P H^{AA}_{AA} \Big) |r) - \epsilon_{\text{GS}} (l|E^B_B|r) \\
&+ \sum_{n=2}^{+\infty} \Big[(l| \left(H^{AA}_{AA}(1-E)^P E^B_B E^A_A + H^{AB}_{AB} E^A_A + H^{BA}_{BA} \right) (E^A_A)^{n-2} H^{AA}_{AA} |r) \\
&\quad + (l|E^B_B \sum_{n'=1}^{n-2} (E^A_A)^{n'-1} H^{AA}_{AA} (E^A_A)^{n-n'-2} H^{AA}_{AA} |r)
\end{aligned}$$

$$+ (l|E_B^B (E_A^A)^{n-2} \left(H_{AA(A)}^{(A)AA} + E_A^A J_{AA}^{AA} + E_A^A H_{(A)AA}^{AA(A)} + E_A^A H_{AA}^{AA} (1-E)^P H_{AA}^{AA} \right) |r)$$
$$- \epsilon_{\text{GS}} (l|E_B^B|r) \Big].$$

The infinite sums on the first and last three lines need to be investigated further. Separating all powers of E_A^A into connected and disconnected parts, the connected parts will yield finite results. This also enables to interchange the sums (with appropriate redefinition of the summation boundaries) in the double sum on the second and second to last line, so as to obtain for e.g. the latter

$$(l|E_B^B (1-E)^P H_{AA}^{AA} (1-E)^P H_{AA}^{AA} |r).$$

The disconnected and potentially diverging contributions that survive in e.g. the last three lines are given by

$$(l|E_B^B|r) \sum_{n=2}^{+\infty} \Big[\sum_{n'=1}^{n-2} (l|H_{AA}^{AA} (E_A^A)^{n-n'-2} H_{AA}^{AA} |r)$$
$$+ (l| \left(H_{AA(A)}^{(A)AA} + J_{AA}^{AA} + H_{(A)AA}^{AA(A)} + H_{AA}^{AA} (1-E)^P H_{AA}^{AA} \right) |r) - \epsilon_{\text{GS}} \Big].$$

By writing the $\sum_{n'=1}^{n-2} = \sum_{n'=-\infty}^{n-2} - \sum_{n'=-\infty}^{0}$ and substituting $n' \to -n'$ in the last sum, we obtain

$$(l|E_B^B|r) \sum_{n=2}^{+\infty} \Big[(l|H_{AA}^{AA} (1-E)^P H_{AA}^{AA} |r)$$
$$+ (l| \left(H_{AA(A)}^{(A)AA} + J_{AA}^{AA} + H_{(A)AA}^{AA(A)} + H_{AA}^{AA} (1-E)^P H_{AA}^{AA} \right) |r) - \epsilon_{\text{GS}} \Big]$$
$$- (l|E_B^B|r)(l|H_{AA}^{AA} (1-E)^P (1-E)^P H_{AA}^{AA} |r).$$

The terms in the remaining infinite sum exactly cancel thanks to presence of ϵ_{GS} and the finite result of the second line is obtained. A similar result is obtained from the disconnected part of the first three lines. Inserting this in the complete expression yields

$$\langle \chi(0)|\hat{H}^2 - \Delta H_{\text{GS}}|\chi(0)\rangle = (l| \left(H_{AA}^{AA} (1-E)^P H_{AA}^{AA} E_B^B + H_{AA(A)}^{(A)AA} E_B^B + J_{AA}^{AA} E_B^B + H_{(A)AB}^{AA(B)} \right.$$
$$\left. + H_{AA}^{AA} H_{BA}^{BA} + H_{AA}^{AA} E_B^B (1-E)^P H_{AA}^{AA} \right) |r)$$
$$+ (l| \left(H_{AA}^{AA} (1-E)^P H_{AB}^{AB} + H_{AA(B)}^{(A)AB} + J_{AB}^{AB} + H_{(A)BA}^{AB(A)} + H_{AB}^{AB} (1-E)^P H_{AA}^{AA} \right) |r)$$
$$+ (l| \left(H_{AA}^{AA} (1-E)^P H_{BA}^{BA} + H_{AB(A)}^{(A)BA} + J_{BA}^{BA} + H_{(B)AA}^{BA(A)} + H_{BA}^{BA} (1-E)^P H_{AA}^{AA} \right) |r)$$
$$+ (l| \left(H_{AA}^{AA} (1-E)^P E_B^B H_{AA}^{AA} + H_{AB}^{AB} H_{AA}^{AA} + H_{BA(A)}^{(B)AA} + E_B^B J_{AA}^{AA} \right.$$
$$\left. + E_B^B H_{(A)AA}^{AA(A)} + E_B^B H_{AA}^{AA} (1-E)^P H_{AA}^{AA} \right) |r)$$

$$
\begin{aligned}
&- 4 \times \epsilon_{\mathrm{GS}} \\
&+ (l| \Big(H^{AA}_{AA}(1-E)^P H^{AA}_{AA} E^A_A + H^{AA}_{AA}(1-E)^P H^{AA}_{AA} + H^{(A)AA}_{AA(A)} E^A_A \\
&\qquad\qquad + J^{AA}_{AA} E^A_A + H^{AA(A)}_{(A)AA} \Big) (1-E)^P E^B_B |r) \\
&+ (l| H^{AA}_{AA}(1-E)^P \Big(H^{AB}_{AB} + E^A_A H^{BA}_{BA} + E^A_A E^B_B (1-E)^P H^{AA}_{AA} \Big) |r) \\
&+ (l| E^B_B (1-E)^P \Big(H^{(A)AA}_{AA(A)} + E^A_A J^{AA}_{AA} + E^A_A H^{AA(A)}_{(A)AA} \\
&\qquad\qquad + H^{AA}_{AA}(1-E)^P H^{AA}_{AA} + E^A_A H^{AA}_{AA}(1-E)^P H^{AA}_{AA} \Big) |r) \\
&+ (l| \Big(H^{AA}_{AA}(1-E)^P E^B_B E^A_A + H^{AB}_{AB} E^A_A + H^{BA}_{BA} \Big) (1-E)^P H^{AA}_{AA} |r) \\
&- 2 \times (l| H^{AA}_{AA}(1-E)^P (1-E)^P H^{AA}_{AA} |r).
\end{aligned}
$$

All other contributions. Next we calculate $\langle \chi(0)|\hat{H}^2|\chi(1)\rangle$. No problems with subtracting an infinite amount of ground state errors is present here, so we have

$$
\begin{aligned}
\langle \chi(1)|\hat{H}^2|\chi(0)\rangle &= (l| \Big(2 \times H^{AA}_{AA}(1-E)^P H^{AA}_{AA}(1-E)^P E^B_A \\
&+ H^{(A)AA}_{AA(A)}(1-E)^P E^B_A + J^{AA}_{AA}(1-E)^P E^B_A \\
&\quad H^{AA(A)}_{(A)AA}(1-E)^P E^B_A + 2 \times H^{AA}_{AA}(1-E)^P H^{AB}_{AA} + H^{AA(B)}_{(A)AA} + H^{(A)AB}_{AA(A)} + J^{AB}_{AA} \Big) E^A_B |r) \\
&+ (l| \Big(2 \times H^{AA}_{AA}(1-E)^P H^{BA}_{AB} + H^{AB(A)}_{(A)AB} + J^{BA}_{AB} + H^{(A)BA}_{AA(B)} \Big) |r) \\
&+ (l| \Big(2 \times H^{AA}_{AA}(1-E)^P E^B_A H^{AA}_{BA} + 2 \times H^{AB}_{AA} H^{AA}_{BA} + H^{BA(A)}_{(A)BA} + H^{(B)AA}_{AB(A)} \Big) |r) \\
&+ 2 \times (l| \Big(H^{AA}_{AA}(1-E)^P E^{BA}_{AB} + H^{AB}_{AA} E^A_B + H^{BA}_{AB} \Big) (1-E)^P H^{AA}_{AA} |r).
\end{aligned}
$$

Analogously,

$$
\begin{aligned}
\langle \chi(2)|\hat{H}^2|\chi(0)\rangle &= (l| \Big(2 \times H^{AA}_{AA}(1-E)^P H^{AA}_{AA}(1-E)^P E^B_A + H^{(A)AA}_{AA(A)}(1-E)^P E^B_A \\
&\quad + J^{AA}_{AA}(1-E)^P E^B_A + H^{AA(A)}_{(A)AA}(1-E)^P E^B_A \\
&\quad + 2 \times H^{AA}_{AA}(1-E)^P H^{AB}_{AA} + H^{AA(B)}_{(A)AA} + H^{(A)AB}_{AA(A)} + J^{AB}_{AA} \Big) (E^A_A) E^A_B |r) \\
&+ (l| \Big(2 \times H^{AA}_{AA}(1-E)^P H^{BA}_{AA} + H^{AB(A)}_{(A)AA} + J^{BA}_{AA} + H^{(A)BA}_{AA(A)} \Big) E^A_B |r) \\
&+ (l| \Big(2 \times H^{AA}_{AA}(1-E)^P E^B_A H^{AA}_{AB} + 2 \times H^{AB}_{AA} H^{AA}_{AB} + H^{BA(A)}_{(A)AB} + H^{(B)AA}_{AA(B)} \Big) |r) \\
&+ 2 \times (l| \Big(H^{AA}_{AA}(1-E)^P E^{BA}_{AA} + H^{AB}_{AA} E^A_A + H^{BA}_{AA} \Big) \Big(H^{AA}_{BA} + E^A_B (1-E)^P H^{AA}_{AA} \Big) |r)
\end{aligned}
$$

and for $n > 2$

$$
\langle \chi(n)|\hat{H}^2|\chi(0)\rangle = (l| \Big(2 \times H^{AA}_{AA}(1-E)^P H^{AA}_{AA}(1-E)^P E^B_A + H^{(A)AA}_{AA(A)}(1-E)^P E^B_A
$$

$$
\begin{aligned}
&+ J^{AA}_{AA}(1-E)^P E^B_A + H^{AA(A)}_{(A)AA}(1-E)^P E^B_A \\
&+ 2\times H^{AA}_{AA}(1-E)^P H^{AB}_{AA} + H^{AA(B)}_{(A)AA} + H^{(A)AB}_{AA(A)} + J^{AB}_{AA}\Big)(E^A_A)^{n-1}E^A_B|r) \\
&+ (l|\Big(2\times H^{AA}_{AA}(1-E)^P H^{BA}_{AA} + H^{AB(A)}_{(A)AA} + J^{AB}_{AA} + H^{(A)BA}_{AA(A)}\Big)(E^A_A)^{n-2}E^A_B|r) \\
&+ (l|\Big(2\times H^{AA}_{AA}(1-E)^P E^B_A H^{AA}_{AA} + 2\times H^{AB}_{AA}H^{AA}_{AA} + H^{BA(A)}_{(A)AA} + H^{(B)AA}_{AA(A)}\Big)(E^A_A)^{n-3}E^A_B|r) \\
&+ 2\times(l|\Big(H^{AA}_{AA}(1-E)^P E^{BA}_{AA} + H^{AB}_{AA}E^A_A + H^{BA}_{AA}\Big) \\
&\qquad\times\Bigg(\sum_{i=0}^{n-4}(E^A_A)^i H^{AA}_{AA}(E^A_A)^{n-4-i}E^A_B + (E^A_A)^{n-3}H^{AA}_{AB} \\
&\qquad\qquad + (E^A_A)^{n-2}H^{AA}_{BA} + (E^A_A)^{n-2}E^A_B(1-E)^P H^{AA}_{AA}\Bigg)|r).
\end{aligned}
$$

We can throw everything together in order to obtain

$$
\begin{aligned}
&\sum_{n=1}^{\infty} e^{-i\kappa n}\langle\chi(n)|\hat{H}^2|\chi(0)\rangle = \\
&\quad \mathrm{e}^{-i\kappa}(l|\Big(2\times H^{AA}_{AA}(1-E)^P H^{AA}_{AA}(1-E)^P + H^{(A)AA}_{AA(A)}(1-E)^P \\
&\qquad + J^{AA}_{AA}(1-E)^P + H^{AA(A)}_{(A)AA}(1-E)^P\Big)E^B_A(1-\mathrm{e}^{-i\kappa}E)^{-1}E^A_B|r) \\
&\quad + \mathrm{e}^{-i\kappa}(l|\Big(2\times H^{AA}_{AA}(1-E)^P H^{AB}_{AA} + H^{AA(B)}_{(A)AA} + H^{(A)AB}_{AA(A)} + J^{AB}_{AA}\Big)(1-\mathrm{e}^{-i\kappa}E)^{-1}E^A_B|r) \\
&\quad + \mathrm{e}^{-i\kappa}(l|\Big(2\times H^{AA}_{AA}(1-E)^P H^{BA}_{AB} + H^{AB(A)}_{(A)AB} + J^{BA}_{AB} + H^{(A)BA}_{AA(B)}\Big)|r) \\
&\quad + \mathrm{e}^{-2i\kappa}(l|\Big(2\times H^{AA}_{AA}(1-E)^P H^{BA}_{AA} + H^{AB(A)}_{(A)AA} + J^{AB}_{AA} + H^{(A)BA}_{AA(A)}\Big)(1-\mathrm{e}^{-i\kappa}E)^{-1}E^A_B|r) \\
&\quad + \mathrm{e}^{-i\kappa}(l|\Big(2\times H^{AA}_{AA}(1-E)^P E^B_A H^{AA}_{BA} + 2\times H^{AB}_{AA}H^{AA}_{BA} + H^{BA(A)}_{(A)BA} + H^{(B)AA}_{AB(A)}\Big)|r) \\
&\quad + \mathrm{e}^{-2i\kappa}(l|\Big(2\times H^{AA}_{AA}(1-E)^P H^{AA}_{AB}E^B_A + 2\times H^{AB}_{AA}H^{AA}_{AB} + H^{BA(A)}_{(A)AB} + H^{(B)AA}_{AA(B)}\Big)|r) \\
&\quad + \mathrm{e}^{-3i\kappa}(l|\Big(2\times H^{AA}_{AA}(1-E)^P E^B_A H^{AA}_{AA} + 2\times H^{AB}_{AA}H^{AA}_{AA} + H^{BA(A)}_{(A)AA} + H^{(B)AA}_{AA(A)}\Big) \\
&\qquad \times(1-\mathrm{e}^{-i\kappa}E)^{-1}E^A_B|r) \\
&\quad + 2\times\mathrm{e}^{-i\kappa}(l|\Big(H^{AA}_{AA}(1-E)^P E^{BA}_{AB} + H^{AB}_{AA}E^A_B + H^{BA}_{AB}\Big)(1-E)^P H^{AA}_{AA}|r) \\
&\quad + 2\times\mathrm{e}^{-2i\kappa}(l|\Big(H^{AA}_{AA}(1-E)^P E^{BA}_{AA} + H^{AB}_{AA}E^A_A + H^{BA}_{AA}\Big) \\
&\qquad \times(1-\mathrm{e}^{-i\kappa}E)^{-1}\Big(H^{AA}_{BA} + E^A_B(1-E)^P H^{AA}_{AA}\Big)|r) \\
&\quad + 2\times\mathrm{e}^{-3i\kappa}(l|\Big(H^{AA}_{AA}(1-E)^P E^{BA}_{AA} + H^{AB}_{AA}E^A_A + H^{BA}_{AA}\Big)(1-\mathrm{e}^{-i\kappa}E)^{-1}H^{AA}_{AB}|r) \\
&\quad + 2\times\mathrm{e}^{-4i\kappa}(l|\Big(H^{AA}_{AA}(1-E)^P E^{BA}_{AA} + H^{AB}_{AA}E^A_A + H^{BA}_{AA}\Big) \\
&\qquad \times(1-\mathrm{e}^{-i\kappa}E)^{-1}H^{AA}_{AA}(1-\mathrm{e}^{-i\kappa}E)^{-1}E^A_B|r).
\end{aligned}
$$

Note that the infinite sum could give rise to one potential divergence coming from the disconnected contribution of the last line of $\langle\chi(n)|H^2|\chi(0)\rangle$ corresponding to

$$\sum_{n=3}^{\infty}\sum_{i=0}^{n-4}(l|\left(H_{AA}^{AA}(1-E)^{P}E_{A}^{B}+H_{AA}^{AB}+H_{AA}^{BA}\right)|r)(l|\left(H_{AA}^{AA}(E_{A}^{A})^{n-4-i}E_{B}^{A}+H_{AB}^{AA}+H_{BA}^{AA}\right)|r).$$

However, the first factor is automatically zero if $|\Psi(A)\rangle$ is a variational minimum within the MPS manifold, as it corresponds exactly to the directional derivative of the energy expectation value in the direction of B.

A.3 Two-Particle Excitations: Technical Details

In this appendix we give all technical details concerning the two-particle ansatz that was defined as

$$|\Upsilon(K,\omega)\rangle=\sum_{n=0}^{+\infty}\sum_{j=1}^{M_n}c_{K,\omega}^{j}(n)|\chi_{j,K}(n)\rangle$$

with

$$|\chi_{j,K}(0)\rangle=\sum_{n=-\infty}^{+\infty}e^{iKn}\sum_{\{s\}=1}^{d}v_L^{\dagger}\left[\prod_{m<n}A^{s_m}\right]B_j^{s_n}\left[\prod_{m>n}A^{s_m}\right]v_R|\{s\}\rangle$$

$$|\chi_{(j_1,j_2),K}(n)\rangle=\sum_{n_1=-\infty}^{+\infty}e^{iKn_1}\sum_{\{s\}=1}^{d}v_L^{\dagger}\left[\prod_{m<n_1}A^{s_m}\right]B_{j_1}^{s_{n_1}}\left[\prod_{n_1<m<n_1+n}A^{s_m}\right]$$
$$\times B_{j_2}^{s_{n_1+n}}\left[\prod_{m>n_1+n}A^{s_m}\right]v_R|\{s\}\rangle.$$

Just as in the case of a one-particle excitation, there is a gauge freedom in this ansatz. We can again choose a left or right gauge fixing condition on the tensors B_j, depending on the situation. We will choose to put all B tensors in the left gauge fixing condition, which has the consequence that the states $|\chi_j(n)\rangle$ are not orthogonal for different n (see further). As was argued in the main body, this choice allows for the strictly local term, for which we keep all variational parameters, to correct for the inability of the other terms to describe the deformation of the particles as they approach. Alternatively, one could choose the left tensor B_{j_1} to be in the left gauge and the right tensor B_{j_2} in the right gauge; this would make the states $|\chi_j(n)\rangle$ orthogonal for different n. When studying bound states with the two-particle ansatz, this might prove to be a better choice.

Similar to the one-particle case, we can enforce the gauge fixing conditions by implementing an effective parametrization in terms of a matrix X with $D^2(d-1)$ elements. As we keep all variational freedom in the strictly local term $|\chi_{j,K}(0)\rangle$, this will correspond to $D^2(d-1)$ variational parameters. In the non-local terms $|\chi_{(j_1,j_2),K}(n)\rangle$

we insert a basis of left-gauged tensors B_{j_1} and B_{j_2} which both describe the (relevant part of the) one-particle spectrum. If we have L particles in the system and we need M basis vectors to describe the dispersion of each, we will have $(L \times M) \times (L \times M)$ basis states $|\chi_{(j_1,j_2),K}(n)\rangle$. The gauge fixing and normalization conditions on all the B tensors can be summarized as

$$(l|E_A^{B_j} = (l|E_A^{B_{j_1}} = (l|E_A^{B_{j_2}} = 0 \quad \text{and} \quad (l|E_{B_{j_2}}^{B_{j_1}}|r) = \delta_{j_1,j_2}.$$

A.3.1 Effective Norm Matrix

The effective norm matrix $(\mathsf{N}_{\text{eff}})_{n'j',nj} = \langle\chi_{j',K}(n')|\chi_{j,K}(n)\rangle$ has matrix elements

$$\langle\chi_{j',K}(0)|\chi_{j,K}(0)\rangle = 2\pi\delta(K-K')(l|E_{B_{j'}}^{B_j}|r) = 2\pi\delta(K-K')\delta_{j,j'}$$

$$\langle\chi_{j',K}(n')|\chi_{j,K}(0)\rangle = 2\pi\delta(K-K')(l|E_{B_{j'_1}}^{B_j}(E_A^A)^{n'-1}E_{B_{j'_2}}^A|r)$$

$$\langle\chi_{j',K}(n)|\chi_{j,K}(n)\rangle = 2\pi\delta(K-K')(l|E_{B_{j'_1}}^{B_{j1}}(E_A^A)^{n-1}E_{B_{j'_2}}^{B_{j2}}|r)$$

$$\langle\chi_{j',K}(n')|\chi_{j,K}(n)\rangle = 2\pi\delta(K-K')(l|E_{B_{j'_1}}^{B_{j1}}(E_A^A)^{n-1}E_A^{B_{j2}}(E_A^A)^{n'-n-1}E_{B_{j'_2}}^A|r) \qquad (n' > n).$$

A.3.2 Effective Hamiltonian Matrix

The effective Hamiltonian matrix $(\mathsf{H}_{\text{eff}})_{n'j',nj} = \langle\chi_{j',K}(n')|\hat{H}|\chi_{j,K}(n)\rangle$ has matrix elements

$$\begin{aligned}
\langle\chi_{j',K'}(0)|\hat{H}|\chi_{j,K}(0)\rangle &= 2\pi\delta(K-K')\Big[\\
&(l|H_{B_{j'}A}^{B_jA}|r) + (l|H_{AB_{j'}}^{AB_j}|r) + \mathrm{e}^{-iK}(l|H_{AB_{j'}}^{B_jA}|r) + \mathrm{e}^{iK}(l|H_{B_{j'}A}^{AB_j}|r) \\
&\quad + (l|E_{B_{j'}}^{B_j}(1-E)^P H_{AA}^{AA}|r) + (l|H_{AA}^{AA}(1-E)^P E_{B_{j'}}^{B_j}|r) \\
&+ \mathrm{e}^{-iK}(l|H_{AA}^{AB_j}(1-\mathrm{e}^{-iK}E)^P E_{B_{j'}}^A|r) + \mathrm{e}^{-2iK}(l|H_{AA}^{B_jA}(1-\mathrm{e}^{-iK}E)^P E_{B_{j'}}^A|r) \\
&+ \mathrm{e}^{iK}(l|H_{AB_{j'}}^{AA}(1-\mathrm{e}^{iK}E)^P E_A^{B_j}|r) + \mathrm{e}^{2iK}(l|H_{B_{j'}A}^{AA}(1-\mathrm{e}^{iK}E)^P E_A^{B_j}|r) \\
&+ \mathrm{e}^{-iK}(l|H_{AA}^{AA}(1-E)^P E_A^{B_j}(1-\mathrm{e}^{-iK}E)^P E_{B_{j'}}^A|r) \\
&+ \mathrm{e}^{iK}(l|H_{AA}^{AA}(1-E)^P E_{B_{j'}}^A(1-\mathrm{e}^{iK}E)^P E_A^{B_j}|r)\Big]
\end{aligned}$$

$$\begin{aligned}
\langle\chi_{(j'_1,j'_2),K'}(1)|\hat{H}|\chi_{j,K}(0)\rangle &= 2\pi\delta(K-K')\Big[\\
&(l|H_{AA}^{AA}(1-E)^P E_{B_{j'_1}B_{j'_2}}^{B_jA}|r) + (l|H_{AB_{j'_1}}^{AB_j}E_{B_{j'_2}}^A|r) + (l|H_{B_{j'_1}B_{j'_2}}^{B_jA}|r)
\end{aligned}$$

$$+ (l|E^{B_j}_{B_{j_1}} H^{AA}_{B_{j'_2}A}|r) + (l|E^{B_jA}_{B_{j'_1}B_{j'_2}}(1-E)^P H^{AA}_{AA}|r)$$

$$+ \mathrm{e}^{-iK}\Big((l|H^{AA}_{AA}(1-E)^P E^{B_j}_{A} E^{AA}_{B_{j'_1}B_{j'_2}}|r) + (l|H^{AB_j}_{AA} E^{AA}_{B_{j'_1}B_{j'_2}}|r) + (l|H^{B_jA}_{AB_{j'_1}} E^{A}_{B_{j'_2}}|r)\Big)$$

$$+ \mathrm{e}^{-2iK}(l|\Big(H^{AA}_{AA}(1-E)^P E^{B_j}_{A} E^{A}_{A} + H^{AB_j}_{AA} E^{A}_{A} + H^{B_jA}_{AA}\Big)(1-\mathrm{e}^{-iK}E)^P E^{AA}_{B_{j'_1}B_{j'_2}}|r)$$

$$+ \mathrm{e}^{iK}\Big((l|H^{AA}_{AA}(1-E)^P E^{AB_j}_{B_{j'_1}B_{j'_2}}|r) + (l|H^{AA}_{AB_{j'_1}} E^{B_j}_{B_{j'_2}}|r) + (l|H^{AB_j}_{B_{j'_1}B_{j'_2}}|r)\Big)$$

$$+ \mathrm{e}^{2iK}(l|\Big(H^{AA}_{AA}(1-E)^P E^{AA}_{B_{j'_1}B_{j'_2}} + H^{AA}_{AB_{j'_1}} E^{A}_{B_{j'_2}} + H^{AA}_{B_{j'_1}B_{j'_2}}\Big)(1-\mathrm{e}^{iK}E)^P E^{B_j}_{A}|r)\Big]$$

$$\langle\chi_{(j'_1,j'_2),K'}(n')|\hat{H}|\chi_{j,K}(0)\rangle = 2\pi\delta(K-K')\Big[$$

$$(l|\Big(H^{AA}_{AA}(1-E)^P E^{B_j}_{B_{j'_1}} E^{A}_{A} + H^{AB_j}_{AB_{j'_1}} E^{A}_{A} + H^{B_jA}_{B_{j'_1}A}\Big)(E^A_A)^{n'-2} E^{A}_{B_{j'_2}}|r)$$

$$+ (l|E^{B_j}_{B_{j'_1}}\Big(\sum_{i=0}^{n'-3}(E^A_A)^i H^{AA}_{AA}(E^A_A)^{n'-3-i}\Big)E^{A}_{B_{j'_2}}|r)$$

$$+ (l|E^{B_j}_{B_{j'_1}}(E^A_A)^{n'-2}\Big(H^{AA}_{AB_{j'_2}} + E^A_A H^{AA}_{B_{j'_2}A} + E^A_A E^{A}_{B_{j'_2}}(1-E)^P H^{AA}_{AA}\Big)|r)$$

$$+ \mathrm{e}^{-iK}(l|\Big(H^{AA}_{AA}(1-E)^P E^{B_jA}_{AB_{j'_1}} + H^{AB_j}_{AA} E^{A}_{B_{j'_1}} + H^{B_jA}_{AB_{j'_1}}\Big)(E^A_A)^{n'-1} E^{A}_{B_{j'_2}}|r)$$

$$+ \mathrm{e}^{-2iK}(l|\Big(H^{AA}_{AA}(1-E)^P E^{B_j}_{A} E^{A}_{A} + H^{AB_j}_{AA} E^{A}_{A} + H^{B_jA}_{AA}\Big)$$

$$(1-\mathrm{e}^{-iK}E)^P E^{A}_{B_{j'_1}}(E^A_A)^{n'-1} E^{A}_{B_{j_2}p}|r)$$

$$+ \mathrm{e}^{iK}(l|\Big(H^{AA}_{AA}(1-E)^P E^{A}_{B_{j'_1}} E^{B_j}_{A} + H^{AA}_{AB_{j'_1}} E^{B_j}_{A} + H^{AB_j}_{B_{j'_1}A}\Big)(E^A_A)^{n'-2} E^{A}_{B_{j'_2}}|r)$$

$$+ (l|\Big(H^{AA}_{AA}(1-E)^P E^{A}_{B_{j'_1}} E^{A}_{A} + H^{AA}_{AB_{j'_1}} E^{A}_{A} + H^{AA}_{B_{j'_1}A}\Big)\Big($$

$$\sum_{j=2}^{n'-1} \mathrm{e}^{ijK}(E^A_A)^{j-2} E^{B_j}_{A}(E^A_A)^{n'-j-1} E^{A}_{B_{j'_2}}$$

$$+ \mathrm{e}^{in'K}(E^A_A)^{n'-2} E^{B_j}_{B_{j'_2}} + \mathrm{e}^{i(n'+1)K}(E^A_A)^{n'-2} E^{A}_{B_{j'_2}}(1-\mathrm{e}^{iK}E)^P E^{B_j}_{A}\Big)|r)\Big]$$

$$\langle\chi_{(j'_1,j'_2),K'}(1)|\hat{H}|\chi_{(j_1,j_2),K}(1)\rangle = 2\pi\delta(K-K')\Big[$$

$$(l|H^{AA}_{AA}(1-E)^P E^{B_{j_1}}_{B_{j'_1}} E^{B_{j_2}}_{B_{j'_2}}|r) + (l|H^{AB_{j_1}}_{AB_{j'_1}} E^{B_{j_2}}_{B_{j'_2}}|r) + (l|H^{B_{j_1}B_{j_2}}_{B_{j'_1}B_{j'_2}}|r)$$

$$+ (l|E^{B_{j_1}}_{B_{j'_1}} H^{B_{j_2}A}_{B_{j'_2}A}|r) + (l|E^{B_{j_1}}_{B_{j'_1}} E^{B_{j_2}}_{B_{j'_2}}(1-E)^P H^{AA}_{AA}|r)$$

$$+ \mathrm{e}^{-iK}(l|\Big(H^{AA}_{AA}(1-E)^P E^{B_{j_1}}_{A} E^{B_{j_2}}_{B_{j'_1}} + H^{AB_{j_1}}_{AA} E^{B_{j_2}}_{B_{j'_1}} + H^{B_{j_1}B_{j_2}}_{AB_{j'_1}}\Big)E^{A}_{B_{j'_2}}|r)$$

$$+ \mathrm{e}^{-i2K}(l|\Big(H^{AA}_{AA}(1-E)^P E^{B_{j_1}}_{A} E^{B_{j_2}}_{A} + H^{AB_{j_1}}_{AA} E^{B_{j_2}}_{A} + H^{B_{j_1}B_{j_2}}_{AA}\Big)(1-\mathrm{e}^{-iK}E)^P E^{A}_{B_{j'_1}} E^{A}_{B_{j'_2}}|r)$$

$$+ \mathrm{e}^{iK}(l|\Big(H^{AA}_{AA}(1-E)^P E^{A}_{B_{j'_1}} E^{B_{j_1}}_{B_{j'_2}} + H^{AA}_{AB_{j'_1}} E^{B_{j_1}}_{B_{j'_2}} + H^{AB_{j_1}}_{B_{j'_1}B_{j'_2}}\Big)E^{B_{j_2}}_{A}|r)$$

$$+ \mathrm{e}^{i2K}(l|\Big(H^{AA}_{AA}(1-E)^P E^{A}_{B_{j'_1}} E^{A}_{B_{j'_2}} + H^{AA}_{AB_{j'_1}} E^{A}_{B_{j'_2}} + H^{AA}_{B_{j'_1}B_{j'_2}}\Big)(1-\mathrm{e}^{iK}E)^P E^{B_{j_1}}_{A} E^{B_{j_2}}_{A}\Big)|r)\Big]$$

$$\langle\chi_{(j'_1,j'_2),K'}(n)|\hat{H}|\chi_{(j_1,j_2),K}(n)\rangle = 2\pi\delta(K-K')\Big[$$

$$(l|\Big(H^{AA}_{AA}(1-E)^P E^{B_{j_1}}_{B_{j'_1}}(E^A_A)^{n-1} + H^{AB_{j_1}}_{AB_{j'_1}}(E^A_A)^{n-1} + H^{B_{j_1}A}_{B_{j'_1}A}(E^A_A)^{n-2}$$

$$+ E_{B_{j_1'}}^{B_{j_1}} \sum_{i=0}^{n-3} (E_A^A)^i H_{AA}^{AA} (E_A^A)^{n-i-3}\Big) E_{B_{j_2'}}^{B_{j_2}} |r)$$
$$+ (l| E_{B_{j_1'}}^{B_{j_1}} (E_A^A)^{n-2} \Big(H_{AB_{j_2'}}^{AB_{j_2}} + E_A^A H_{B_{j_2'}A}^{B_{j_2}A} + E_A^A E_{B_{j_2'}}^{B_{j_2}} (1-E)^P H_{AA}^{AA} \Big) |r)$$
$$+ \mathrm{e}^{-iK} (l| \Big(H_{AA}^{AA} (1-E)^P E_A^{B_{j_1}} E_{B_{j_1'}}^A + H_{AA}^{AB_{j_2}} E_{B_{j_1'}}^A + H_{AB_{j_1'}}^{B_{j_1}A} \Big) (E_A^A)^{n-2} E_A^{B_{j_2}} E_{B_{j_2'}}^A |r)$$
$$+ (l| \Big(H_{AA}^{AA} (1-E)^P E_A^{B_{j_1}} E_A^A + H_{AA}^{AB_{j_1}} E_A^A + H_{AA}^{B_{j_1}A} \Big) \Big($$
$$\sum_{j=2}^{n-1} \mathrm{e}^{-ijK} (E_A^A)^{j-2} E_{B_{j_1'}}^A (E_A^A)^{n-j-1} E_A^{B_{j_2}} (E_A^A)^{j-1}$$
$$+ \mathrm{e}^{-inK} (E_A^A)^{n-2} E_{B_{j_1'}}^{B_{j_2}} (E_A^A)^{n-1}$$
$$+ \mathrm{e}^{-i(n+1)K} (E_A^A)^{n-2} E_A^{B_{j_2}} (1 - \mathrm{e}^{-iK} E)^P E_{B_{j_1'}}^A (E_A^A)^{n-1} \Big) E_{B_{j_2'}}^A |r)$$
$$+ \mathrm{e}^{iK} (l| \Big(H_{AA}^{AA} (1-E)^P E_{B_{j_1'}}^A E_A^{B_{j_1}} + H_{AB_{j_1'}}^{AA} E_A^{B_{j_1}} + H_{B_{j_1'}A}^{AB_{j_1}} \Big) (E_A^A)^{n-2} E_{B_{j_2'}}^A E_A^{B_{j_2}} |r)$$
$$+ (l| \Big(H_{AA}^{AA} (1-E)^P E_{B_{j_1'}}^A E_A^A + H_{AB_{j_1'}}^{AA} E_A^A + H_{B_{j_1'}A}^{AA} \Big) \Big($$
$$\sum_{j=2}^{n-1} \mathrm{e}^{ijK} (E_A^A)^{j-2} E_A^{B_{j_1}} (E_A^A)^{n-j-1} E_{B_{j_2'}}^A (E_A^A)^{j-1}$$
$$+ \mathrm{e}^{inK} (E_A^A)^{n-2} E_{B_{j_2'}}^{B_{j_1}} (E_A^A)^{n-1}$$
$$+ \mathrm{e}^{i(n+1)K} (E_A^A)^{n-2} E_{B_{j_2'}}^A (1 - \mathrm{e}^{iK} E)^P E_A^{B_{j_1}} (E_A^A)^{n-1} \Big) E_A^{B_{j_2}} |r) \Big]$$

$$\langle \chi_{(j_1',j_2'),K'}(2) | \hat{H} | \chi_{(j_1,j_2),K}(1) \rangle = 2\pi\delta(K-K') \Big[$$
$$(l| \Big(H_{AA}^{AA} (1-E)^P E_{B_{j_1'}}^{B_{j_1}} E_A^{B_{j_2}} + H_{AB_{j_1'}}^{AB_{j_1}} E_A^{B_{j_2}} + H_{B_{j_1'}A}^{B_{j_1}B_{j_2}} \Big) E_{B_{j_2'}}^A |r)$$
$$+ (l| E_{B_{j_1'}}^{B_{j_1}} \Big(H_{AB_{j_2'}}^{B_{j_2}A} + E_A^{B_{j_2}} H_{B_{j_2'}A}^{AA} + E_A^{B_{j_2}} E_{B_{j_2'}}^A (1-E)^P H_{AA}^{AA} \Big) |r)$$
$$+ \mathrm{e}^{-iK} (l| \Big(H_{AA}^{AA} (1-E)^P E_A^{B_{j_1}} E_{B_{j_1'}}^{B_{j_2}} + H_{AA}^{AB_{j_1}} E_{B_{j_1'}}^{B_{j_2}} + H_{AB_{j_1'}}^{B_{j_1}B_{j_2}} \Big) E_A^A E_{B_{j_2'}}^A |r)$$
$$+ \mathrm{e}^{-i2K} (l| \Big(H_{AA}^{AA} (1-E)^P E_A^{B_{j_1}} E_A^{B_{j_2}} + H_{AA}^{AB_{j_1}} E_A^{B_{j_2}} + H_{AA}^{B_{j_1}B_{j_2}} \Big)$$
$$\times (1 - \mathrm{e}^{-iK} E)^P E_{B_{j_1'}}^A E_A^A E_{B_{j_2'}}^A |r)$$
$$+ \mathrm{e}^{iK} (l| \Big(H_{AA}^{AA} (1-E)^P E_{B_{j_1'}}^A E_A^{B_{j_1}} + H_{AB_{j_1'}}^{AA} E_A^{B_{j_1}} + H_{B_{j_1'}A}^{AB_{j_1}} \Big) E_{B_{j_2'}}^{B_{j_2}} |r)$$
$$+ (l| \Big(H_{AA}^{AA} (1-E)^P E_{B_{j_1'}}^A E_A^A + H_{AB_{j_1'}}^{AA} E_A^A + H_{B_{j_1'}A}^{AA} \Big) \Big($$
$$\mathrm{e}^{i2K} E_{B_{j_2}P}^{B_{j_1}} E_A^{B_{j_2}} + \mathrm{e}^{i3K} E_{B_{j_2}P}^A (1 - \mathrm{e}^{iK} E)^P E_A^{B_{j_1}} E_A^{B_{j_2}} \Big) |r) \Big]$$

$$\langle \chi_{(j_1',j_2'),K'}(n+1) | \hat{H} | \chi_{(j_1,j_2),K}(1) \rangle = 2\pi\delta(K-K') \Big[$$
$$(l| \Big(H_{AA}^{AA} (1-E)^P E_{B_{j_1'}}^{B_{j_1}} (E_A^A)^{n-1} + H_{AB_{j_1'}}^{AB_{j_1}} (E_A^A)^{n-1} + H_{B_{j_1'}A}^{B_{j_1}A} (E_A^A)^{n-2} \Big) E_A^{B_{j_2}} E_{B_{j_2'}}^A |r)$$
$$+ (l| E_{B_{j_1'}}^{B_{j_1}} \Big(\sum_{i=0}^{n-3} (E_A^A)^i H_{AA}^{AA} (E_A^A)^{n-i-3} \Big) E_A^{B_{j_2}} E_{B_{j_2'}}^A |r)$$

$$+ (l|E^{B_{j_1}}_{B_{j'_1}} (E^A_A)^{n-2} \Big(H^{AB_{j_2}}_{AA} E^A_{B_{j'_2}} + E^A_A H^{B_{j_2}A}_{AB_{j'_2}}$$

$$+ E^A_A E^{B_{j_2}}_A H^{AA}_{B_{j'_2}A} + E^A_A E^{B_{j_2}}_A E^A_{B_{j'_2}} (1-E)^P H^{AA}_{AA} \Big) |r)$$

$$+ \mathrm{e}^{-iK} (l| \Big(H^{AA}_{AA} (1-E)^P E^{B_{j_1}}_A E^A_{B_{j'_1}} + H^{AB_{j_1}}_{AA} E^A_{B_{j'_1}} + H^{B_{j_1}A}_{AB_{j'_1}} \Big) (E^A_A)^{n-2} E^{B_{j_2}}_A E^A_A E^A_{B_{j_2}p} |r)$$

$$+ (l| \Big(H^{AA}_{AA} (1-E)^P E^{B_{j_1}}_A E^A_A + H^{AB_{j_1}}_{AA} E^A_A + H^{B_{j_1}A}_{AA} \Big) \Big($$

$$\sum_{j=2}^{n-1} \mathrm{e}^{-ijK} (E^A_A)^{j-2} E^A_{B_{j'_1}} (E^A_A)^{n-j-1} E^{B_{j_2}}_A (E^A_A)^{n'-n+j-1} E^A_{B_{j'_2}}$$

$$+ \mathrm{e}^{-inK} (E^A_A)^{n-2} E^{B_{j_2}}_{B_{j'_1}} (E^A_A)^{n'-1} E^A_{B_{j'_2}}$$

$$+ \mathrm{e}^{-i(n+1)K} (E^A_A)^{n-2} E^{B_{j_2}}_A (1 - \mathrm{e}^{-iK} E)^P E^A_{B_{j'_1}} (E^A_A)^{n'-1} E^A_{B_{j'_2}} \Big) |r)$$

$$+ \mathrm{e}^{iK} (l| \Big(H^{AA}_{AA} (1-E)^P E^A_{B_{j'_1}} E^{B_{j_1}}_A + H^{AA}_{AB_{j'_1}} E^{B_{j_1}}_A + H^{AB_{j_1}}_{B_{j'_1}A} \Big) (E^A_A)^{n-1} E^{B_{j_2}}_{B_{j'_2}} |r)$$

$$+ (l| \Big(H^{AA}_{AA} (1-E)^P E^A_{B_{j'_1}} E^A_A + H^{AA}_{AB_{j'_1}} E^A_A + H^{AA}_{B_{j'_1}A} \Big) \Big($$

$$\sum_{j=2}^{n'-1} \mathrm{e}^{ijK} (E^A_A)^{j-2} E^{B_{j_1}}_A (E^A_A)^{n'-j-1} E^A_{B_{j'_2}} (E^A_A)^{n-n'+j-1} E^{B_{j_2}}_A$$

$$+ \mathrm{e}^{in'K} (E^A_A)^{n'-2} E^{B_{j_1}}_{B_{j'_2}} (E^A_A)^{n-1} E^{B_{j_2}}_A$$

$$+ \mathrm{e}^{i(n'+1)K} (E^A_A)^{n'-2} E^A_{B_{j'_2}} (1 - \mathrm{e}^{iK} E)^P E^{B_{j_1}}_A (E^A_A)^{n-1} E^{B_{j_2}}_A \Big) |r) \Big]$$

$$\langle \chi_{(j'_1, j'_2), K'}(n') | \hat{H} | \chi_{(j_1, j_2), K}(1) \rangle = 2\pi\delta(K - K') \Big[$$

$$(l| \Big(H^{AA}_{AA} (1-E)^P E^{B_{j_1}}_{B_{j'_1}} E^{B_{j_2}}_A + H^{AB_{j_1}}_{AB_{j'_1}} E^{B_{j_2}}_A + H^{B_{j_1}B_{j_2}}_{B_{j'_1}A} \Big) (E^A_A)^{n'-2} E^A_{B_{j'_2}} |r)$$

$$+ (l|E^{B_{j_1}}_{B_{j'_1}} H^{B_{j_2}A}_{AA} (E^A_A)^{n'-3} E^A_{B_{j'_2}} |r)$$

$$+ (l|E^{B_{j_1}}_{B_{j'_1}} E^{B_{j_2}}_A \sum_{i=0}^{n'-4} (E^A_A)^i H^{AA}_{AA} (E^A_A)^{n'-4-i} E^A_{B'_{j_2}} |r)$$

$$+ (l|E^{B_{j_1}}_{B_{j'_1}} E^{B_{j_2}}_A (E^A_A)^{n'-3} \Big(H^{AA}_{AB_{j'_2}} + E^A_A H^{AA}_{B_{j'_2}A} + E^A_A E^A_{B_{j'_2}} (1-E)^P H^{AA}_{AA} \Big) |r)$$

$$+ \mathrm{e}^{-iK} (l| \Big(H^{AA}_{AA} (1-E)^P E^{B_{j_1}}_A E^{B_{j_2}}_{B_{j'_1}} + H^{AB_{j_1}}_{AA} E^{B_{j_2}}_{B_{j'_1}} + H^{B_{j_1}B_{j_2}}_{AB_{j'_1}} \Big) (E^A_A)^{n'-1} E^A_{B_{j'_2}} |r)$$

$$+ \mathrm{e}^{-i2K} (l| \Big(H^{AA}_{AA} (1-E)^P E^{B_{j_1}}_A E^{B_{j_2}}_A + H^{AB_{j_1}}_{AA} E^{B_{j_2}}_A + H^{B_{j_1}B_{j_2}}_{AA} \Big)$$

$$\times (1 - \mathrm{e}^{-iK} E)^P E^A_{B_{j'_1}} (E^A_A)^{n'-1} E^A_{B_{j'_2}} |r)$$

$$+ \mathrm{e}^{iK} (l| \Big(H^{AA}_{AA} (1-E)^P E^A_{B_{j'_1}} E^{B_{j_1}}_A + H^{AA}_{AB_{j'_1}} E^{B_{j_1}}_A + H^{AB_{j_1}}_{B_{j'_1}A} \Big) E^{B_{j_2}}_A (E^A_A)^{n'-3} E^A_{B_{j'_2}} |r)$$

$$+ (l| \Big(H^{AA}_{AA} (1-E)^P E^A_{B_{j'_1}} E^A_A + H^{AA}_{AB_{j'_1}} E^A_A + H^{AA}_{B_{j'_1}A} \Big) \Big($$

$$\sum_{j=2}^{n'-2} \mathrm{e}^{ijK} (E^A_A)^{j-2} E^{B_{j_1}}_A E^{B_{j_2}}_A (E^A_A)^{n'-j-2} E^A_{B'_{j_2}}$$

$$+ \mathrm{e}^{i(n'-1)K} (E^A_A)^{n'-3} E^{B_{j_1}}_A E^{B_{j_2}}_{B_{j'_2}}$$

$$+\,e^{in'K}(E_A^A)^{n'-2}E_{B_{j_2'}}^{B_{j_1}}E_A^{B_{j_2}}$$
$$+\,e^{i(n'+1)K}(E_A^A)^{n'-2}E_{B_{j_2'}}^A(1-e^{iK}E)^P E_A^{B_{j_1}}E_A^{B_{j_2}}\Big)|r)\Big]$$

$$\begin{aligned}
\langle\chi_{(j_1',j_2'),K'}(n')|\hat{H}|\chi_{(j_1,j_2),K}(n)\rangle &= 2\pi\delta(K-K')\Big[\\
&(l|\Big(H_{AA}^{AA}(1-E)^P E_{B_{j_1'}}^{B_{j_1}}E_A^A + H_{AB_{j_1'}}^{AB_{j_1}}E_A^A + H_{B_{j_1'}A}^{B_{j_1}A}\Big)(E_A^A)^{n-2}E_A^{B_{j_2}}E_A^A(E_A^A)^{n'-n-2}E_{B_{j_2'}}^A|r)\\
&+(l|E_{B_{j_1'}}^{B_{j_1}}\Big(\sum_{i=0}^{n-3}(E_A^A)^i H_{AA}^{AA}(E_A^A)^{n-i-3}\Big)E_A^{B_{j_2}}E_A^A(E_A^A)^{n'-n-2}E_{B_{j_2'}}^A|r)\\
&+(l|E_{B_{j_1'}}^{B_{j_1}}(E_A^A)^{n-2}\Big(H_{AA}^{AB_{j_2}}E_A^A + E_A^A H_{AA}^{B_{j_2}A}\Big)(E_A^A)^{n'-n-2}E_{B_{j_2'}}^A|r)\\
&+(l|E_{B_{j_1'}}^{B_{j_1}}(E_A^A)^{n-1}E_A^{B_{j_2}}\Big(\sum_{i=0}^{n'-n-3}(E_A^A)^i H_{AA}^{AA}(E_A^A)^{n'-n-3-i}E_{B_{j_2'}}^A + (E_A^A)^{n'-n-2}H_{AB_{j_2'}}^{AA}\\
&\qquad + (E_A^A)^{n'-n-1}H_{B_{j_2'}A}^{AA} + (E_A^A)^{n'-n-1}E_{B_{j_2'}}^A(1-E)^P H_{AA}^{AA}\Big)|r)\\
&+e^{-iK}(l|\Big(H_{AA}^{AA}(1-E)^P E_A^{B_{j_1}}E_{B_{j_1'}}^A + H_{AA}^{AB_{j_1}}E_{B_{j_1'}}^A + H_{AB_{j_1'}}^{B_{j_1}A}\Big)\\
&\qquad\times(E_A^A)^{n-2}E_A^{B_{j_2}}(E_A^A)^{n'-n}E_{B_{j_2'}}^A|r)\\
&+(l|\Big(H_{AA}^{AA}(1-E)^P E_A^{B_{j_1}}E_A^A + H_{AA}^{AB_{j_1}}E_A^A + H_{AA}^{B_{j_1}A}\Big)\Big(\\
&\quad\Big(\sum_{j=2}^{n-1}e^{-ijK}(E_A^A)^{j-2}E_{B_{j_1'}}^A(E_A^A)^{n-j-1}E_A^{B_{j_2}}(E_A^A)^j\Big)(E_A^A)^{n'-n-1}E_{B_{j_2'}}^A\\
&\quad+e^{-inK}(E_A^A)^{n-2}E_{B_{j_1'}}^{B_{j_2}}(E_A^A)^{n'-1}E_{B_{j_2'}}^A\\
&\quad+e^{-i(n+1)K}(E_A^A)^{n-2}E_A^{B_{j_2}}(1-e^{-iK}E)^P E_{B_{j_1'}}^A(E_A^A)^{n'-1}E_{B_{j_2'}}^A\Big)|r)\\
&+e^{iK}(l|\Big(H_{AA}^{AA}(1-E)^P E_{B_{j_1'}}^A E_A^{B_{j_1}} + H_{AB_{j_1'}}^{AA}E_A^{B_{j_1}} + H_{B_{j_1'}A}^{AB_{j_1}}\Big)\\
&\qquad\times(E_A^A)^{n-1}E_A^{B_{j_2}}(E_A^A)^{n'-n-2}E_{B_{j_2'}}^A|r)\\
&+(l|\Big(H_{AA}^{AA}(1-E)^P E_{B_{j_1'}}^A E_A^A + H_{AB_{j_1'}}^{AA}E_A^A + H_{B_{j_1'}A}^{AA}\Big)\Big(\\
&\quad\sum_{j=2}^{n'-n-1}e^{ijK}(E_A^A)^{j-2}E_A^{B_{j_1}}(E_A^A)^{n-1}E_A^{B_{j_2}}(E_A^A)^{n'-n-j-1}E_{B_{j_2'}}^A\\
&\quad+e^{i(n'-n)K}(E_A^A)^{n'-n-2}E_A^{B_{j_1}}(E_A^A)^{n-1}E_{B_{j_2'}}^{B_{j_2}}\\
&\quad+\sum_{j=n'-n+1}^{n'-1}e^{ijK}(E_A^A)^{j-2}E_A^{B_{j_1}}(E_A^A)^{n'-j-1}E_{B_{j_2'}}^A(E_A^A)^{n-n'+j-1}E_A^{B_{j_2}}\\
&\quad+e^{in'K}(E_A^A)^{n'-2}E_{B_{j_2'}}^{B_{j_1}}(E_A^A)^{n-1}E_A^{B_{j_2}}\\
&\quad+e^{i(n'+1)K}(E_A^A)^{n'-2}E_{B_{j_2'}}^A(1-e^{iK}E)^P E_A^{B_{j_1}}(E_A^A)^{n-1}E_A^{B_{j_2}}\Big)|r)\Big].
\end{aligned}$$

A.3.3 Asymptotic Regime

The expressions for the effective norm and Hamiltonian matrices above are largely determined by powers of the transfer matrices. The power of the transfer matrices behaves as

$$(E^A_A)^n = |r)(l| + \mathcal{O}\left(\mathrm{e}^{-n/\xi}\right) \qquad \text{as} \quad n \to \infty,$$

where the correlation length ξ of the MPS was defined in Eq. (3.3). The asymptotic regime in N_{eff} and H_{eff} is reached when the corrections can be safely neglected, i.e. $n > \xi \times \log(1/\epsilon)$ where ϵ is the allowed error.

The effective norm matrix reduces to the unit matrix in this regime

$$\langle \chi_{(j_1',j_2'),K'}(n') | \chi_{(j_1,j_2),K}(n) \rangle = 2\pi\delta(K-K')\delta_{n',n}\delta_{j_1',j_1}\delta_{j_2',j_2},$$

and the effective Hamiltonian matrix is greatly simplified:

$$\begin{aligned}
&\langle \chi_{(j_1',j_2'),K'}(n) | \hat{H} | \chi_{(j_1,j_2),K}(n) \rangle = 2\pi\delta(K-K')\Big[\\
&\quad \delta_{j_1,j_1'}(l|\Big(H^{AB_{j_2}}_{AB_{j_2'}} + H^{B_{j_2}A}_{B_{j_2'}A} + E^{B_{j_2}}_{B_{j_2'}}(1-E)^P H^{AA}_{AA} + H^{AA}_{AA}(1-E)^P E^{B_{j_2}}_{B_{j_2'}}\Big)|r) \\
&\quad + \delta_{j_2,j_2'}(l|\Big(H^{AA}_{AA}(1-E)^P E^{B_{j_1}}_{B_{j_1'}} + H^{AB_{j_1}}_{AB_{j_1'}} + H^{B_{j_1}A}_{B_{j_1'}A} + E^{B_{j_1}}_{B_{j_1'}}(1-E)^P H^{AA}_{AA}\Big)|r)\Big] \\
&\langle \chi_{(j_1',j_2'),K'}(n+1) | \hat{H} | \chi_{(j_1,j_2),K}(n) \rangle = 2\pi\delta(K-K')\Big[\\
&\quad \delta_{j_1,j_1'}(l|\Big(H^{AA}_{AA}(1-E)^P E^{B_{j_2}}_{A} E^{A}_{B_{j_2'}} + H^{AB_{j_2}}_{AA} E^{A}_{B_{j_2'}} + H^{B_{j_2}A}_{AB_{j_2'}}\Big)|r) \\
&\quad + \delta_{j_2,j_2'}\mathrm{e}^{iK}(l|\Big(H^{AA}_{AA}(1-E)^P E^{A}_{B_{j_1'}} E^{B_{j_1}}_{A} + H^{AA}_{AB_{j_1'}} E^{B_{j_1}}_{A} + H^{AB_{j_1}}_{B_{j_1'}A}\Big)|r)\Big] \\
&\langle \chi_{(j_1',j_2'),K'}(n') | \hat{H} | \chi_{(j_1,j_2),K}(n) \rangle = 2\pi\delta(K-K')\Big[\\
&\quad \delta_{j_1,j_1'}(l|\Big(H^{AA}_{AA}(1-E)^P E^{B_{j_2}}_{A} E^{A}_{A} + H^{AB_{j_2}}_{AA} E^{A}_{A} + H^{B_{j_2}A}_{AA}\Big)(E^A_A)^{(n'-n-2)} E^{A}_{B_{j_2'}}|r) \\
&\quad + \mathrm{e}^{iK(n'-n)}\delta_{j_2,j_2'}(l|\Big(H^{AA}_{AA}(1-E)^P E^{A}_{B_{j_1'}} E^{A}_{A} + H^{AA}_{AB_{j_1'}} E^{A}_{A} + H^{AA}_{B_{j_1'}A}\Big)(E^A_A)^{(n'-n-2)} E^{B_{j_1}}_{A}|r)\Big].
\end{aligned}$$

One can observe that the matrix elements indeed form a repeating row of block matrices, centered around the diagonal and exponentially decaying

$$\begin{aligned}
(\mathsf{H}_{\text{eff}})_{n'j_1'j_2',nj_1j_2} &= \langle \chi_{(j_1',j_2'),K'}(n') | \hat{H} | \chi_{(j_1,j_2),K}(n) \rangle \\
&= (A_{n'-n})_{j_1'j_2',j_1j_2} \\
&= \mathcal{O}\left(\mathrm{e}^{-|n'-n|/\xi}\right) \qquad \text{as} \quad |n-n'| \to \infty.
\end{aligned}$$

A.3.4 Two-Particle Form Factors

Again we start from the spectral function

$$S(\kappa,\omega)=\sum_{n=-\infty}^{+\infty}\int_{-\infty}^{+\infty}\mathrm{d}t\,\mathrm{e}^{i(\omega t-\kappa n)}\langle\Psi_0|O_n^\dagger(t)O_0(0)|\Psi_0\rangle.$$

Inserting a projector on the two-particle subspace, the two-particle contribution to this function can be written as ($\Gamma_2(\kappa,\omega)$ is the set of all two-particle states at that momentum-energy combination)

$$S(\kappa,\omega)_{2\mathrm{p}}=\sum_{i\in\Gamma_2(K,\omega)}\left|\langle\Upsilon_\gamma(\kappa,\omega)|\hat{O}_0|\Psi(A)\rangle\right|^2.$$

If we denote the coefficients $c^j(n)$ of the two-particle states as

$$c^j(n)=c^j_{\mathrm{local}}(n)+\sum_{\gamma=1}^{2\Gamma}q^\gamma\mathrm{e}^{i\kappa_\gamma n}v^j_\gamma$$

such that $c^j_{\mathrm{local}}(n)\approx 0$ if $n>R$ for some value of R. The overlap appearing in the spectral functions can be calculated as

$$\langle\Psi(A)|\hat{O}_0|\Upsilon(K,\omega)\rangle=\sum_{n=0}^{\infty}\sum_j c^j(n)\langle\Psi(A)|\hat{O}_0|\chi_{j,K}(n)\rangle$$

where

$$\begin{aligned}\langle\Psi(A)|\hat{O}_0|\chi_{j,K}(0)\rangle&=(l|O_A^{B_j}|r)+e^{iK}(l|O_A^A(1-\mathrm{e}^{iK}E)^PE_A^{B_j}|r)\\ \langle\Psi(A)|\hat{O}_0|\chi_{(j_1,j_2),K}\rangle&=(l|O_A^{B_{j_1}}(E_A^A)^{n-1}E_A^{B_{j_2}}|r)+e^{iK}(l|O_A^A(1-\mathrm{e}^{iK}E)^PE_A^{B_{j_1}}(E_A^A)^{n-1}E_A^{B_{j_2}}|r)\\ &=(l|\left(O_A^{B_{j_1}}+e^{iK}O_A^A(1-\mathrm{e}^{iK}E)^PE_A^{B_{j_1}}\right)(E_A^A)^{n-1}E_A^{B_{j_2}}|r).\end{aligned}$$

We have

$$\begin{aligned}\langle\Psi(A)|\hat{O}_0|\Upsilon(K,\omega)\rangle&=\sum_j c^j(0)\left((l|O_A^{B_j}|r)+\mathrm{e}^{iK}(l|O_A^A(1-\mathrm{e}^{iK}E)^PE_A^{B_j}|r)\right)\\ &+\sum_{n=1}^{R}\sum_{j_1,j_2}c_{\mathrm{local}}^{(j_1,j_2)}(n)(l|\left(O_A^{B_{j_1}}+\mathrm{e}^{iK}O_A^A(1-\mathrm{e}^{iK}E)^PE_A^{B_{j_1}}\right)(E_A^A)^{n-1}E_A^{B_{j_2}}|r)\\ &+\sum_{\gamma=1}^{2\Gamma}q^\gamma\sum_{j_1,j_2}v_\gamma^{(j_1,j_2)}(l|\left(E_A^{B_{j_1}}+\mathrm{e}^{iK}O_A^A(1-\mathrm{e}^{iK}E)^PE_A^{B_{j_1}}\right)(1-\mathrm{e}^{i\kappa_\gamma n}E)^PE_A^{B_{j_2}}|r).\end{aligned}$$

A.4 Proof of Eq. (3.26)

Let us start with the polynomial eigenvalue equation for the asymptotic solutions of the scattering problem, Eq. (3.20)

$$\sum_{m=-M}^{+M} \mu^m \mathsf{A}_m \boldsymbol{v} = \omega \boldsymbol{v}.$$

For a given $\mu \in \mathbb{C}$, this equation is an ordinary eigenvalue problem with eigenvalue ω and eigenvector $\boldsymbol{v}$. Given the property $\mathsf{A}_m^\dagger = \mathsf{A}_{-m}$, we are only assured of a Hermitian eigenvalue problem (and thus of real eigenvalues ω) if $\mu^* = \mu^{-1}$, i.e. if μ is on the unit circle. So for any $\mu_\gamma = \exp(i\kappa)$, let there be eigenvalues $\omega_\gamma(\kappa)$ and corresponding normalized eigenvectors $\boldsymbol{v}_\gamma(\kappa)$. The functions $\omega_\gamma(\kappa)$ and $\boldsymbol{v}_\gamma(\kappa)$ are assumed to be smooth such that at least the first derivatives are well defined. By taking the derivative of the eigenvalue equation with respect to κ we obtain

$$\sum_{m=-M}^{+M} im\mathsf{A}_m \mathrm{e}^{i\kappa m} \boldsymbol{v}_\gamma(\kappa) + \sum_{m=-M}^{+M} \mathsf{A}_m \mathrm{e}^{i\kappa m} \frac{\mathrm{d}\boldsymbol{v}_\gamma}{\mathrm{d}\kappa}(\kappa) = \frac{\mathrm{d}\omega_\gamma}{\mathrm{d}\kappa}(\kappa) \boldsymbol{v}_\gamma(\kappa) + \omega_\gamma(\kappa) \frac{\mathrm{d}\boldsymbol{v}_\gamma}{\mathrm{d}\kappa}(\kappa).$$

By multiplying this equation with $\boldsymbol{v}_{\gamma'}(\kappa)^\dagger$ and using the normalization $\boldsymbol{v}_{\gamma'}(\kappa)^\dagger \boldsymbol{v}_\gamma (\kappa) = \delta_{\gamma'\gamma}$, we tobtain the following relation for later use

$$\sum_{m=-M}^{+M} im \boldsymbol{v}_{\gamma'}(\kappa)^\dagger \mathsf{A}_m \mathrm{e}^{i\kappa m} \boldsymbol{v}_\gamma(\kappa) = \delta_{\gamma'\gamma} \frac{\mathrm{d}\omega_\gamma}{\mathrm{d}\kappa}(\kappa). \tag{A.1}$$

Now consider a two particle eigenstate $|\Upsilon(K, \omega)\rangle$, which has the asymptotic form

$$\boldsymbol{c}(K, \omega) = \sum_{\gamma=1}^{2\Gamma} q_\gamma \mathrm{e}^{i\kappa_\gamma n} \boldsymbol{v}_\gamma.$$

We can introduce the projectors (we will omit all dependencies on the total momentum K)

$$P_R = \sum_{n=0}^{R} \sum_{j=1}^{L_n} |\chi_j(n)\rangle\langle\chi_j(n)| \quad \text{and} \quad P_R^\perp = \sum_{n>R} \sum_{j=1}^{L_n} |\chi_j(n)\rangle\langle\chi_j(n)|$$

so that we have

$$\langle\Upsilon(\omega)|P_R^\perp H P_R|\Upsilon(\omega)\rangle = \langle\Upsilon(\omega)|P_R H P_R^\perp|\Upsilon(\omega)\rangle$$

upon the condition that $|\Upsilon(\omega)\rangle$ is an eigenstate. If we choose $R > M$, we can insert the asymptotic form for the effective Hamiltonian

$$\sum_{n=0}^{R}\sum_{n'>R} \boldsymbol{c}(n')^\dagger \mathsf{A}_{n-n'}\boldsymbol{c}(n) - \boldsymbol{c}(n)^\dagger \mathsf{A}_{n'-n}\boldsymbol{c}(n') = 0.$$

Since $\mathsf{A}_m = 0$ for $|m| > M$, this allows to restrict the summations and rewrite this equality as

$$\mathrm{Im}\left[\sum_{m=1}^{M}\sum_{n=R+1-m}^{R} \boldsymbol{c}(n)^\dagger \mathsf{A}_m \boldsymbol{c}(n+m)\right] = 0.$$

We can insert the asymptotic form for $\boldsymbol{c}(n)$ to obtain for the "diagonal terms" where γ has the same value for both sums

$$\begin{aligned}
\mathrm{Im}\left[\sum_{\gamma=1}^{2\Gamma}|q_\gamma|^2\sum_{m=1}^{M}\sum_{n=R+1-m}^{R} \boldsymbol{v}_\gamma^\dagger \mathsf{A}_\mathrm{m}\boldsymbol{v}_\gamma \mathrm{e}^{i\kappa_\gamma m}\right] &= \mathrm{Im}\left[\sum_{\gamma=1}^{2\Gamma}|q_\gamma|^2\sum_{m=1}^{M} m\boldsymbol{v}_\gamma^\dagger \mathsf{A}_\mathrm{m}\boldsymbol{v}_\gamma \mathrm{e}^{i\kappa_\gamma m}\right] \\
&= -\sum_{\gamma=1}^{2\Gamma}|q_\gamma|^2\sum_{m=1}^{M}\left(im\boldsymbol{v}_\gamma^\dagger \mathsf{A}_\mathrm{m}\boldsymbol{v}_\gamma \mathrm{e}^{i\kappa_\gamma m} - im\boldsymbol{v}_\gamma^\dagger {\mathsf{A}_\mathrm{m}}^\dagger\boldsymbol{v}_\gamma \mathrm{e}^{-i\kappa_\gamma m}\right) \\
&= -\sum_{\gamma=1}^{2\Gamma}|q_\gamma|^2\sum_{m=-M}^{M} im\boldsymbol{v}_\gamma^\dagger \mathsf{A}_\mathrm{m}\boldsymbol{v}_\gamma \mathrm{e}^{i\kappa_\gamma m} \\
&= -\sum_{\gamma=1}^{2\Gamma}|q_\gamma|^2\frac{\mathrm{d}\omega_\gamma}{\mathrm{d}\kappa}(\kappa_\gamma)
\end{aligned}$$

and this expression has to equal zero if we can show that the contribution of all "non-diagonal terms" ($\gamma \neq \gamma'$ in the two sums) vanish. We look at a single contribution with $\gamma \neq \gamma'$ and the corresponding term with γ and γ' interchanged, and first assume $\kappa_\gamma \neq \kappa_{\gamma'}$.

$$\begin{aligned}
&\mathrm{Im}\Bigg[q_{\gamma'}^* q_\gamma \sum_{m=1}^{M}\sum_{n=R+1-m}^{R} \boldsymbol{v}_{\gamma'}^\dagger \mathsf{A}_\mathrm{m}\boldsymbol{v}_\gamma \mathrm{e}^{i\kappa_\gamma m}\mathrm{e}^{i(\kappa_\gamma-\kappa_{\gamma'})n} \\
&\qquad + q_\gamma^* q_{\gamma'} \sum_{m=1}^{M}\sum_{n=R+1-m}^{R} \boldsymbol{v}_{\gamma}^\dagger \mathsf{A}_\mathrm{m}\boldsymbol{v}_{\gamma'} \mathrm{e}^{i\kappa_{\gamma'} m}\mathrm{e}^{i(\kappa_{\gamma'}-\kappa_{\gamma})n}\Bigg] \\
&= \mathrm{Im}\left[\frac{\mathrm{e}^{i(\kappa_\gamma-\kappa_{\gamma'})(R+1)}}{\mathrm{e}^{i(\kappa_\gamma-\kappa_{\gamma'})}-1} q_{\gamma'}^* q_\gamma \sum_{m=1}^{M} \boldsymbol{v}_{\gamma'}^\dagger \mathsf{A}_\mathrm{m}\boldsymbol{v}_\gamma \left(\mathrm{e}^{i\kappa_\gamma m} - \mathrm{e}^{i\kappa_{\gamma'} m}\right)\right.
\end{aligned}$$

$$
\begin{aligned}
&\quad + \frac{\mathrm{e}^{i(\kappa_{\gamma'}-\kappa_\gamma)(R+1)}}{\mathrm{e}^{i(\kappa_{\gamma'}-\kappa_\gamma)}-1} q_\gamma^* q_{\gamma'} \sum_{m=1}^{M} \boldsymbol{v}_\gamma^\dagger \mathsf{A}_\mathsf{m} \boldsymbol{v}_{\gamma'} \left(\mathrm{e}^{i\kappa_{\gamma'} m} - \mathrm{e}^{i\kappa_\gamma m}\right)\Bigg] \\
&= \frac{1}{2\sin((\kappa_\gamma-\kappa_{\gamma'})/2)} \mathrm{Re}\Bigg[\mathrm{e}^{i(\kappa_\gamma-\kappa_{\gamma'})(R+1/2)} q_{\gamma'}^* q_\gamma \sum_{m=1}^{M} \boldsymbol{v}_{\gamma'}^\dagger \mathsf{A}_\mathsf{m} \boldsymbol{v}_\gamma \left(\mathrm{e}^{i\kappa_\gamma m} - \mathrm{e}^{i\kappa_{\gamma'} m}\right) \\
&\quad - \mathrm{e}^{i(\kappa_{\gamma'}-\kappa_\gamma)(R+1/2)} q_\gamma^* q_{\gamma'} \sum_{m=1}^{M} \boldsymbol{v}_\gamma^\dagger \mathsf{A}_\mathsf{m} \boldsymbol{v}_{\gamma'} \left(\mathrm{e}^{i\kappa_{\gamma'} m} - \mathrm{e}^{i\kappa_\gamma m}\right)\Bigg] \\
&= \frac{1}{2\sin((\kappa_\gamma-\kappa_{\gamma'})/2)} \mathrm{Re}\Bigg[\mathrm{e}^{i(\kappa_\gamma-\kappa_{\gamma'})(R+1/2)} q_{\gamma'}^* q_\gamma \sum_{m=1}^{M} \boldsymbol{v}_{\gamma'}^\dagger \mathsf{A}_\mathsf{m} \boldsymbol{v}_\gamma \left(\mathrm{e}^{i\kappa_\gamma m} - \mathrm{e}^{i\kappa_{\gamma'} m}\right) \\
&\quad - \mathrm{e}^{i(\kappa_\gamma-\kappa_{\gamma'})(R+1/2)} q_{\gamma'}^* q_\gamma \sum_{m=1}^{M} \boldsymbol{v}_{\gamma'}^\dagger \mathsf{A}_\mathsf{m}^\dagger \boldsymbol{v}_\gamma \left(\mathrm{e}^{-i\kappa_{\gamma'} m} - \mathrm{e}^{-i\kappa_\gamma m}\right)\Bigg] \\
&= \frac{1}{2\sin((\kappa_\gamma-\kappa_{\gamma'})/2)} \mathrm{Re}\Bigg[\mathrm{e}^{i(\kappa_\gamma-\kappa_{\gamma'})(R+1/2)} q_{\gamma'}^* q_\gamma \sum_{m=-M}^{M} \boldsymbol{v}_{\gamma'}^\dagger \mathsf{A}_\mathsf{m} \boldsymbol{v}_\gamma \left(\mathrm{e}^{i\kappa_\gamma m} - \mathrm{e}^{i\kappa_{\gamma'} m}\right)\Bigg].
\end{aligned}
$$

Note that we are missing the term for $m = 0$, but that this term is zero anyway because of the factor $(\mathrm{e}^{i\kappa_\gamma m} - \mathrm{e}^{i\kappa_{\gamma'} m})$. Finally noting that $\sum_{m=-M}^{M} \mathsf{A}_\mathsf{m} \boldsymbol{v}_\gamma \mathrm{e}^{i\kappa_\gamma m} = \omega \boldsymbol{v}_\gamma$ and $\sum_{m=-M}^{M} \boldsymbol{v}_{\gamma'}^\dagger \mathsf{A}_\mathsf{m} \mathrm{e}^{i\kappa_{\gamma'} m} = \omega \boldsymbol{v}_{\gamma'}^\dagger$, it is clear that both contributions cancel and the total expression evaluates to zero. Finally, we consider the case that $\kappa_\gamma = \kappa_{\gamma'} = \kappa$. We obtain

$$
\begin{aligned}
&\mathrm{Im}\left[q_{\gamma'}^* q_\gamma \sum_{m=1}^{M} \sum_{n=R+1-m}^{R} \boldsymbol{v}_{\gamma'}^\dagger \mathsf{A}_\mathsf{m} \boldsymbol{v}_\gamma \mathrm{e}^{i\kappa m} + q_\gamma^* q_{\gamma'} \sum_{m=1}^{M} \sum_{n=R+1-m}^{R} \boldsymbol{v}_\gamma^\dagger \mathsf{A}_\mathsf{m} \boldsymbol{v}_{\gamma'} \mathrm{e}^{i\kappa m}\right] \\
&\qquad = \mathrm{Im}\left[q_{\gamma'}^* q_\gamma \sum_{m=1}^{M} m \boldsymbol{v}_{\gamma'}^\dagger \mathsf{A}_\mathsf{m} \boldsymbol{v}_\gamma \mathrm{e}^{i\kappa m} + q_\gamma^* q_{\gamma'} \sum_{m=1}^{M} m \boldsymbol{v}_\gamma^\dagger \mathsf{A}_\mathsf{m} \boldsymbol{v}_{\gamma'} \mathrm{e}^{i\kappa m}\right] \\
&\qquad = \mathrm{Im}\left[q_{\gamma'}^* q_\gamma \sum_{m=-M}^{M} m \boldsymbol{v}_{\gamma'}^\dagger \mathsf{A}_\mathsf{m} \boldsymbol{v}_\gamma \mathrm{e}^{i\kappa m}\right].
\end{aligned}
$$

In the last line, we replaced the second term of the line before by the negative of its complex conjugate, since we are taking the imaginary part of the whole expression anyway. Using that $\boldsymbol{v}_\gamma$ and $\boldsymbol{v}_{\gamma'}$ correspond to some $\boldsymbol{v}_\gamma(\kappa)$ and $\boldsymbol{v}_{\gamma'}(\kappa)$ with different $\gamma \neq \gamma'$ but equal $\omega_\gamma(\kappa) = \omega_{\gamma'}(\kappa)$, we can employ Eq. (A.1) to conclude that this term is zero.

MIX
Papier aus verantwortungsvollen Quellen
Paper from responsible sources
FSC® C105338

If you have any concerns about our products,
you can contact us on
ProductSafety@springernature.com

In case Publisher is established outside the EU,
the EU authorized representative is:
Springer Nature Customer Service Center GmbH
Europaplatz 3, 69115 Heidelberg, Germany

Printed by Libri Plureos GmbH
in Hamburg, Germany